Granivorous birds
in ecosystems

The International Biological Programme was established by the International Council of Scientific Unions in 1964 as a counterpart of the International Geophysical Year (IGY). The subject of the IBP was defined as 'The Biological Basis of Productivity and Human Welfare', and the reason for its establishment was recognition that the rapidly increasing human population called for a better understanding of the environment as a basis for the rational management of natural resources. This could be achieved only on the basis of scientific knowledge, which in many fields of biology and in many parts of the world was felt to be inadequate. At the same time it was recognised that human activities were creating rapid and comprehensive changes in the environment. Thus, in terms of human welfare, the reason for the IBP lay in its promotion of basic knowledge relevant to the needs of man.

The IBP provided the first occasion on which biologists throughout the world were challenged to work together for a common cause. It involved an integrated and concerted examination of a wide range of problems. The Programme was coordinated through a series of seven sections representing the major subject areas of research. Four of these sections were concerned with the study of biological productivity on land, in freshwater, and in the seas, together with the processes of photosynthesis and nitrogen-fixation. Three sections were concerned with adaptability of human populations, conservation of ecosystems and the use of biological resources.

After a decade of work, the Programme terminated in June 1974 and this series of volumes brings together, in the form of syntheses, the results of national and international activities.

INTERNATIONAL BIOLOGICAL PROGRAMME 12

Granivorous birds
in ecosystems

Their evolution, populations, energetics, adaptations, impact and control

Edited by

Jan Pinowski
Assistant Professor, Institute of Ecology, Polish Academy of Sciences

S. Charles Kendeigh
Professor Emeritus of Zoology, University of Illinois

CAMBRIDGE UNIVERSITY PRESS

CAMBRIDGE
LONDON · NEW YORK · MELBOURNE

Published by the Syndics of the Cambridge University Press
The Pitt Building, Trumpington Street, Cambridge CB2 1RP
Bentley House, 200 Euston Road, London NW1 2DB
32 East 57th Street, New York, NY 10022, USA
296 Beaconsfield Parade, Middle Park, Melbourne 3206, Australia

First published 1977

Printed in Great Britain at the
University Press, Cambridge

Library of Congress Cataloguing in Publication Data
Main entry under title:
Granivorous birds in ecosystems, their evolution, popula-
tions, energetics, adaptations, impact and control.
(International biological programme; 12)
Bibliography: p.
Includes index.
1. Birds–Ecology. 2. Granivores. I. Pinowski, Jan
II. Kendeigh, Samuel Charles, 1904– III. Series.
ISBN 0 521 21504 8

Contents

Contents

Table des matières

Table des matières

Содержание

Содержание

Contenido

Contenido

List of contributors

J. B. Cragg, Faculty of Environmental Design, University of Calgary, Calgary 44, Alberta, Canada

Victor R. Dol'nik, Zoological Institute, Soviet Academy of Sciences, 199164, Leningrad, USSR

*M. I. Dyer, Natural Resource Ecology Laboratory, Colorado State University, Fort Collins, CO 80523, USA

Valery M. Gavrilov, Cathedra of Vertebrate Zoology, Biological Faculty, Moscow State University, 117234, Moscow, USSR

*Richard F. Johnston, Museum of Natural History and Department of Systematics and Ecology, University of Kansas, Lawrence, KS 66045, USA

*S. Charles Kendeigh, Vivarium Building, University of Illinois, Wright and Healey Streets, Champaign, Illinois 61820, USA

William J. Klitz, Department of Zoology, University of California, Berkeley, California 94730, USA

Andrzej Myrcha, Institute of Ecology, Polish Academy of Sciences, Dziekanów Leśny near Warsaw, 05–150 Łomianki, Poland

Barbara Pinowska, Institute of Ecology, Polish Academy of Sciences, Dziekanów Leśny near Warsaw, 05–150 Łomianki, Poland

*Jan Pinowski, Institute of Ecology, Polish Academy of Sciences, Dziekanów Leśny near Warsaw, 05–150 Łomianki, Poland

Peter Ward, Institute of Terrestrial Ecology, Monks Wood Experimental Station, Abbots Ripton, Huntingdon PE17 2LS, England

*John A. Wiens, Department of Zoology, Oregon State University, Corvallis, Oregon 97331, USA

* Senior authors.

Preface

Investigations in the IBP Section concerned with the Productivity of Terrestrial Ecosystems (PT) have taken two forms: *Major Biome Studies* concerned in the main with the mode of operation of total ecosystems, and *Special Studies* in which the ecology of important groups of consumers has been examined in considerable detail. It was as part of the Special Studies approach that scientists concentrated on the role of certain species of granivorous birds in ecosystems. The topic was clearly related to the general aim of IBP, namely *Biological Productivity and Human Welfare*. Granivorous birds are of worldwide occurrence, they are a component of man-made ecosystems and, because of their association with man's main food crops, their study was a matter of urgency and required international collaboration.

It is worthwhile reminding readers of this volume of the original aims of what was described in IBP News No. 13 (1969) as Theme 10. Granivorous Birds: 'Though this theme includes studies on energy flow through populations of a number of bird species, the major IBP objective is the integrated, cooperative study of the genus *Passer*, especially the house sparrow (*P. domesticus* (L.)) and the tree sparrow (*P. montanus* (L.)). These two species enjoy a world-wide distribution and live in a variety of habitat types under a wide range of different climatic conditions. The global study of the population dynamics, morphology and bioenergetics of this genus will provide information on the influence of a number of environmental factors on secondary production.'

This synthesis volume, in which full use is made of the systems approach, is based on a large body of data much of which is from non-IBP sources. The major IBP input has been concerned with the welding together of this background material. The quality of this achievement in synthesis is a resultant of very close collaboration between the many workers who willingly exchanged unpublished results, rough drafts of papers, chapters and guidelines to provide opportunities for maximum interaction.

During the last decade, ecological and evolutionary studies have tended to go their separate ways. This volume recognizes the need to unite these two highly interdependent biological disciplines. Furthermore, in linking them with fundamental work on bioenergetics, new insights are provided into the behavioural and applied aspects of granivory. Studies which encompass archaeological and anthropological investigations form a backcloth for the interpretation of detailed investigations of the population biology and bioenergetics of birds. Taken together, they provide a basis for assessing an organism's ability to exploit its adaptations for granivory. It is a measure of the confidence that can be placed in our present knowledge that Johnston & Klitz (Chapter 2, this volume) are, as a result of a thorough analysis of the adaptations of the house sparrow for commensal granivory, prepared to predict that other species '. . . will not make the same impress on man's agriculture or his life in general'. It is said that there are no problem children, only problem parents. This volume indicates that it is man who must accept responsibility for the extent to which granivorous species have become an obvious feature of his environment and it provides guidelines for the development of rational systems of management.

In spite of the wealth of detailed knowledge now available, many gaps exist in the data base. These will certainly be filled in by future studies but if we lack fundamental ecological information on species within the genus *Passer*, then how much do we know about other less numerous and less obvious species which are components of ecosystems altered by man? The chapters which touch on management highlight some of the unknowns. Of particular concern is our lack of knowledge of the behaviour of populations. Without adequate quantitative measures of behaviour under field conditions, full use cannot be made of the advanced and sophisticated information on the bioenergetics of birds which is given a prominent place in this volume.

In spite of their importance, the investigations on granivorous birds might be described as the Forgotten Theme of IBP(PT). Some countries, whose scientists have made major contributions to the development of the theme, have either not mentioned it at all or have only touched upon it in their National Reports. More to the point, many of the scientists found it difficult to obtain support from IBP National Funds. They received some help from the small budget of the Special Committee of IBP but the completion of this volume owes more to the dedication of the individuals involved than to the availability of adequate grants. Their cooperative effort is not to stop with the official end of IBP. Their *Newsletter* will continue to appear, thanks to the efforts of Dr Jan Pinowski and the financial support promised by the Polish Academy of Sciences. Furthermore, the group, which has now been formally recognized by the International Association for Ecology, has issued an invitation to other scientists who are studying granivorous birds to join in its activities.

The continued existence of collaborative endeavours of this type is necessary if biologists are to tackle problems of global concern and, in particular, problems which have pronounced cultural and social components. Apart from the scientific impact, IBP has provided scientists with many opportunities to become familiar with other cultures and other points of view. In this regard, those who were present at the final editorial meeting of the Granivorous Birds Theme will long remember one of Czechoslovakia's scientists, a specialist on the population dynamics of

Passer domesticus and *Passer montanus*, giving a recital which included selections from Dvořák and Chopin, on a violin constructed by one of the staff of the Polish Research Station from wood grown in the grounds of the station.

J. B. Cragg
Killam Memorial Professor,
Faculty of Environmental Design,
University of Calgary,
Calgary, Canada

1. Introduction

S. C. KENDEIGH & J. PINOWSKI

The present volume is a synthesis of the studies and researches of the Working Group on Granivorous Birds, organized within the Section on Terrestrial Productivity (PT) of the IBP. The studies of the Working Group have centered on weed seed and grain-eating species because of their abundance and importance in both natural and man-made ecosystems, the ease with which they may be worked in both the field and the laboratory and their importance to man.

Development of the project

The idea for such a Working Group originated as part of the program in secondary productivity of terrestrial ecosystems at the Institute of Ecology of the Polish Academy of Science under the directorship of Dr K. Petrusewicz. An international conference covering the general program and including several papers dealing with birds was held in Jabłonna, near Warsaw, Poland, from 31 August to 6 September 1966.

Dr Jan Pinowski, principal ornithologist of the Polish Institute of Ecology, has headed the Working Group throughout the period of its activities. In the autumn of 1965 he sent letters to over 100 ornithologists around the world proposing the organization of the Group and on 31 May 1966, obtained its official approval as a project of the IBP. There was considerable interest expressed in the proposal, and a central steering committee was organized at the Fourteenth International Ornithological Congress, 27 July 1966, in Oxford, England. Besides Dr Pinowski as chairman, this committee included Dr D. Summers-Smith of England, Dr F. J. Turček of Czechoslovakia, and Drs R. F. Johnston and S. C. Kendeigh of USA.

Close cooperation between Drs Pinowski and Kendeigh in the planning and execution of the program on granivorous birds began in 1966 when the latter was on a Cultural Exchange Fellowship arranged between the USA and USSR Academies of Science. *En route* to the Soviet Union he spent several days at the Institute of Ecology, Polish Academy of Science, and later worked for two and a half weeks with Dr Victor Dol'nik in Leningrad. Dr Dol'nik is a co-author of Chapter 5.

In order to develop and encourage the work of the Group and to serve as a medium for the exchange of ideas and reports, Dr Pinowski began issuing a periodical from the Polish Institute of Ecology in 1967 entitled, 'International Studies on Sparrows'. Twenty-two numbers in nine volumes had appeared by 1976. The third number in 1967 listed the names of 79 co-investigators in 25 different countries. Since that time, new participants have been added while others have become inactive. Later numbers contained an extensive bibliography of the genus *Passer*.

Interest in the program has also been maintained by a number of national and international conferences. On 3 September 1969, Dr Kendeigh chaired a half-day symposium at the meeting of the American Ornithologists' Union (AOU) at Fayetteville, Arkansas. Fourteen papers or abstracts from this session were published as Ornithological Monographs No. 14 by the AOU under the title: 'A symposium of the house sparrow *Passer domesticus* (L.) and European tree sparrow *P. montanus* (L.) in North America'.

The first general session of the Working Group was held on 6–8 September 1970, at the Hague and at Arnhem in the Netherlands. This meeting was under the sponsorship of Dr J. A. L. Mertens and Dr J. H. van Balen of the Netherlands Institute of Ecological Research, Research Institute for Nature Management. The proceedings were published in book form by the Polish Institute of Ecology, under the editorship of Dr Kendeigh and Dr Pinowski and the title: 'Productivity, population dynamics and systematics of granivorous birds'. Thirty-one papers are included in this volume. There were 52 participants from 17 countries attending the meeting.

A second general meeting of the Working Group was held at the Polish Institute of Ecology at Dziekanów Leśny, near Warsaw, on 3–7 September 1973, sponsored by Dr Pinowski. Twenty-two persons, representing 11 different countries, gave reports. The purpose of the session was to organize and begin work on this synthesis volume covering the research findings of the Working Group over the seven-year span in which the IBP had been active. Preliminary outlines of chapters were prepared, chapter editors selected and chapter contents discussed. Following these meetings, a delightful trip by train, bus and boat through north-central Poland to the Baltic Sea and back allowed further informal discussions until the sessions adjourned on 10 September.

Dr John A. Wiens organized the next working session at Oregon State University, Corvallis, Oregon, USA, on 10–12 July 1974. This meeting was intended to consolidate and integrate the thinking of North American collaborators and was participated in by nine persons from the USA and Canada. Considerable progress was made toward finalizing the arrangement, authorship and contents of the chapters.

The above meeting was followed by one arranged by Dr M. I. Dyer at Colorado State University, Fort Collins, Colorado, USA, on 7–12 October 1974, participated in by 13 collaborators from Yugoslavia, the Netherlands, German Federal Republic (GFR), Poland, Nigeria, Canada and the USA.

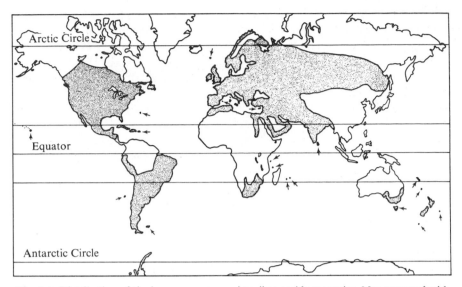

Fig. 1.1. Distribution of the house sparrow, primarily a resident species. Map prepared with the help of Richard F. Johnston.

The last meeting of chapter authors to prepare and coordinate the book manuscript was held at Szymbark, Poland, on 17–21 March 1975. Consultants from the Netherlands, Czechoslovakia and Poland also attended.

In addition to the organized conferences, Dr Wiens consulted with a number of collaborators on a special trip to Europe in the summer of 1974 and Dr Dyer did likewise both in the winter of 1974–1975 and in January 1976. Dr Pinowski made two trips to the USA in 1975 for work with Drs Dyer, Wiens, Kendeigh and others. Dr Peter Ward, a co-author of Chapter 7, also came to the USA in October 1975 to work with Drs Dyer and Wiens.

Scope and approach

The IBP and this series of volumes are concerned with productivity and human welfare, with the structure and function of ecosystems, with the analysis of natural resources and how they may be utilized on a sustained yield basis, and with the health and happiness of mankind as he seeks to occupy and utilize the various parts of the world. The agricultural, industrial and cultural ecosystems which man occupies are inherently unstable and would quickly revert to Nature were man to lose his dominance over them. Man is beginning to learn that there are limits to his capacity to exert such dominance which are set in part by the attributes of natural systems and in part by man's ignorance of how to manipulate

3

S. C. Kendeigh & J. Pinowski

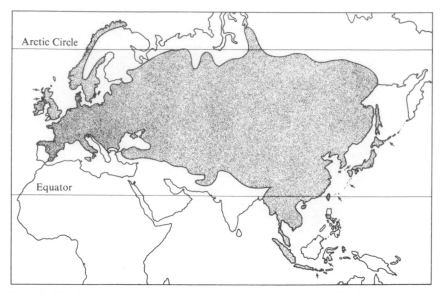

Fig. 1.2. Distribution of the European tree sparrow, primarily a resident species. It also occurs in the region around St. Louis, Missouri, USA, and in southeast Australia.

these attributes. Granivorous birds may be direct competitors of man for food and may also harbor zoonotic disease organisms that affect his health. To control man-dominated ecosystems, it is necessary to understand the role played by these species and their relationships with man. This requires not only the direct study of population ecology in the field but also the acquisition of experimental information in the laboratory about how birds are made, how they function and how they behave.

The house sparrow (Fig. 1.1) has been given most attention in this volume because of its wide dispersal and long and close association with man, followed by the European tree sparrow (Fig. 1.2), red-winged blackbird, common grackle, brown-headed cowbird, red-billed quelea, dickcissel and horned lark. All of these species belong to the order Passeriformes, sub-order Oscines. While other granivores, such as columbids, galliforms and some waterfowl have been largely ignored, this concentration on a few well-studied species will hopefully provide insights which may apply to granivores in general.

The house sparrow *Passer domesticus*, European tree sparrow *P. montanus*, Spanish sparrow *P. hispaniolensis* (Temminck) and red-billed quelea (dioch) *Quelea quelea* (L.), are all in the family of weaver finches Ploceidae. The first three are separated into the sub-family Passerinae, the last in the sub-family Ploceinae. The red-winged blackbird *Agelaius phoeniceus* (L.), common grackle *Quiscalus quiscula* (L.), and brown-

Fig. 1.3. Distribution of the breeding range (stippled) and wintering range (cross-hatched) of the red-winged blackbird.

headed cowbird *Molothrus ater* (Boddaert) are in the family Icteridae and sub-family Icterinae. The dickcissel *Spiza americana* (Gmelin), belongs to the family Emberizidae and sub-family Cardinalinae, and the horned lark *Eremophila alpestris* (L.) to the family Alaudidae. Classification and nomenclature throughout this volume follows Peters' *Check-list of birds of the world*, published by the Museum of Comparative Zoology, Cambridge, Massachusetts, USA. Common names will be generally used, but

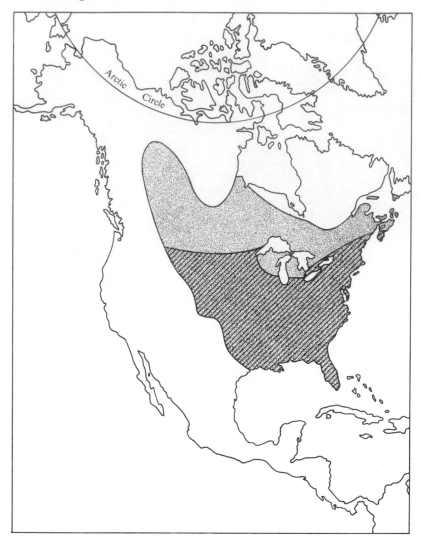

Fig. 1.4. Distribution of the breeding range (stippled) and wintering range (cross-hatched) of the common grackle.

these will be followed by the scientific name the first time that the species is cited in the chapter.

The three species of icterids are largely confined to North America (Figs. 1.3 to 1.5). They are generally resident where they occur, except in the northern portions of their ranges where there is movement southward in the winter. The dickcissel is a grassland species that nests in mid-North America and is highly migratory (Fig. 1.6). The genus *Quelea* occurs south

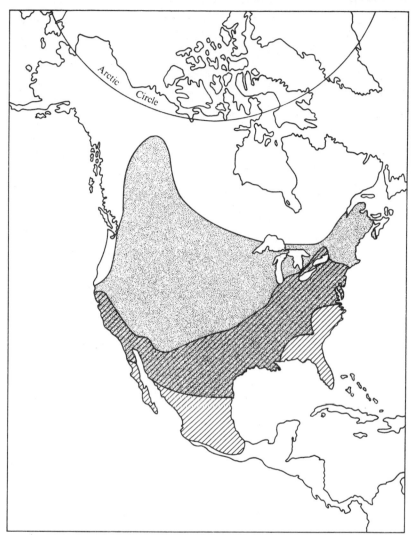

Fig. 1.5. Distribution of the breeding range (stippled) and winter range (cross-hatched) of the brown-headed cowbird.

of the Sahara in Africa. *Quelea quelea* is a migratory species inhabiting the dry savannah and grasslands (Fig. 1.7). Flocks travel over considerable distances, stop to form breeding colonies whenever and wherever they find conditions favorable, and thus the species may be abundant in an area one year and almost absent the next. The horned lark is a grassland and tundra species, widely distributed over the northern hemisphere in both the Old and New Worlds (not mapped). There is a shift of individuals

7

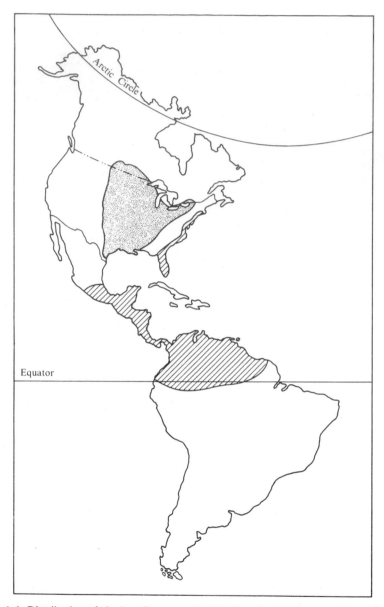

Fig. 1.6. Distribution of the breeding range (stippled) and wintering range (cross-hatched) of the dickcissel. Map prepared by J. W. Tatschl, Kansas State University, USA.

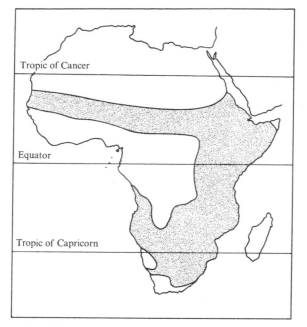

Fig. 1.7. Distribution of the red-billed quelea, a nomadic species. Map prepared with the help of Peter Ward.

southward to overwinter, although in many regions the species is represented throughout the year.

The genus *Passer* probably originated in the savannah biotope of tropical Africa and dispersed northward down the Nile Valley in late Miocene or Pliocene times. From here, beginning in the Pleistocene, it spread widely through northern Africa, Europe and Asia. With the help of man, particularly in the nineteenth century, *P. domesticus* was introduced into North America, Australia, New Zealand, South America, the Hawaiian Islands and South Africa (Fig. 1.1). The spread of the species at known times into widely different localities having different terrain, climates, vegetation and biotic associates provides a unique opportunity for the study of evolutionary adjustments and adaptations. *P. montanus* (Fig. 1.2) was brought to North America in 1870 and became established in the vicinity of St. Louis, Missouri. Its dispersal from this center has been very limited. *P. hispaniolensis* occurs in Spain, Portugal, North Africa, the Mediterranean islands, the Balkan Peninsula and eastward to beyond the Caspian Sea and into Iran and Afghanistan. Populations in the northern part of the species' range migrate south for the winter or disperse as vagrants, but southern populations are largely resident. Chapter 2 deals with dispersal of the house sparrow, how the species came into close association with man, its morphological variations and adaptations in different geographic areas, and the mechanics of its genetic and evolutionary development.

Once a species invades a new area and finds a favorable environment, it multiplies in numbers and establishes a population. The population grows to the carrying capacity of the area with respect to food, cover, competitors, predators and climate. The population density varies locally and temporally as these environmental factors dictate. At equilibrium, natality balances mortality, but the way these two processes become effective, and whether in fact equilibrium is attained, varies with conditions, with time and with the species. Analysis of population dynamics is the subject of Chapter 3.

Size is commonly measured as weight, and fluctuations in weight among adults in a species are an index of energy balance. Likewise, increase in weight of growing young, together with the increase in number of individuals, give a measure of productivity of the species, as Chapter 4 explains. The impact of a species depends not only on numbers of individuals but also on the size of the individuals. A few large birds may consume as much food, monopolize as much space and exert as much effect as a large number of small birds. Biomass tends to equate numbers and size both in adults and in growing young.

In order to exist and to function, a bird, as all organisms, requires energy. Whether or not a bird can exhibit a specific behavior depends, in large part, on the amount of energy it can mobilize. The occurrence of a species in a region, its population size, its need to migrate and the timing and conditioning of all events in its yearly cycle, therefore, depend on fluctuations in its energy resources. Chapter 5 analyzes the components in the daily energy budget throughout the year and attempts to provide equations of general application.

Current efforts to document and understand the structure and function of ecosystems are founded on population dynamics, on energy flow patterns and rates and on relevant environmental parameters. These factors are put into a computer model which makes possible the quantification of energy demands, food consumption and the potential impact of avian consumers in ecosystems. Chapter 6 thus draws upon and integrates the material of the three preceding chapters.

Chapter 7 is concerned with the practical application of these studies. Emphasis is placed on methods of evaluating the economic impact of birds on cereal grains, conditions under which bird species become destructive, management techniques and control strategies.

Although almost any bird species may at times be destructive to man's interests, the most important bird pests are granivorous. Chapter 8 analyzes why this is so, both because of adaptive strategies in morphology, physiology and behavior and because of the environmental conditions to which they are evolutionarily adjusted.

This volume by no means provides all the answers to questions that may arise. Rather, it is intended to be innovative and to introduce new approaches, which may eventually provide a fuller insight into the importance of birds in ecosystems and on man's livelihood. Chapter 9 summarizes these approaches and points out what needs to be done and what kinds of information need to be collected.

Although this book centers around granivorous birds, the concepts developed and the procedures outlined are equally applicable to species with other food habits. The population dynamics of other species vary only in details, energy requirements are the same regardless of how they are met; the computer model needs adjustment only of the input variables, and then management is concerned only with that particular crop and pest species involved. In perusing the chapters that follow, the reader might well give attention to how the material may be applied to his own interests and problems.

Acknowledgements

We wish to acknowledge first and foremost the great contribution to the work of this Group by the Institute of Ecology of the Polish Academy of Science, including its former director, Dr K. Petrusewicz and present director, Dr R. Klekowski. Without the initiation of the program and the promotion, publications, meetings and financial support furnished by this Institute, the project may never have been developed, let alone brought to its present successful conclusion with the publication of this volume. We are also grateful to the Netherlands Institute of Ecological Research, particularly Dr J. H. van Balen and Dr J. A. L. Mertens, for sponsoring the first general meeting of the Working Group and to the AOU for supporting a symposium and its publication, especially to Dr John William Hardy and Dr Martin L. Morton who edited Ornithological Monograph No. 14.

The National Science Foundation of the USA provided a substantial grant in support of the Corvallis and Fort Collins meetings and for salaries, travel, subsistence and other expenses. This grant was solicited and administered by Dr M. I. Dyer and Dr J. A. Wiens. The Central Office of the IBP in London, England, the Foreign Currency Program administered by the Smithsonian Institution of Washington D.C., and the Commonwealth Fund of the British Commonwealth all provided funds to participants for international travel.

Special mention needs to be made for the considerable help furnished

Table 1.1. *Participants in symposia and work sessions*

Name	Country	AOU symposium, USA	The Hague–Arnhem, Netherlands	Dziekanów Leśny, Poland	Corvallis, USA	Fort Collins, USA	Szymbark, Poland
Ted. R. Anderson	USA	X	—	X	X	—	—
Frantisek Balát	Czechoslovakia	—	—	X	—	—	X
Paul H. Baldwin	USA	—	X	—	—	—	—
J. H. van Balen	Netherlands	—	X	—	—	X	—
Jon C. Barlow	Canada	X	—	—	—	—	—
L. Bruce Barnett	USA	X	—	—	—	—	—
Zdenek Bauer	Czechoslovakia	—	—	—	—	—	X
Floyd H. Blackmore	USA	X	—	—	—	—	—
Charles R. Blem	USA	X	—	X	—	—	—
L. Bortoli	Chad	—	—	X	—	—	—
E. J. Buyckx	Italy	—	X	—	—	—	—
P. Clausing	German Democratic Republic (GDR)	—	—	X	—	—	—
Mary Heimerdinger Clench	USA	X	—	—	—	—	—
James B. Cragg	Canada	—	—	X	X	X	X
D. G. Dawson	New Zealand	—	X	—	—	—	—
A. F. DeBont	Republic democratique du Congo	—	X	—	—	—	—
A. Dhondt	Belgium	—	X	—	—	—	—
G. Diagne	Senegal	—	X	—	—	—	—
James A. Dick	Canada	X	—	—	—	—	—
Victor R. Dol'nik	USSR	—	—	X	—	—	X
H. Dominas	Poland	—	X	X	—	—	—
Melvin I. Dyer	USA	—	X	X	X	X	X
C. C. H. Elliott	South Africa	—	X	—	—	—	—
P. R. Evans	England	—	X	—	—	—	—
G. Folk	Czechoslovakia	—	—	X	—	—	—
V. M. Galushin	USSR	—	X	—	—	—	—
James A. Gessaman	USA	—	—	—	—	X	—
Ph. Gramet	France	—	X	X	—	—	—
G. Grün	GDR	—	—	X	—	—	—
G. P. Hekstra	Netherlands	—	X	—	—	—	—
J. Hulscher	Netherlands	—	X	—	—	—	—
Richard F. Johnston	USA	X	X	—	X	X	X
P. J. Jones	Botswana	—	X	—	—	—	—
V. Javanovic	Yugoslavia	—	—	X	—	X	—
Werner Keil	GFR	—	X	X	—	—	—
S. Charles Kendeigh	USA	X	X	X	X	X	X
William J. Klitz	USA	X	—	—	—	—	—
H. Klomp	Netherlands	—	X	—	—	—	—
J. M. Lienart	Chad	—	X	—	—	—	—
R. Liversidge	South Africa	—	X	—	—	—	—
H. Löhrl	GFR	—	—	X	—	—	—
W. J. Maher	Canada	—	X	—	—	—	—
Elden W. Martin	USA	X	X	—	—	—	—
J. A. L. Mertens	Netherlands	—	X	X	—	X	—
Z. B. Mirza	Pakistan	—	X	—	—	—	—

Table 1.1. (*cont.*)

Name	Country	AOU sym-posium, USA	The Hague–Arnhem, Nether-lands	Dzieka-nów Leśny, Poland	Cor-vallis, USA	Fort Collins, USA	Szym-bark, Poland
Carl J. Mitchell	USA	X	—	—	—	—	—
G. J. Morel	Senegal	—	X	—	—	—	—
M. Y. Morel	Senegal	—	X	—	—	—	—
A. Myrcha	Poland	—	X	—	—	—	—
R. M. Naik	India	—	X	X	—	—	—
Charles A. North	USA	X	X	—	X	—	—
R. J. O'Connor	England	—	X	—	—	—	—
H. Oelke	GFR	—	—	X	—	X	—
T. K. Palmer	USA	—	X	—	—	—	—
K. Petrusewicz	Poland	—	—	—	—	—	X
Barbara Pinowska	Poland	—	—	X	—	—	X
Jan Pinowski	Poland	—	X	X	—	X	X
J. D. Rising	Canada	—	X	—	—	—	—
Chandler S. Robbins	USA	X	X	—	—	—	—
R. Rossbach	GFR	—	X	—	—	—	—
J. Roy	Chad	—	—	X	—	—	—
R. A. Ryder	USA	—	X	—	—	—	—
D. C. Seel	Wales	—	X	—	—	—	—
Charles G. Sibley	USA	—	X	—	—	—	—
W. R. Siegfried	South Africa	—	X	—	—	—	—
D. D. B. Summers	England	—	X	—	—	—	—
J. Tahon	Belgium	—	X	—	—	—	—
Carol L. Votava	USA	X	—	—	—	—	—
Peter Ward	Nigeria	—	X	—	—	X	—
January Weiner	Poland	—	X	X	—	X	X
George C. West	USA	—	—	X	X	—	—
K. Westerterp	Netherlands	—	—	—	—	—	X
M. Wieloch	Poland	—	X	X	—	—	—
John A. Wiens	USA	—	—	X	X	X	X
Raymond L. Will	USA	X	—	—	—	—	—
E. N. Wright	England	—	X	—	—	—	—
W. B. Yapp	England	—	X	—	—	—	—
N. J. Yterberg	Norway	—	X	—	—	—	—
John L. Zimmerman	USA	—	—	—	X	—	—

by Dr James B. Cragg, Convener of Section PT, IBP, for advice concerning the preparation of this volume and for the acquisition of travel funds. He went out of his way to attend several of our working sessions and to participate in the discussions.

We are very appreciative to a number of persons for technical information and evaluation of reports. Some of this came through correspondence, but especially important was the give and take discussion of principles and details that developed at the sessions. The participants gave freely of their knowledge and time. Although they will be personally

13

S. C. Kendeigh & J. Pinowski

acknowledged in the individual chapters, we believe it desirable also to recognize their contribution at this point by listing in Table 1.1 their attendance at particular sessions. The authors hope that the interchange of ideas has been of as much interest and value to the other participants as it has been to them.

Finally, but by no means least, we humbly acknowledge the great mass of unpublished data contributed unstintingly by many collaborators without which our thesis could not have been developed to the degree that it has. Their names will be listed in the chapters to which they apply.

— — — — — — — — — — — — — — —

Conversion of energy units

We will use calories for units of energy in this book. The following equivalents, however, will make easy conversion into the joules and watts of the International System of Units.

1 kcal = 4.184 kilo joules

1 kcal bird^{-1} hour^{-1} = 1.163 watts

1 kcal bird^{-1} day^{-1} = 0.048 watts

1 Mcal bird^{-1} year^{-1} = 0.133 watts

2. Variation and evolution in a granivorous bird: the house sparrow

R. F. JOHNSTON & W. J. KLITZ

Granivorous birds can hardly have been important competitors of man for grass seeds until man developed the culture of grasses and became a sedentary grain producer. Avian and human seed specialists of 10 000 to 100 000 years ago would have evolved to respond to seed crops of reasonable and predictable abundance, irrespective of whether the grass species was found in seral or stable ecologic conditions. But it is difficult to conceive of man and birds coming into serious competition until man increased the predictability both to birds and himself of finding large quantities of large-sized seeds of annual grasses in one place for a persistent period of time. This man did by instituting monoculture, subsequent long-term storage, and overwinter feeding of himself and livestock at farmsteads. As Wiens & Johnston (Chapter 8, this volume) have noted, there were several species of birds ready to fit into a food niche of this kind.

As a consequence, today each region of the world has a set of birds that variably exploits food opportunities around granivorous man, and the total of inadvertent biomass thus maintained is considerable. In some places the locally most abundant birds are these commensals. But only a few species are persistently committed to commensalism, and only one, the house sparrow *Passer domesticus* (L.), seems to be an obligate commensal. Accordingly, it is appropriate in a general survey of the ecology of granivorous birds to examine the possible lines of development followed by the house sparrow in becoming the prototypical avian commensal granivore. The remainder of this chapter develops certain lines exploring the history and current status of adaptation in house sparrows of a number of geographic regions.

To do this we first provide a largely intuitive account of the evolutionary history of *P. domesticus*, using both the fossil record of *Passer* and the archeological record of man at the time he was developing sedentary agricultural habits as a tiller of the primordial wheats and barley of the Near East. This narrative is followed by an examination of the more recent circumstances surrounding the introduction of house sparrows to North America in the 1850s, along with indications of the many ways in which the birds have very rapidly adjusted to life in a new continental setting.

15

R. F. Johnston & W. J. Klitz

History of the house sparrow

The genus Passer

The evolutionary history of the house sparrow, either as a granivorous bird or as an obligate commensal of man, is partly a reflection of the evolutionary history of all the species of the genus *Passer*. Fifteen species are referred to the genus today (Table 2.1); morphological relationships of the 15 are depicted in Fig. 2.1, a tree diagram based on a matrix of taxonomic distances between the species computed from raw variables of wing length, tail length, bill length, bill width and tarsus length. Each species is represented by average sizes of the five variables, obtained by summing dimensions of at least five females and five males and dividing by 10, except for *P. pyrrhonotus* (two females and two males were used) and *P. domesticus* (several thousand specimens were used).

The branching dendrogram of Fig. 2.1 has two main stems, one including *P. domesticus* and the other *P. montanus*. Each branch contains replicates, three of *P. domesticus* and two of *P. montanus*; the replications are included partly because significant differentiation has occurred within *P. domesticus* and *P. montanus*, and partly because the degrees of apparent relationship suggested within *P. domesticus* and *P. montanus* can serve as gauges of the reliability of the way in which information is summarized in the diagram.

The three continental samples of *P. domesticus* cluster tightly together and are joined at slightly greater distances by *P. hispaniolensis* and *P. griseus*. This group is joined by a quartet of *P. simplex*, *P. ammodendri*, *P. melanurus* and *P. flaveolus*, and the two stems are in turn united with *P. iagoensis*. This set of eight species is linked at a relatively great distance to the remaining seven species of the genus. The seven are dominated by the replicate samples of *P. montanus*, nearly spanning the enormous reach of Eurasia; *P. montanus* is joined by the pair *P. rutilans* and *P. castanopterus*, this set by a trio, *P. moabiticus*, *P. luteus* and *P. pyrrhonotus*, and the six by *P. eminibey* at slightly greater distance.

It is important to note that the species of greatest phenetic affinity have distinct geographic distributions, save in the case of *P. domesticus* and *P. hispaniolensis*, species of recent common ancestry and not yet fully isolated reproductively. Consistent with this observation is the fact that the two species of greatest phenetic distance of all the others are *P. iagoensis* and *P. eminibey*, each of which is sympatric (in part) with four other species of *Passer*. Thus, the diagram, wholly phenetically-based, has important ecological implications concerning the degree to which morphologically similar species are enjoined from living close to one another. However, for present purposes, the most important conclusion drawn from the dendrogram is that, since granivorous commensals are to

16

Table 2.1. *The species of the genus* Passer, *their primary zoogeographic affinity, and their putative sympatry with primitive agricultural man*[a]

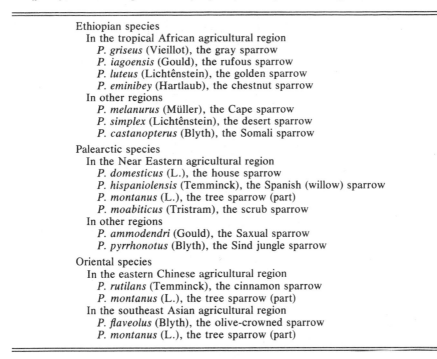

Ethiopian species
  In the tropical African agricultural region
    P. griseus (Vieillot), the gray sparrow
    P. iagoensis (Gould), the rufous sparrow
    P. luteus (Lichtênstein), the golden sparrow
    P. eminibey (Hartlaub), the chestnut sparrow
  In other regions
    P. melanurus (Müller), the Cape sparrow
    P. simplex (Lichtênstein), the desert sparrow
    P. castanopterus (Blyth), the Somali sparrow

Palearctic species
  In the Near Eastern agricultural region
    P. domesticus (L.), the house sparrow
    P. hispaniolensis (Temminck), the Spanish (willow) sparrow
    P. montanus (L.), the tree sparrow (part)
    P. moabiticus (Tristram), the scrub sparrow
  In other regions
    P. ammodendri (Gould), the Saxual sparrow
    P. pyrrhonotus (Blyth), the Sind jungle sparrow

Oriental species
  In the eastern Chinese agricultural region
    P. rutilans (Temminck), the cinnamon sparrow
    P. montanus (L.), the tree sparrow (part)
  In the southeast Asian agricultural region
    P. flaveolus (Blyth), the olive-crowned sparrow
    P. montanus (L.), the tree sparrow (part)

[a] 'Centers' of indigenous development of agriculture (Harlan, 1971).

be found in both the major branches of relationship within *Passer*, we can expect commensal tendencies from virtually any species of the genus. In fact *P. domesticus*, *P. hispaniolensis*, *P. griseus*, *P. melanurus*, *P. iagoensis*, *P. montanus*, *P. castanopterus*, *P. rutilans*, *P. luteus* and *P. eminibey* are known to nest around man's dwellings, or to be granivorous commensals of man or both.

Summers-Smith (1963) believes the group to have had an African origin, perhaps from an ancestor like *P. griseus*. He envisions the black-bibbed species evolving and spreading into north Africa using the Nile River as an avenue to temperate areas, and then easterly into Asia. Summers-Smith speaks in terms of one such radiation out from Africa, but if we are to follow the suggestions in Fig. 2.1, two Asiatic colonizations seem to have occurred, the first reflecting a *P. montanus*-like species and the second a *P. hispaniolensis/domesticus*-like species. In any event, our thinking is that several members of the genus have long occurred in or near (1) tropical Africa, (2) the Near East, (3) eastern China, and (4) southeast Asia,

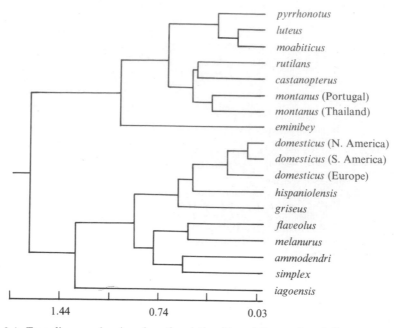

Fig. 2.1. Tree diagram showing phenetic relationships of the species of *Passer*, computed from a matrix of distance coefficients over five variables, using the unweighted pair-group method with arithmetic averages. Correlation of the tree back to the distance matrix is 0.774.

four of the six 'centers' and related foci of development of the earliest agricultural men (cf. Harlan, 1971).

Tree sparrows had an evolutionary history distinct from house and willow sparrows, probably originating in eastern Asia in complete isolation from the house–willow sparrow distributional centrum in the Mediterranean basin. The course of developing commensalism in tree sparrows may have had similarities to that which we shall put forth for house sparrows, but the story lacks the antecedents of an interpretable fossil record; distributional sympatry with sedentary, agricultural human populations certainly occurred (cf. Harlan, 1971), and a series of datable events of possible consequence for developing commensalism may be presumed for the present to parallel those available for the house sparrow.

The house sparrow as an almost obligate commensal of man does not occur permanently away from man's activities (Summers-Smith, 1963), and by definition, it probably never did. The house sparrow could have existed in its present morphological form prior to the origin of agricultural man if we assume *P. domesticus bactrianus* is primitive rather than derived. Or, if not, perhaps a species such as *P. hispaniolensis* can serve in the mind's eye as the prototype of the house aprrow, though this does

18

not assume that present-day house and Spanish sparrows shared anything more than a common ancestor. We should note well that the Spanish sparrow is osteologically indistinguishable from the house sparrow, so available Pleistocene fossils cannot help to assign a date range for the origin of the house sparrow.

Fossil record

The oldest fossils of house sparrow-like birds come from the Paleolithic strata of Ouum-Qatafa Cave, in Wadi Khareitoun, near Bethlehem, Israel. Several long bones and a coracoid were assigned by Tchernov (1962) to either *P. domesticus* or *P. hispaniolensis*; since we cannot reliably distinguish bones of one species from the other in the large series of recent specimens in our collection, we can only say that the fossils represent either species or their common ancestor of perhaps 12 000 B.P. Two premaxillae from older strata were described as *P. predomesticus* by Tchernov (cf. Markus, 1964), the specimens coming from middle Acheulian layers within the Mindel-Riss interglacial, perhaps prior to 400 000 B.P. The later specimens assigned by Tchernov to both *P. domesticus* and *P. hispaniolensis* were found in Mousterian (Würm glaciation) layers in the same cave.

Other fossils of *Passer*, likened to *P. domesticus*, have been found in cave sites near Galilee in layers around 10 000 to 15 000 B.P. (Bar-Yosef & Tchernov, 1966). Whatever the specific affinities of these fossils, the species they represent could not have been an ecologically real 'house sparrow', given that no house sparrow niche existed. Nevertheless, populations that could have given rise to *P. domesticus* and *P. hispaniolensis* did indeed occur in suitable places in the Near East for a period of time well anticipating the development of sedentary man there.

The beginnings of agricultural man in the Near East

Sedentary human groups existed possibly before agriculture was developed, but the archaeological record shows that hunters and food gatherers occupied sites only temporarily. In fact, few non-agricultural sites have been found in the most intensively studied regions of Europe and the Near East (Piggott, 1965); these are dated to about 11 000 B.P. More than temporary occupancy seems to have occurred at communities at Nahal Oren and Eynan, Palestine, where the economy included husbandry of grazing animals and artifacts included flint-bladed knives or straight sickles that could have been used for cutting some crop, perhaps a wild cereal (Braidwood & Howe, 1960, from Piggott, 1965). The settlements have a radiocarbon date of 10 850 B.P.

19

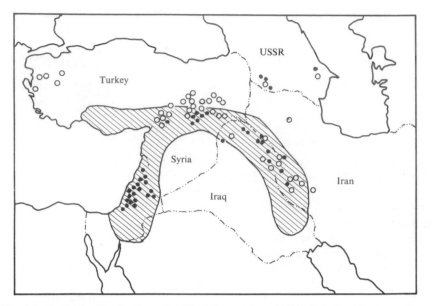

Fig. 2.2. Probable distribution of wild einkorn wheat (solid circles) and emmer (open circles), with general distribution of early farming villages indicated by cross-hatching. Source: Harlan & Zohary (1966).

A date of 10000 B.P. can be used for the general appearance of food-producing villages (Flannery 1965; Harlan & Zohary, 1966; Hole, 1966; Renfrew, 1973). At around 9000 B.P. at Jericho a settlement of about 4 ha existed, at which a cereal crop and domestic goats were important. This town was of mud and brick construction, having walls with towers up to 10 m high. The community apparently was in use by man for about 1000 years, and Piggott (1965) supposes that it and others like it formed the basis for the development of later economies and larger towns of the fertile crescent region (Fig. 2.2).

People at Argissa, Thessaly, grew wheat, barley, flax and probably millet, tended sheep, pigs and cattle, and had dogs at a date before 8000 B.P. Slightly later still, in many localities in Macedonia stone-using agriculturalists were able to set up and maintain permanent communities of mud and brick-walled houses; this kind of settlement became widely distributed in Mesopotamia, eastern Europe, the lowland Germanic region and around the Mediterranean Sea in Italy, Sicily, France and the Iberian peninsula by 7000 B.P. (Table 2.2). Some of these settlements persisted for 3000 years.

Construction by man of walls for housing was much easier in the Near East than in more mesic regions, because simple, mud bricks and mud mortar were satisfactory for long-term use in places having hot summers

Table 2.2. *Chronology of agricultural communities in Europe and the Near East: earliest records*[a]

Locality	Years before present
Palestine	10850
Jericho	9000
Thessaly	8300
Macedonia	8220
Eastern Europe in general	7000
Mesopotamia: fertile crescent	7000
Italy (coastal)	7000
Sicily	7000
France (south coastal)	7000
Iberia (coastal)	7000
Malta	6000
Italy (northern lake district)	5000
France (central region)	5000
British Isles	5000
Northern Europe in general	5000
Scandinavia	5000

[a] Mostly from Piggott (1965).

and relatively dry winters. More complicated building was required for humid central and northern European sites, and it is likely that appropriate techniques for such building developed later than those used for building in the xeric regions of the Near East. The point is well taken, for the archaeological record makes relatively little distinction in dates for Mesopotamian and east–central European sites (Table 2.2). In any event, in European settlements a framing of poles and wattles was used for support of walls, and roofs were gabled and pitched, a more complicated construction than that in Mesopotamia.

The background for development of sedentary agriculture in the Near East is almost certainly the primordial restriction of wild barley *Hordeum* and the wild wheats *Triticum aegipiloides* Thurb. and *T. dicoccoides* Koern. to Anatolia, Palestine, the northern Tigris and Euphrates watersheds, and the region south and southeast of the Caspian Sea (Harlan & Zohary, 1966; see Fig. 2.2). Local adaptation of barley and wheat populations to those regions would have been important in limiting practice of early agriculture of these important grains precisely to those regions. The adaptive limitations of the wild cereal species would ultimately be broached, but for an unknown number of centuries sedentary man could live only where his grains would let him. Consequently, the earliest bird commensals would have had to have been those opportunistic or pre-

21

R. F. Johnston & W. J. Klitz

adapted species with the Near East in their distributional ranges. As we have noted, antecedents of *P. domesticus* lived exactly there.

Possible evolutionary development of the house sparrow

Initial steps taken by presumptive house sparrows in association with Neolithic man would have stemmed from pre-adaptations toward such associations, and we may conceive of the sparrow–man commensalism developing repeatedly relatively early and as the development of man's settlements occurred. The full benefit of commensalism would have remained unavailable to sparrows as long as they were migratory, leaving the Near East in winter. The ancestor of the house sparrow is, of course, not known to have been migratory, but today no mid-latitude continental population of any *Passer* species but the house sparrow is permanently resident, so it is likely that the ancestor of the house sparrow was migratory also. In addition, the climate of the Near Eastern region of our concern was considerably cooler prior to 11 000 B.P., to judge by the pollen record (Wright, 1968). This increases the likelihood that the ancestors of the house sparrow were migrants. And, of course, the Spanish sparrow, a very close relative with whom the house sparrow shares a common genetic pool in many places, is today largely migratory, being a permanent resident only on Sardinia.

Overwintering in the Near East would have been under very strong positive selection for the following reasons: (1) a regular supply of grain would have been available to birds all winter; (2) hazards of migration, including moving to unfamiliar areas, setting up winter quarters and moving back in spring, would have been avoided; (3) optimal nest sites and nest-building materials for nesting around man would have been available to overwintering birds before returning migrants. Overwintering, or non-migratory, individuals would perhaps have lived under lower risk of death and would have had a reproductive advantage over migrants; if migrant and non-migrant phenotypes were genetically-based (which is likely), non-migratory commensals would have become predominant among birds around man's settlements. Such non-migratory individuals also would have tended to mate with others of like behavior and perhaps of similar genetic constitution, which could have been responsible for establishing partial genetic discontinuity between old-line migrants and the developing house-type sparrows. Old-line migrants naturally would have continued their old ways, probably nesting in riparian groves in a fashion like Spanish sparrows today. These points are summarized in Table 8.2 (p. 335) and Fig. 2.3.

Developing house sparrows, essentially in genetic contact with old-line birds but showing behavioral and physiologic differences from them, could

22

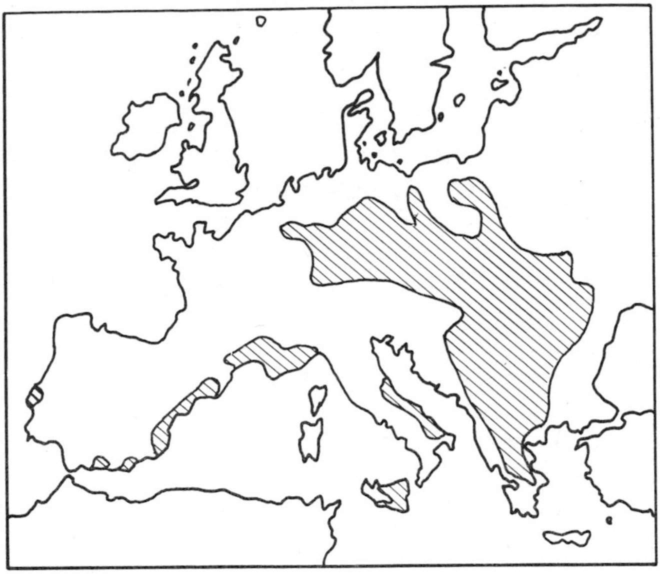

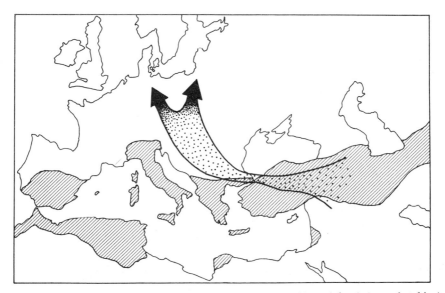

Fig. 2.3. Geographic distribution of agricultural communities in Europe, approximately 7000 B.P. Source: Piggott (1965).

Fig. 2.4. Current distribution of the Spanish sparrow *Passer hispaniolensis* (cross-hatching) and possible avenue of colonization of northern Europe by non-migratory populations of *P. hispaniolensis*-like stock in the period 10000–7000 B.P.

23

have existed for 2000 years before opportunity for further differentiation opened up to them. The opportunity came when man moved from the Near East into eastern and central Europe after the considerable climatic improvement in the period 10 000–5000 B.P. (Moberg, 1966). Then, for the first time the presumptive house sparrows would have met climatic conditions different from those to which they had previously been exposed (Fig. 2.4). Moving into central Europe would have thus intensified the relationship with man, at least as it concerned finding shelter in winter. It would also have rendered more tenuous the genetic link with the old-line sparrows back in Near Eastern and Mediterranean regions.

It may be questioned whether the developing house sparrows had the capacity to make such adjustments in periods of time that were in the order of a relatively few centuries. If we judge their capacity for evolutionary change by the response of English house sparrows introduced in 1852 into North America, they could well have made the adjustments. We anticipate certain details to be presented below in noting that North American house sparrows have elaborated considerable size and color differentiation in a period not greatly exceeding 100 years (Selander & Johnston, 1967; Johnston & Selander, 1971, 1973 a; Johnston, 1973 a; Kluge & Kerfoot, 1973). The differentiates are so organized relative to climate and geography that there is no question but that the differentiation is adaptive. We think, if essentially contemporary populations can have elaborated such a complex picture of evolutionary adjustment in a trivial number of generations, that the Mediterranean sparrow forebears of 10 000 B.P. could have done a similar job in following the migration of agricultural man both into northern Europe and into western India from his own origins in the Near East.

Evolutionary biology

Man brought the house sparrow *P. domesticus* to North America with the expectation that the species could maintain its human–commensal relationships and duplicate its European ecology in the New World. The species has been notably successful in realizing such expectation (Barrows, 1889; Summers-Smith, 1963), and today, some 120 years after the initial introductions of sparrows into New York City, breeding populations can be found (American Ornithologists' Union Check List, 1957) from sub-arctic Canada (around 61° N) to southern Mexico, with incipient colonization of sub-tropical Guatemala now occurring (around 15° N). The birds successfully live at Furnace Creek, Death Valley, below sea level, and at Leadville, Colorado, around 3100 m in elevation, as well as from the Atlantic to the Pacific. Such a distributional range in North America puts house sparrows into a great variety of temperature, precipitation and biotic regimes, which the birds' close association with man only partly ameliorates.

Ecologic success in North America is thus a reflection of countless local successes, each of which had to be evolved from the range of available variation in morphology, physiology and behavior present in the phenotypes of their predominantly English ancestors 12 decades ago. That this could be done in the grand manner is a trenchant comment on phenetic variation in a bird species, irrespective of whether such variation is to any degree genetically-based in a direct sense (as distinguished from being ecophenotypic). A trenchant comment demands examination, and this chapter accordingly provides a summary of information on phenetics and genetics of North American and European house sparrows. Much of what is to be summarized is work done only in the past 10 years, even though there has been persistent interest in house sparrows, at least of North America, ever since their introduction (Barrows, 1889; Bumpus, 1899; Kalmbach, 1940; Lack, 1940; Kendeigh, 1944; Calhoun, 1947). Interest owes partly to the bird's ubiquity as a human commensal, and thus to its potentially great economic importance in agricultural settings. But, interest also derives from the facts of introduction to North America and subsequent colonization of the continent – house sparrows afford us the avian paradigm of the evolutionary process.

This is possible largely because the apparent adaptation has occurred within the context of a natural experiment, which began with the inoculation of the species into New York in 1852, and wherein the European (= English) populations may serve as controls. Aside from this circumstance, house sparrows are not completely ideal subjects for evolutionary study. Firstly, they are not easy to breed in captivity, so genetic information is difficult to get except by electrophoretic means. Secondly, they live so close to man that it is not always clear precisely what environment the birds experience at a given locality; it is unlikely that sparrows living at Peace River, Canada, experience temperatures of $-50\,°C$ for more than a few minutes at a time, to judge from laboratory studies of Kendeigh (1944), but no one really knows. Nevertheless, the birds were treated by man in such a way as to make them now extremely useful for studies in descriptive and experimental evolutionary biology.

Dispersal

Barrows (1889) has details about the several introductions of house sparrows in eastern North America in the period 1850–1869. The most engaging anecdote is that of Mr Nicolas Pike with his 200 dollars, securing perhaps 100 birds that were ultimately liberated in Brooklyn and which, in all probability, were the first successful house sparrows in North

Table 2.3. *Representative distances of dispersal by house sparrows in North America in the nineteenth century*

Probable source	Distance and direction moved
New York City, New York[a]	Southerly, 48 km year⁻¹
	Westerly, 64 km year⁻¹
	Northerly, 120 km year⁻¹
San Francisco, California[a]	Eureka, California, 120 km year⁻¹
	Hollister, California, 30 km year⁻¹
Galveston, Texas[a]	Jefferson, Texas, 21 km year⁻¹
Topeka, Kansas[a]	Lawrence, Kansas, 14 km year⁻¹
Salt Lake City, Utah[b]	Cache Valley, Utah, 8 km year⁻¹
	Manti, Utah, 14 km year⁻¹
	Veyo, Utah, 14 km year⁻¹
Tucson, Arizona[c]	Yuma, Arizona, 48 km year⁻¹

[a] Barrows (1889).
[b] Selander (1950).
[c] Phillips, Marshall & Monson (1964).

America. This was in 1852; later inoculations are too numerous for our consideration here.

There is no satisfactory record of an earlier introduction, but the observations of Vik (1962) should indicate the potential of the species in undertaking dispersal movement in association with man. Vik tells of the journey of the M/S *Stavangfjord* from Oslo, Norway, to New York City, New York, in July 1960. Six house sparrows, all apparently birds-of-the-year, appeared on the ship on 10 July after leaving Kristiansand, Norway. Passengers fed and watered the birds. Three sparrows flew in the direction of nearby Cape Race and Sable Island on 16 July, and three flew toward Long Island on 17 July. Vik says the chief officer of the ship found the accompanying sparrows in no way unusual, although no mention was made of whether other sparrows had made an entire voyage at any earlier time.

Dispersal by the exploiting of man is not unusual for sparrows (summary in Summers-Smith, 1963) and it appears that such techniques may be necessary for continuing occupancy of marginal habitat. Lund (1956) thinks that the species could not live in Norwegian Finnmark (roughly 70° N) without using grain ships to cover the great distances of the wholly inhospitable environment that lie between man's settlements there.

Nineteenth-century North American data on movements away from inoculation sites (Barrows, 1889) suggest that house sparrows can maintain dispersal movement on the average of about 16 to 24 km year⁻¹ (Table 2.3). When dispersal was aided by man, average distances are less meaningful, but appear to be 48 km year⁻¹ southward from New York, 64 km year⁻¹ westward from New York, and 120 km year northward, within the first 34 years of their occupancy of North America (Barrows, 1889). The

Variation and evolution in the house sparrow

Table 2.4. *Representative distances of dispersal by house sparrows in Kansas, from 1874–1886*[a]

Place	Date	km from Topeka	km year^{-1}
Abilene	1884	128	12.8
Bronson	1884	136	13.6
Cherryvale	1884	192	19.2
Eldorado	1882	160	20.0
Eureka	1884	144	14.4
Fort Scott	1885	152	13.8
Iola	1885	128	11.6
Manhattan	1880	72	12.0
Oswego	1885	208	18.9
Parsons	1886	192	17.0
Stafford	1886	272	22.6
Wichita	1885	200	18.1

[a] Introduced to Topeka in 1874 from New York (Barrows, 1889).

estimate of around 16 km year^{-1} when unaided by man is supported by independent data of Selander (1950) for the sparrows of Utah, but in equally difficult country sparrows averaged 48 km year^{-1} in Arizona (Phillips, Marshall & Monson, 1964). Details for early dispersal in Kansas are shown in Table 2.4.

Continental occupancy of North America is generally considered to have been accomplished by 1900, but filling in and expansion of range has continued to occur. Sparrows reached Death Valley, California, in 1917 (Grinnell, 1919) and Mexico City, Mexico, by 1935. Movement southward is still occurring at temperate elevations; the species has recently been recorded at Quetzaltenango, Guatemala, 106 km south of Mexico, at an elevation of 2350 m (Thurber, 1972). Occupancy of settlements in northern North America is tenuous and erratic. A small colony persists at Churchill, Manitoba (Godfrey, 1966), but colonies in southern Northwest Territories, as at Hay River, are not permanent. Those on the southern fringes of James Bay in northern Ontario are also unable to maintain themselves every year, requiring dispersing birds from more southerly sites to maintain a population (J. C. Barlow, personal communication).

Observations by field naturalists at the Hastings Reservation of the University of California provide another view of dispersal capability by house sparrows (Davis, 1973). There is no colony of house sparrows at the Reservation and the nearest breeding group is 24 km away. Over a period of 34 years house sparrows are known to have appeared at the Reservation seven different times: 1939, 1962, 1967, 1968, 1971, 1973 and 1974. Four occurrences were in March or April (three by females), and four were in October (two by females, one by a male, and one by an unknown bird). All the visiting sparrows were wary of human observers and could not be closely approached; they spent their time perched in conspicuous places calling loudly and persistently. Twice the visiting individual stayed for two days, but the others were not seen beyond the first day. Such observations suggest a relatively regular, low-level potential for dispersal under present conditions in coastal California. In most instances there was just a single bird on the move. Taking only the records for the 1960s and 70s, there were eight individuals in the six years; if conditions for colonization of the Reservation by sparrows had been adequate, they would certainly be breeding there now.

This picture of movements presents both faces of adaptive dispersal. Firstly, sufficient long-distance (20–30 km/year) dispersal occurs that potential habitat is regularly monitored. Secondly, most sparrows actually do not move much (Summers-Smith, 1963). Such population structure fits readily into the conceptual frame provided by Sewall Wright long ago when he spoke of proliferation of 'adaptive peaks'.

Intralocality variation in size

Samples from Europe and the Americas are close in mean sizes for skin variables (Table 2.5), and this is true also for skeletal variables (Johnston, 1973 *a*). At any one locality the means for all variables show differences between years and between young and old individuals; this has been summarized for age and sex classes by Johnston (1973 *b*) for one locality sample from eastern Kansas. Intralocality variation in samples of fully-grown birds between years tends to be relatively small, especially when analyzed one character at a time, but it is not infrequently meaningful when assessed by multivariate statistics (Johnston, Niles & Rohwer, 1972). Storm-induced mortality can cause minor shifts in mean sizes of bones and other body parts, and in the samples examined by Bumpus (1899) surviving males were generally larger than those that perished, a distinction that is missed when single variables are examined one by one (Grant, 1972). The surviving females, on the other hand, showed less character variance for four size variables than did those that perished, whether assessed by univariate or multivariate techniques. This agrees with the classical textbook example of stabilizing selection in wild populations, in that individuals of either large or small extreme sizes perish more readily than individuals at or near mean size.

We should expect that size proportions of parts as well as overall size will be important in dictating survival in times of temperature stress, such as occurred with the Bumpus sparrows. One aspect of bodily proportions,

Table 2.5. *Unweighted character means (mm) for three continental samples of house sparrows* [a]

	Males			Females		
Character	Europe	North America	South America	Europe	North America	South America
Wing	76.90	77.37	77.29	74.13	74.64	74.62
Tail	58.66	57.85	58.65	56.86	55.98	56.99
Bill length	9.18	9.32	8.99	9.10	9.30	8.96
Bill width	7.34	7.14	7.04	7.33	7.15	7.05
Tarsus	18.66	17.95	18.19	17.82	17.82	17.52

[a] Sample sizes. Europe, 793 ♂♂, 558 ♀♀; North America, 2255 ♂♂, 1534 ♀♀; South America, 721 ♂♂, 468 ♀♀.

that between the size of body core and the length of limbs, will be further examined below, under a consideration of character correlations.

Variability in characters

Coefficients of variability for size characters tend to be similar within any character for all localities, irrespective of continental source. Thus, specimen samples from South America show mean coefficients of variability about equal to those for either North American or European samples (Table 2.6). There is a suggestion of slightly greater variability in tarsus length for American birds, but this is restricted to males and is in any event only on the order of 0.5%.

Character correlations

Correlations between pairs of variables within one sex at a locality vary with the age of birds in the sample (Table 2.7). This is especially true for samples of very young birds (cf. Ruprecht, 1968), but it is also the case for mature birds of different ages. The specimens on which Table 2.7 is based were all fully-grown individuals; those labelled 'adult' were about 17 months or more old, and those labelled 'sub-adults' were five to eight months old. The differences in the pairwise correlations for any character are trivial. However, in these samples the sub-adult birds tended to have slightly longer leg bones and slightly smaller sternal and coracoid dimensions than adult birds, and the univariate covariation between character sets, such as that for core and for limbs, tended to be different in the two samples. Presumably, the sub-adults represent a range of phenotypes turned out by adults the previous spring and summer and had faced

Table 2.6. *Mean coefficients of variability for five size characters in house sparrows from different continental sources*

Sex and character	Sample sources and sample sizes			
	Europe (35)	England– Germany (7)	North America (53)	South America (11)
Males				
Wing	2.06	2.08	1.94	1.88
Tail	2.81	2.85	2.71	2.58
Bill length	4.15	4.24	3.78	3.75
Bill width	3.41	2.86	3.47	3.85
Tarsus	3.59	3.61	3.84	4.17
Females				
Wing	2.04	1.94	1.91	1.98
Tail	2.71	2.65	2.78	2.84
Bill length	3.72	3.79	3.84	3.73
Bill width	3.52	3.24	3.46	3.62
Tarsus	3.94	3.94	3.82	4.07

Table 2.7. *Intercharacter correlation coefficients for variables of body core in two age classes of male house sparrows from eastern Kansas*[a]

		Adults (N = 44)			
		Coracoid	Sternal length	Keel length	Sternal depth
	Coracoid	—	0.491	0.477	0.423
sub-adults	Sternal length	0.552	—	0.823	0.313
(N = 76)	Keel length	0.436	0.792	—	0.229
	Sternal depth	0.459	0.543	0.367	—

[a] All coefficients exceed 0.300 for adults and 0.228 for sub-adults and show a significant correlation at the level of $P = 0.05$; sample taken October 1971 at Manhattan, Riley County, Kansas.

no wintertime size selection. Adults supposedly have been selected for the most adaptive proportions – large core but short limbs. The simple difference in covariation, which is evident in a number of bony character-sets for both sexes, is examined for core and limb variables of males in Tables 2.8 and 2.9. In the adult sample, limb bones and keel length showed significant correlations, and the only lack of correlation was between leg bones and depth of sternum. The adults thus seem to show a slightly greater amount of character covariation than sub-adults, but it is difficult to make a strong case on the basis of these data. We cannot be sure that surviving individuals of the sub-adult set would show, a year later,

Table 2.8. *Correlation coefficients between body core variables and limb bone variables in sub-adult male house sparrows from eastern Kansas* [a]

	Coracoid	Sternal length	Keel length	Sternal depth
Humerus length	0.644	0.391	0.206	0.354
Ulna length	0.600	0.380	0.214	0.308
Femur length	0.498	0.326	0.197	0.335
Tibiotarsus length	0.590	0.270	0.116	0.335
Tarsometatarsus length	0.617	0.241	0.139	0.281

[a] All coefficients except those for keel length exceed 0.228 and show a significant correlation at the level of $P = 0.05$; sample ($N = 76$) taken November 1971 at Manhattan, Riley County, Kansas.

Table 2.9. *Correlation coefficients between body core variables and limb bone variables in adult male house sparrows from eastern Kansas* [a]

	Coracoid	Sternal length	Keel length	Sternal depth
Humerus length	0.684	0.588	0.436	0.349
Ulna length	0.681	0.573	0.471	0.312
Femur length	0.711	0.498	0.359	0.256*
Tibiotarsus length	0.733	0.556	0.383	0.287*
Tarsometatarsus length	0.628	0.483	0.326	0.175*

[a] All coefficients except for those marked * exceed 0.300 and show a significant correlation at the level of $P = 0.05$; sample ($N = 44$) taken November 1971 at Manhattan, Riley County, Kansas.

correlations like those of the present adult sample. But our case here is only that adults and sub-adults show different character correlations between elements of the body core and the limbs. To examine the contention that the different sets of character correlations are significantly different between sub-adult and adult samples, and that the increase in adult covariation is somehow adaptive, we must employ a different analytic technique.

Canonical correlation analysis strongly suggests that the differences in correlations seen at a univariate level are not just occurring at random and may well be evidence of an increase in adaptive covariation in the adults. The canonical correlations given in Table 2.10 represent a form of general regression used to extract factors which are uncorrelated with each system but where pairs of factors have maximum correlations between systems, in this case systems of core and limb variables. The first variate maximizes the canonical correlation between the two data sets. The resulting coefficient will be the largest product-moment correlation

31

R. F. Johnston & W. J. Klitz

Table 2.10. *The first two canonical correlations between body core variables and limb bone variables in two age classes of male house sparrows from eastern Kansas* [a]

Sample	Canonical correlation coefficient (R)	R^2	% determination	
Sub-adults	1 = 0.697	0.486	48.6	52.9%
	2 = 0.288	0.083	4.3	
Adults	1 = 0.816	0.665	66.5	72.1%
	2 = 0.410	0.168	5.6	

[a] Variables as in Tables 2.5 and 2.6; sample taken November 1971 at Manhattan, Riley County, Kansas.

that can be developed between linear functions of the two systems. Then a second factor orthogonal to (uncorrelated with) the first is located (Cooley & Lohnes, 1971; Karr & James, 1975). The first two canonical correlations (of which only the first represents a statistically significant correlation) between core and limbs for the sub-adults account for only 53% of the total determination in the two sets of variables, whereas for adults the figure is 72%. Hence, fewer individuals in the adult sample depart from a certain relationship between core and limbs than in the sub-adult sample, bringing to mind what occurred in the Bumpus (1899) sparrows subjected to severe winter weather. Our present samples do not show significant differences in mean sizes of the bones or in their variances as measured at a univariate level, but there is a major decrease in the bivariate scatter of individuals about the line representing covariation between several core and limb variables. This may be taken as an index to reduction in overall variance in the adults relative to the sub-adults.

Sexual dimorphism

At any locality the sexes differ in all dimensional (Table 2.11) and most coloristic variables, but some of these differences are not significant between the sexes. A discriminant function analysis (Cooley & Lohnes, 1971) of specimens from North America (Fig. 2.5) shows a very highly significant difference in overall size between the sexes. The most important variables showing small mean differences are associated with the bill and the skull (Selander & Johnston, 1967: 228). But, body core and the flight structures always show relatively strong sexual dimorphism (Selander & Johnston, 1967; Johnston & Selander, 1971, 1973a; Johnston, 1973b). The various linear variables have slightly different relative differences in size between males and females. These dimensional differences in the aggregate amount to about 2.8% difference in body size between the large

32

Table 2.11. *Sexual dimorphism in sizes of five skin characters: percent of male mean sizes reached by female means*[a]

| | Females (% of male size) | | |
	Europe	North America	South America
Wing	96.39	96.47	96.54
Tail	96.93	96.77	97.16
Bill length	99.15	99.79	99.60
Bill width	99.85	100.14	100.07
Tarsus	99.09	99.28	96.35

[a] Sample sizes: Europe, 793 ♂♂, 558 ♀♀; North America, 2255 ♂♂, 1534 ♀♀; South America, 712 ♂♂, 468 ♀♀.

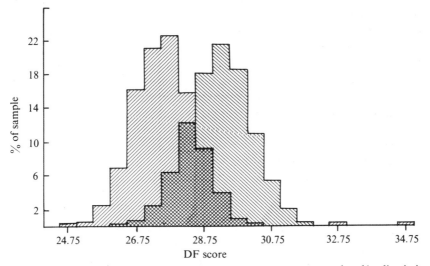

Fig. 2.5. Frequency histograms of overall size of house sparrows, as rendered by discriminant function (DF) scores on five variables of size from skin specimen samples from North America; left histogram, females, right histogram, males. Degree of overlap in histograms is indicated by double-hatching, and provides a visual measure of sexual-size overlap. Degree of sexual dimorphism in size shows a Mahalanobis distance of $D^2 = 6.854$ ($P \leqslant 0.001$). Males, $N = 2234$; females, $N = 1534$.

males relative to the smaller females. It would appear that at least some part of this difference is owing to classical Darwinian sexual selection (Selander & Johnston, 1967), and some part to strictly survival aspects of size in relation to fitness (Johnston *et al.*, 1972).

Young individuals in a sparrow population tend to be less sexually dimorphic in size than adults. Young females seem to grow less rapidly than young males; differential growth rates in the sexes could be related

33

to the advantages of rapid assumption of large (= adult) size by males, without a parallel advantage in females (Johnston, 1973 *a*). Such a differential growth rate between the sexes leads to a small number of variables (sternum length and depth, and humerus, coracoid and keel lengths) being sexually dimorphic in size when the birds are perhaps four months old. The same picture is obtained by looking at differences in sizes in adults versus young birds at a locality – most of the variables assessed in females are of different sizes in the two ages, but only about half are different in males (Johnston, 1973 *b*).

Interlocality variation

Interlocality variation refers to significant differences between mean values of variables in locality samples. For house sparrows we know that such variation includes both genetic and phenetic variables, the former of which is discussed elsewhere in this chapter. Phenetic variables known to vary geographically in sparrows include single-character size and color variables (Calhoun, 1947; Johnston & Selander, 1964, 1971, 1973 *b*; Packard, 1967; Johnston, 1969, 1973 *c*), multivariate assessments of size and proportions in dimensional variables (Johnston & Selander, 1973 *a*), frequency of albinotic feathers, of schizochroic rufous in the bib of males, and of assumption of definitive male plumage in first-year birds (Selander & Johnston, 1967). It is reasonable to expect any measurable quality in house sparrows to show geographic variation. Our account here will emphasize multivariate geographic variation.

Basis of multivariate study

As S. J. Gould (Gould & Johnston, 1972) has remarked, parts tend to be big in big animals. There is a strong tendency for covariation to occur in this sense as well as in the sense of morphological symmetry, which leads to high correlations between, say, the bony elements of the wing, as a reflection of their relationships to each other in the physiology and mechanics of flight. There is, in short, a great degree of correlation between bodily parts in individuals and in samples, owing presumably to both these sorts of adaptive covariation.

Such correlation or covariation entails a large amount of redundant information. If we look at variables one at a time the redundancy may be bothersome, but such redundancy is the very basis of some of the most powerful multivariate statistical assessments, such as principal component (PC) analysis (Sneath & Sokal, 1973; Cooley & Lohnes, 1971). Partly because high character correlations are found in house sparrows, population samples are nearly ideal for multivariate analysis of this sort.

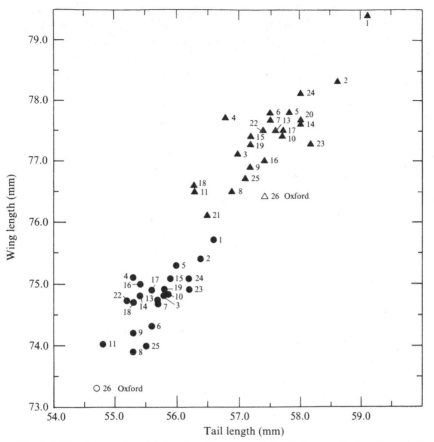

Fig. 2.6. Bivariate scatter plot showing a linear correlation between wing length and tail length in locality samples of house sparrows, mostly from North America; circles, females, triangles, males. Specimens from Oxford, England, are specially labelled and represent a stock of sparrows having close geographic ties with populations lineally ancestral to those in North America today. Locality code: (1) Edmonton, Canada; (2) Montreal, Canada; (3) Detroit, Michigan; (4) Roodhouse, Illinois; (5) Salt Lake City, Utah; (6) Lawrence, Kansas; (7) Vancouver, Canada; (8) Oakland, California; (9) Sacramento, California; (10) Los Angeles, California; (11) Death Valley, California; (13) Las Cruces, New Mexico; (14) Bastrop, Texas; (15) Austin, Texas; (16) Devine, Texas; (17) Houston, Texas; (18) Bishop, Texas; (19) Progreso, Texas; (20) Zachary, Louisiana; (21) Baton Rouge, Louisiana; (22) Gainesville, Florida; (23) Mexico City, Mexico; (24) Oaxaca City, Mexico; (25) Oahu, Hawaii; (26) Oxford, England. Males ($N = 25$), $r = 0.845$ ($P \leqslant 0.001$); females ($N = 23$), $r = 0.834$ ($P \leqslant 0.001$). Source: R. K. Selander & R. F. Johnston (unpublished).

Information on single-character correlations over large sets of specimens of house sparrows has been presented earlier in the form of phenograms (Johnston, 1969; Johnston & Selander, 1971), but here we wish to use a simpler technique, the bivariate scatter plot (Fig. 2.6). In addition to crudely showing geographic variation in feather length, the diagram shows a highly significant correlation between wing and tail lengths for each sex, and, additionally, a bivariate linear size continuum running from the point representing the smallest females (from Oxford, England) to that of the largest males (from Edmonton, Alberta). Knowing sample mean wing length, irrespective of sex, enables one to predict with great certainty the length of tail for that sample.

To rotate the axes and shift the origin of such a plot, as is entailed in PC analysis, would provide a new variable encompassing most of the variation in general 'feather length', and it might be worth the computational effort. But it would be a trivial exercise because the bivariate case is readily assessed by eye. However, such rotation and relocation, as that done in PC analysis, is not trivial for n-dimensional cases where $n > 2$. We emphasize these matters, firstly, to show that the nature of character covariation in house sparrows is ideal for examination by PC analysis and, secondly, to suggest the great amount of information that can be subsumed by the first few new axes in a PC analysis over one or two dozen variables. In house sparrows we regularly find the first principal component (PC I) representing 50% of the trace of variance over 16 variables, which means PC I has eight times the information content of any one of the sixteen.

By way of general explanation, the PC analysis (Pearson, 1901) extracts an axis (PC I) to give the linear combination of variables that accounts for the greatest amount of variation in the original data matrix. Subsequent principal components (II, III, etc.) statistically orthogonal to one another, account for progressively smaller amounts of variation. With this multivariate technique it is possible to consider simultaneously all the original variables.

Geographic variation in size

Sizes of most parts of house sparrows as well as their body weights vary geographically in highly predictable ways, depending on the continental source of specimen samples (Johnston, 1973 a). Variation in single variables has been depicted by projections of pie-diagrams onto maps of western Europe (Johnston, 1969) and of North America (Johnston & Selander, 1971), by isolines on maps of North America (Johnston & Selander, 1971, 1973 b), by linear regression plots against some environmental variables (Johnston & Selander, 1964; Packard, 1967; Johnston, 1969), and by tabular mean summaries with indications of statistically homogeneous subsets ('STP') (Johnston, 1969; Johnston & Selander, 1971). These show that there is a general increase in size of parts to the north, or with a decrease in winter temperatures, in North America, and they show the reverse in Europe.

Overall size, as is given by PC I in a PC analysis, has been depicted by isolines on maps and by linear regression plots (Johnston & Selander, 1971, 1973 a; Gould & Johnston, 1972; Johnston, 1973 a). Information content of such plots or maps can be as much as 71 % over six variables (Johnston *et al.*, 1972), 60% over 16 variables (Johnston, 1973 a) and 35% over five variables (Johnston & Selander, 1973 a). PC I shows that general size increases with latitude in North America and decreases with latitude in Europe, a peculiarity briefly discussed elsewhere (Johnston & Selander, 1973 a: 377–8; see also page 120, this volume).

Geographic variation in proportions

Proportional relationships in skeletal and skin variables have been ex-amined through PC analysis, and the graphic and tabular data summarizing the nature of PC II and PC III may be found in the publications dealing with PC I cited above. The most informative proportional relationship thus far demonstrated is that in PC II for skeletal variables. In PC II for house sparrows of any continental source, we see the sharp rendition of a contrast between core and limb dimensions: in any sample or set of samples proportional dimensions of core and limbs vary inversely to each other. Thus, either large or small individuals at high latitudes tend to have relatively short limb bones in view of their body core dimensions, and birds at low latitudes have relatively enormous limb bone lengths compared with their body cores. Geographic variation in PC II is the same in Europe and North America in spite of the fact that the size vectors run counter to another in the two continental populations.

Another aspect of proportions is that between the sizes of the sexes in house sparrows. Sexual dimorphism in size can be depicted in a number of ways, but plotting scores along a discriminant axis is very nearly optimal. Fig. 2.5 presents such a plot, derived from a discriminant function analysis over the five skin variables for 3768 specimens from North America (Johnston & Selander, 1973 a). D^2, the conventional Mahalanobis measure of distance between taxonomic units over the five skin variables, varies predictably in North America and Europe – sexual distance in-creases with increases in latitude or with decreases in locality mean winter temperatures. Recent assessments of sexual distance over 14 skeletal variables of the samples from North America and Europe show the same pattern. In the skeletal samples it is particularly clear that, since head and beak sizes vary completely in parallel with overall body sizes, the foraging tools of sparrows in the far north are more different in sizes than those of birds in the south: it is inescapable that house sparrows are, in fact, engaged in enlarging the food niche at high latitudes, and the five-variable data base used earlier simply was not powerful enough to support the

37

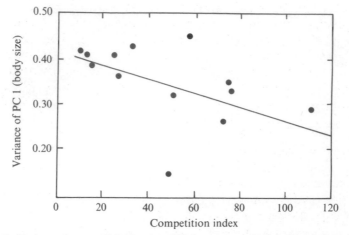

Fig. 2.7. Variance in overall body size (PC 1, sexes pooled) in North American house sparrows as a function of interspecific competition for seeds in winter. Spearman's rank correlation coefficient is $r_s \leqslant -0.761$ ($P = 0.01$, $N = 13$). The competition index is based on winter densities of competing bird granivores, the fraction of the diet in a competitor identical to that of the house sparrow, and mean body weight of the competitor species (Hamilton, 1974).

conclusion, attractive though it was at the time (Johnston & Selander, 1973 a).

Generalized character variance

Within each sex for each single variable or for any principal component there is homogeneity of variance (Selander & Johnston, 1967; Hamilton, 1974). However, variance of PC I locality mean scores for the pooled sexes increases with an increase in latitude, as might have been expected knowing the pattern of geographic variation in sexual D^2. As Hamilton points out, for both European and North American house sparrows the increase in pooled generalized variance at high latitudes provides an exception to Fretwell's (1972 a) theory of body size in seed-eating sparrows. Our explanation of increased character variance at high latitudes involves resource partitioning by birds of different sizes and is dependent on a progressive release from interspecific competition by other granivorous birds at high latitudes in winter (Fig. 2.7).

Geographic variation in plumages

Fresh fall plumage in house sparrows varies in a predictable fashion over gradients aligned on geography. Spectrophotometric analysis shows brightness (CIE Y values: Judd, 1950) to vary clinally as follows: nape and flank for females in Italy are darker to the north and under conditions of high precipitation, paler to the south and where precipitation is light, with a parallel cline for brightness in pileum of males. For brightness in breast plumage of females in North America, the cline runs from north-eastern stations to southwestern ones, showing the same relationship to precipitation schedules as did the Italian birds (Johnston, 1973 c). The overall pattern of variation conforms to Gloger's ecogeographic rule (Bowers, 1956) for fresh fall plumage colors.

As the feathers wear and are subjected to sunlight, their colors change. In samples from plains grassland localities from eastern Kansas to eastern Colorado, a strong regression relationship is seen for nape or flank brightness Y on elevation, if specimens are secured in autumn, with unabraded and unfaded feathers. Specimen samples taken in February, May and July show progressive deterioration of the regression relationship. Color adaptation thus seems to occur mainly at the time new feathers appear in autumn and through early winter.

Genetic variation in Europe and North America

House sparrows are not readily bred in captivity (Mitchell & Hayes, 1973), so few genetic data are available for the species. Empiric phenetic data from analysis of electrophoretically-mobile enzymes and proteins have been gathered, however, and these phenotypic data fit simple models of genetic inheritance (Klitz, 1972). We have little doubt that if sparrows were test-mated the results would support contentions that stained products on gels are phenotypic features bearing one-to-one relationships to allelic pairs at genetic loci. This would both conform to expectation (Smithies, 1955; Markert & Moller, 1959; Lewontin & Hubby, 1966) and provide us with a technique less complex than cross-mating for getting genetic information on this important animal.

It is therefore one of the ironies of nature that house sparrows prove to be among the more genetically monomorphic organisms that have been examined electrophoretically. Of the 23 genetic loci that have been examined and for which some electrophoretic mobility has been found (Table 2.12), only 17% have proven polymorphic (cf. Selander & Johnston, 1973); this is less than in such organisms as mice *Mus musculus* L. (Selander, Hunt & Yang, 1969), or fruit flies *Drosophila pseudoobscura* Frolowa (Lewontin & Hubby, 1966), in which around 30–35% polymorphism is usual.

From one locality to the next the range in polymorphisms is anywhere from 13 to 22%, with higher percentages relatively common in European samples and the lower percentage characteristic of those from North America. Mean heterozygosity per individual in 10 samples from Europe

Table 2.12. *Allelic frequencies at esterase and transferrin loci in house sparrows from Europe and North America, sexes combined*

Locus	Allele	Ten European localities ($N = 122$)	Lawrence, Kansas ($N = 36$)	Manhattan, Kansas ($N = 225$)	Point Reyes, California ($N = 157$)
Esterase	$Es-1^f$	0.59	0.54	0.533	0.484
	$Es-1^s$	0.41	0.46	0.434	0.433
	$Es-1^c$	—	—	0.033	0.083
		($N = 140$)	($N = 36$)	($N = 223$)	($N = 114$)
Transferrin	$Tr-1^a$	0.60	0.58	0.610	0.380
	$Tr-1^b$	0.30	0.29	0.305	0.350
	$Tr-1^c$	0.10	0.13	0.085	0.270

Table 2.13. *Occurrence of esterase and transferrin genotypes in house sparrows from Manhattan, Kansas, October 1971*

	Genotypes	Females		Males	
		SNCO[a]	SCO[b]	SNCO	SCO
Esterase	$Es-1^{ff}$	20	3	22	15
	$Es-1^{fs}$	36	18	38	20
	$Es-1^{ss}$	9	5	16	8
	$Es-1^{fc}$	3	3	1	0
	$Es-1^{sc}$	3	0	3	2
	$Es-1^{cc}$	0	0	0	0
Total		71	29	80	45
Transferrin	$Tr-1^{aa}$	9	2	9	6
	$Tr-1^{ab}$	26	11	25	12
	$Tr-1^{bb}$	28	14	30	17
	$Tr-1^{ac}$	4	1	0	0
	$Tr-1^{bc}$	3	0	13	6
	$Tr-1^{cc}$	1	1	1	1
Total		71	29	78	42

[a] SNCO – skull not completely ossified.
[b] SCO – skull completely ossified.

was 0.059, ranging from 0.047 to 0.075. For three North American samples only 0.045 of the loci of an average individual were estimated to be heterozygous.

An earlier study (Klitz, unpublished) using a smaller and slightly different set of loci provided similar results. For 20 loci in 400 specimens from 13 North American localities, less than 10% of the loci were, on the average, polymorphic. At that time only one clearly polymorphic enzyme was detectable, anodally-migrating esterases leaving several closely-spaced bands on gels. The mode of inheritance could not be deduced from the patterns, but they were nonetheless repeatable and probably represented the products of one or two polymorphic loci. This study is in no way inconsistent with the one mentioned just above, and we may say that house sparrows have a tendency toward genic monomorphism.

In this respect they are like some species of the rodent genus *Dipodomys* (Johnson & Selander, 1971) and another passerine bird *Zonotrichia capensis* (Müller) (Nottebohm & Selander, 1972). It would be useful to have a second estimate of the amount of genic variability in house sparrows, and one that would be largely orthogonal to the electrophoretic approach, since it may be that house sparrows are to some extent intractable to electrophoretic techniques.

Geographic variation in gene frequencies

Of the five genes found to be polymorphic in Europe, three were segregating at comparable and intermediate frequencies between localities, but the other two had near-fixation of the common allele (Table 2.13). A 6-phosphogluconate dehydrogenase (PGD) locus segregated three alleles at most localities in Europe (Fig. 2.8), the exceptions occurring only in small samples. The Italian sample, although of identical allelic composition, showed a somewhat different pattern from other localities, the two most common alleles elsewhere becoming the least and second most common in Italy. A phosphoglucomutase (PGM) locus from liver revealed two different rare alleles, although never more than one of these occurred at a given locality. One of these rare alleles was highest in Italy ($f = 0.07$) and dropped to 0.02 and 0.01 in Geneva, Switzerland, and Tour du Valat, France, respectively, the nearest adjacent localities. A malic dehydrogenase (MDH) locus sporadically segregated a rare allele at four of the eight localities at which this locus was recorded. The PGM, PGD and MDH loci were monomorphic in North America.

A kidney esterase and a transferrin polymorphism are the only ones known from North American samples. Since the samples are from three localities and are fairly robust in size (Table 2.12), it is not likely that the three other polymorphisms characteristic of European samples were missed. The American esterase and transferrin seem to have the same alleles as the European material, with the exception of a rare esterase allele. The frequencies of the alleles at both loci show no significant differences between the Kansan and the pooled European samples. However, at Point Reyes, California, the esterase *c* allele (*Es-1ᶜ*) shows an increase in frequency over that at Manhattan, Kansas, and all three transferrin allelic frequencies at Point Reyes are different from the others.

41

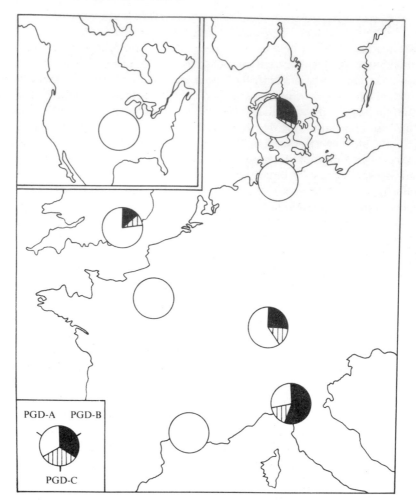

Fig. 2.8. Geographic distribution of frequencies of three alleles of the 6-phosphogluconate dehydrogenase (PGD) locus segregating in Europe. Localities from north to south are Aarhus, Denmark ($N = 27$); Hamburg, Germany ($N = 27$); London, England ($N = 40$); Paris, France ($N = 36$); Radolfzell, Germany ($N = 30$); Parma, Italy ($N = 21$); and Tour du Valat, France ($N = 42$).

Diverse aspects of evolutionary potential

Preceding material shows that house sparrows in North America are phenetically and genetically different from their European ancestors. Two aspects of such differentiation are important to consider now – firstly, is the evident change small or large and, secondly, how much of the change is actually evolutionary descent with modification and how much is ecophenotypic? There are some fragmentary and largely indirect answers to these questions, which will be summarized herewith.

The amount of phenetic differentiation

North American populations of house sparrows are judged to have the same degree of intrapopulation character variance as have their European forebears from western and southern sectors of Europe; but measures of interlocality variability are less for the American than for the European birds (Johnston & Selander, 1971, 1973 *a*). Wing length, tail length, bill length and bill width are especially constrained in geographic variation in the American samples. A possible cause for such modest interlocality variability is that there was a restriction on available variability in the birds making up the inoculum samples of the 1850s and later. Such a restriction could conceivably be owing to the founder effect (low variance in the inocula, due to chance trapping effort), or to restricted variation in the source itself (low variance in the populations from which the inocula originated). Expectation *a priori* favors the latter, since the number of sparrows introduced in the 1850s was large, in the thousands, which very likely rules out the founder effect.

An analytical approach to this problem is that taken by Kluge & Kerfoot (1973), which compares the amount of intersample variation with the amount of intrasample variation. Kluge & Kerfoot assume that a large intrasample variance will permit development of large intersample variability, and that small intrasample variance will constrain intersample variability. The necessary data for the examination are plotted in Fig. 2.9. Both the North American and European samples provide a reasonably similar relationship between ranges in character-state means and the standard deviations of the appropriate characters. However, the isolated English and German data show restricted range in character-state means for a given standard deviation, using the other samples as a gauge.

An interpretation of the information in Fig. 2.9 is that there was relatively little interpopulation variability in the English and German birds making up the source stock of American sparrows in the last century. They probably had 'normal' intralocality variance, and it is presumably out of this, rather more restricted, variability that the American birds have elaborated their character differentiation. Thus, the evident restriction on variability seen in North American samples is probably a consequence of the considerable homogeneity in size of the English–German sources: the European populations were not randomly assorted and the selection of inocula was only from one segment of them. Perhaps it is surprising that the American birds show as much variation as they do, and in this regard it is important to note that the American samples come from a geographic area approximately four times the size of the area sampled in Europe, as well as from climatic environments significantly more diverse than the European. In any event, it is now critical to know how much of this evident variation is genetically-based and how much is ecophenotypic.

43

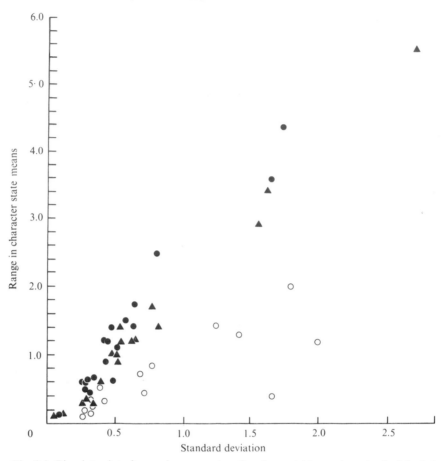

Fig. 2.9. Bivariate plot of range in mean values for size variables against standard deviation for each such mean, for 18 variables in North American specimens (solid triangles), 19 variables in European specimens (solid circles), and 15 variables in English–German specimens (open circles), of house sparrows.

The relationship between phenetic and genetic differentiation

A thorough assessment of this relationship requires a robust program in quantitative genetics, which is not feasible because the birds are difficult to breed in captivity. This difficulty results partly from lack of attention by aviculturalists to a species that is largely considered a pest and partly from the birds not being ideally suited to cage life; Mitchell & Hayes (1973) reported very great nestling mortality and an overall breeding success of 12% in their aviary studies in Texas. No genetic studies in which test-matings are a critical feature have been attempted.

Two approaches have been used to estimate the likelihood of the phenetics of house sparrows being appreciably influenced by direct genetic coding. The first employed six replications of PC analyses on large specimen sets over 16 skeletal size variables (Johnston, 1973 *a*). The specimen sources were both North American (four samples) and European (two samples). The assumption was that if character loadings on the first few PCs were similar or identical, such similarity would have been achieved through some means other than environmental induction of an ecophenotypic nature, for the samples originated in three distinct climatic, physiographic and biotic environments.

The character loadings were clearly similar for PC I and II but were vaguely alike or different for PC III. Consensus of opinion is that since PC I summates the size variables uniformly (Blackith & Reyment, 1971), no special significance pertains to identity in PC I for the six replicates; and PC III refers to such a small fraction of the trace of variance, it is not significant if we find variation in character loads. But PC II, which uniformly summarized the inverse relationship between body core and limb proportions, in both sexes and in samples from both continental sources, must certainly represent results of products of a common genetic pool. So, some part of the measured phenetics was reasonably directly genetically grounded, and some part of the phenetic variation seen in the North American samples is interpreted to be a consequence of genetic variation.

Another approach to the problem is through multivariate statistical techniques, employed on specimen samples for which there is some biochemical genotypic information. The assumptions in all the exercises undertaken were that specimens sorted to one genotype of one genetic locus would be to some extent more phenetically homogeneous than a sample composed of specimens unsorted for genotypes of that locus; and, that if there is linkage between a known gene and others coding for gross morphology, then less gross morphological heterogeneity should also be evident in a specimen sample sorted to one biochemical genotype.

Specimens used were part of those described earlier, from one flock at Manhattan, Kansas (Table 2.13). Genotypic information on a polymorphic serum transferrin and a polymorphic kidney esterase was obtained, along with data on size of 33 morphological variables. Biometric techniques employed were multiple discriminant analysis, PC analysis, and canonical correlation analysis, in all instances looking for those specimen sets showing greatest covariation among genetic and phenetic variables.

Classification analysis

The three most common esterase genotypes were used as input groups and some combination of the 33 size variables were used as discriminators.

45

R. F. Johnston & W. J. Klitz

Table 2.14. *Classification analysis of female house sparrows from Manhattan, Kansas: gross morphology on esterase genotypes* [a]

Original input group	Number of specimens classified into esterase genotypes			
	$Es-1^{ff}$	$Es-1^{fs}$	$Es-1^{ss}$	% correct
$Es-1^{ff}$	17	5	1	74
$Es-1^{fs}$	9	37	8	69
$Es-1^{ss}$	2	2	10	71

[a] After entry number 14 of a step-wise classification analysis using principal components (PCs) I–XXXIII of 33 variables of gross morphology as discriminators; the F-ratio of the U-statistic is 2.54 (d.f. 28 and 150). The variables loading on the first five PCs to enter are bill length versus foramen magnum diameter; wing length; gut and liver weight versus foramen magnum diameter; hallux length; scapula length.

Raw variables enabled up to 60% of the specimens to be correctly classified back to input groups; variables transformed to principal components enabled classifications to be from 69% to 74% correct (Table 2.14). Tests employing 'fake genotypes' gave inadequate discrimination and assigned specimens essentially at random to original input groups.

Canonical correlation analysis

Specimen sets, either sorted to a genotype or consisting of undifferentiated samples of one sex and age, were investigated for relative degree of morphological covariation shown between two sets of size variables. These variables were four from the body core skeleton and five from the appendicular skeleton, known to vary inversely to one another in size over a very broad latitudinal range.

The largest canonical correlations were obtained for genotypically homogeneous specimens of adult birds. Sub-adults, whether of a given genotype or unsorted, had lesser degrees of correlation between core and limb elements than adults; but genotypic groups again tended to have higher canonical correlations between core and limbs than undifferentiated specimen sets (Table 2.15).

Probabilities that a genetic locus coding for a biochemical character will be linked otherwise with a specific set of phenetic variables of size and proportion are no doubt relatively small, and yet this seems to have been found in the Manhattan sparrow sample at an esterase locus. The relationship seems to be a real one, and not fortuitous, on the basis of the following considerations.

(1) Classification of specimens back to original input groups by dis-

Variation and evolution in the house sparrow

Table 2.15. *The canonical correlation between body core and limb variables in samples of male house sparrows from Manhattan, Kansas*

Age and sample size	Canonical correlation R		R^2	Cumulative % determination	Chi-square d.f.	
Adults						
All (44)	I	0.818**	0.669	66.9	55.7	20
	II	0.482	0.232	74.6	13.7	12
$Es\text{-}1^{fs}$ (20)	I	0.823*	0.678	67.8	33.9	20
	II	0.781	0.610	87.4	18.0	12
Sub-adults						
All (75)	I	0.669**	0.489	48.9	56.0	20
	II	0.263	0.069	52.4	9.8	12
$Es\text{-}1^{fs}$	I	0.631	0.398	39.8	23.9	20
	II	0.359	0.129	47.5	7.7	12

* $P \leqslant 0.05$;   ** $P \leqslant 0.005$.

criminant analysis was as much as 74% accurate, if the input groups were real genotypes.

(2) Deliberate destruction of organization, by randomly assigning specimens to 'fake genotypes', resulted in inept and inaccurate classifications.

(3) The transferrin genotypes were not adequately classified by any analysis used, which is to some extent reassuring in that it would have been entirely beyond any reasonable expectation to have found two biochemical loci bearing indirect linkage to gross morphology, when the two loci are the only ones presently known to be polymorphic.

(4) Expectation was fully realized in the tests undertaken by canonical correlation analysis – a genotypic sample tended to show greater morphologic homogeneity than an unsorted sample of the same provenance.

Some fraction of the genic elements that elaborate sizes and proportions in house sparrows thus seems to be monitored by a genetic locus coding directly for an esterase. Such a relationship is independent support for the idea that a significant part of the variation in sizes and proportions of house sparrows is under genetic control. This means that modifications in sizes and proportions of the birds subsequent to their introduction into North America must certainly have been instances of evolutionary, rather than ecophenotypic, change.

Significance of character variation to commensal granivory

We have now reached a position from which we may ask what the preceding material has to say about the house sparrow as a commensal granivore and about commensal granivory in general. There are a number of distinct lines to examine.

Basic evolutionary adjustment

The house sparrow seems clearly a product of interaction between sparrows and sedentary, agricultural man. Hence, there are on the order of 10000 to 11000 years of successful association between man and sparrow. This is a long-term relationship, sufficient to allow fine-grain adaptation of the birds to man's way. A number of other bird species have had similar temporal opportunities to make adjustments to man, and we need especially to note *Passer hispaniolensis, P. montanus, P. iagoensis, P. griseus, P. castanopterus, P. melanurus, P. rutilans, P. luteus, Streptopelia decaocto* (Friv.), *Agelaius* spp. and *Columba livia* Gmelin. In no sense could the complexity of adaptation shown by *P. domesticus* have developed without antecedent behavioral and ecologic trends in this direction, nor is it likely that so many members of *Passer* could have come to a similar ecology by sheer chance or by convergence: the *Passer* commensals share a pre-adapted common ancestor.

Capacity for complex size adaptation

Significant north–south clinal size variation occurs in both North American and European populations of house sparrows. The selective agencies are complex and by no means understood, for the size clines run counter to each other on the two continents. Likely selective pressures must include a set of variable levels of competition from other granivores, and most importantly must include periodic wintertime interspecific competitive release on both continents. It is very likely the occurrence at low and mid-latitudes in Europe of a relatively large congener (*P. hispaniolensis*) and a relatively small one (*P. montanus*) generated a need for solutions by European house sparrows different from those of the Americas, and it is a mark of characteristic flexibility that the different solutions seem equally viable.

Capacity for rapid character modification

Large variance in character variables permits rapid response to selection for change in size, color, rates, etc. European house sparrows generally have about twice as much interlocality variance as the North and South American house sparrows, but the latter, nevertheless, have the capacity of elaborating at least a first-order approximation of adaptive geographic variation in an absolutely short time. Calhoun's (1947) indication of around a 1 mm increase in wing length in the first 55 years the sparrows lived in the northeastern sector of the USA provides one estimate of capacity for change, and Fig. 2.6 can be used to obtain an estimate of 2.5–3 mm increase in wing length for sparrows anywhere in North America over the first 110 years of their occupancy there. At least some part of the size modifications discussed here seems to be conventionally based in the genotype, but environmental induction probably also is significant.

Capacity for variable niche-partitioning

Outside constraints on the dimensions of the house sparrow niche vary and permit sparrows to adjust to such lability if they can. House sparrows at high latitudes in both Europe and North America respond to intraspecific pressures, seemingly on the food resource, by increasing variance of sizes of bill and head, which increases the potential range in seed sizes and overall amount of food available to the population. This opportunity is not allowed at low latitudes, and intraspecific competition for food must be met in other ways, as yet not discerned but clearly not dependent on consequences of large character variance.

Capacity for long-distance dispersal

Current populations of house sparrows are known to produce individuals capable of moving away from the home colony (cf. discussion in Summers-Smith, 1963). The periodic low-level appearance of house sparrows at the Hastings Reservation, California, is instructive in a quantitative way, and the experience of actual colonization by sparrows of offshore islands cultivated during World War II by the British is qualitatively impressive (Summers-Smith, 1963). Furthermore, nineteenth-century rates of colonization of the American west had a low-level average of about 15 km year^{-1}, a rate we know *a posteriori* to have been thoroughly adequate for exploring every available human commensal opportunity at that time.

It seems clear from this examination that house sparrows are adjusted in a number of ways toward being a successful commensal granivore. In that sense it is irrelevant that they may not serve as the model on which we might focus thinking about other species practising commensal granivory – they are one apex of such granivory and have an enormous potential for maintaining such adaptation. But other species will not make the same impress on man's agriculture or his life in general. The rock dove apparently was not 'wild' enough and now, although it can exploit grain and nesting sites in a similar way to the house sparrow (see Chapter 8, this volume), it really is a domestic species, more exploited by man than it exploits man. And, from one species to another, the story is always one in which some aspect of commensalism is less than that recorded for house sparrows. But if sparrows do not actually serve as the paradigmatic commensal granivore, they show how complicated and thoroughgoing the set of adaptations can be, or, indeed, needs to be.

Summary

(1) Man's cereal monoculture, grain storage and feeding of livestock are such to encourage commensal granivory by other animals. The house sparrow *P. domesticus* is examined as the most thoroughly evolved bird commensal, a status achieved in the past 10000 years. Although all current species of *Passer* seem to be capable of grain-eating commensalism, some are not commensals, and only the house sparrow has an interpretable fossil record. The species is known to have occurred 10000 to 15000 years ago in one region of man's development of a sedentary agriculture based on wheat and barley. To judge by their behavior today, house sparrows were likely to have adapted to the new commensal niche almost as soon as man created it.

(2) The house sparrow has relatively small effective population sizes but is capable each generation of undertaking considerable dispersal movement. Such capability allowed exploration and finding of all available colony sites in a wholly adequate fashion in nineteenth-century North America.

(3) Complex size adaptation, part of it genetic, is characteristic of house sparrows. The well-known north–south size cline and the inverse relationship of body core to limb length in North America are only parts of the complexity in size adjustment.

(4) Size and other variables are modified rapidly, in relatively few generations. The basis for this is seemingly permissive character variance within localities. As an instance, wing length has increased 3 mm in mean size in no more than 110 generations in North American house sparrows relative to the mean for English birds.

(5) Heavy interspecific competition, at least from other kinds of birds for food, at southern latitudes constrains character variance and allows little secondary sexual size dimorphism, a situation clearly relaxed at higher latitudes, supporting the niche-variation hypothesis.

(6) Coincidences between variation in enzyme biochemistry and gross variation in size strongly suggest linkage disequilibrium between a locus for a kidney esterase and loci for general body size. This is evidence for a genetic basis for size variation; thus we know that size changes in North American house sparrows in the past 120 generations have been evolutionary changes. This conclusion is related to the significance of points (3) and (4) above.

(7) Granivorous commensalism in house sparrows therefore clearly requires a complex set of adaptations, important behavioral components of which have not been examined here. Other species are likely also to require reasonably complex adaptive character complexes should they attempt to move completely to commensal granivory.

Drafts of this paper were read at different stages of completion by most members of the several conferences of the Working Group, and the comments generated this way usually resulted in improvements in the manuscript. Frances C. James, Ned K. Johnson and Robert K. Selander read late drafts of the paper and provided helpful criticism. Fiscal support was provided by the National Science Foundation (GB 35906X) and the General Research Fund of The University of Kansas. We very much appreciate the several kinds of assistance provided by our colleagues and the public agencies.

3. Population dynamics

M. I. DYER, J. PINOWSKI & B. PINOWSKA

The behavioral evolution and dispersal of the house sparrow *Passer domesticus* (L.), has been shown to parallel the cultural evolution and dispersal of man (Chapter 2, this volume) as the bird species developed a close association for food and shelter. It is, therefore, important that the basic functions pertaining to population dynamics of the house sparrow and other granivorous species of economic importance be analysed so that we assemble information that may lend support to theoretical considerations of bird populations and also aid understanding of resource relatives. While there are few or no data to consider in the same light of the house sparrow–man association for these other bird species, certain important attributes that allow for an association with man are present (see Chapter 8, this volume). It is possible that remote links have existed for the other species, and thus it is of heuristic interest to document as much as we can about population dynamics of a few other granivorous species. We give special attention, therefore, to the house sparrow; the European tree sparrow *Passer montanus* (L.); the Spanish sparrow *P. hispaniolensis* (Temminck); the red-winged blackbird *Agelaius phoeniceus* (L.); the common grackle *Quiscalus quiscula* (L.); and the brown-headed cowbird *Molothrus ater* (Boddaert). The house sparrow is nearly cosmopolitan in distribution; the other species are largely restricted to single continents (Chapter 1, this volume).

The parameters to be developed deal primarily with natality and mortality functions as they affect the number of young raised to the fledging stage and their rate of survival thereafter. The ultimate objectives are quantitative data on the number of young produced per year (productivity), the rate of population turnover, how the parameters vary geographically and temporally and explanations of the phenomena.

The genus *Passer*

Social structure

The house sparrow is much more attached to its nest and nesting area throughout the year than is the European tree sparrow. *P. domesticus* leaves its nesting sites only during late summer, when flocks range into the local countryside to feed. After this flocking phase, house sparrows return to their individual nesting areas in September and October. At this time they often rebuild their nests; this activity can occur at any time throughout the year (Summers-Smith, 1963). As house sparrow young leave the nest, they congregate into flocks of increasing size throughout the summer and feed in agricultural fields containing barley, wheat, grain sorghum and oats. Later in autumn, flocks start to decrease in size. Rarely do they range more than 1 to 2 km from their roosting centers (Gersdorf, 1955; Speyer, 1956; Summers-Smith, 1956; Fallet, 1958 a; North, 1968, 1973 b).

In late summer the roosting centers are concentrated in deciduous trees in the countryside; during the fall and winter, village populations roost in houses and farm buildings whereas birds that live in larger cities concentrate in bare deciduous trees in the center of the city, often numbering in the thousands in these latter locations (Moreau, 1931; Fallet, 1958 a; Graczyk, 1961; Dawson, 1967; Will, 1969). One race, *P. domesticus bactrianus* Zar. and Kud., routinely undertakes typical migration between central and south-central Asia and the southeastern part of the continent. It enters its breeding grounds in April and early May, raises one brood, gathers in large flocks for the rest of the summer, and leaves in September and October. During its stay on the breeding grounds it roosts in large numbers in deciduous trees and forages in grain fields and open steppe (Rustamov, 1958; Gavrilov, É. I., 1974).

As adult tree sparrows complete their breeding cycle and as juveniles mature, flocks slowly increase in size. At this time the flocks invade grain fields when seeds are ripening. From September to early November the flocks decrease in size as the tree sparrow returns to its nesting sites and participates in early morning sexual displays. This autumnal sexual behavior continues until cold weather ensues (Pinowski, 1965 b, 1966, 1967 a). Eggs are occasionally laid during this period (Hasse, 1962). At the end of this autumnal period of sexual activity, the flocks are reformed and the birds return to roosting in nest boxes in trees nearby. This roosting activity lasts until spring when the breeding season recommences. During this period, flocks leave the colony at sunrise to forage in surrounding fields and return to the roost at sunset. The autumnal sexual display period is important in that it provides an opportunity for young birds of the year to select a nesting site to be used the following breeding season.

In the spring, tree sparrows re-enact the sexual displays of the previous autumn and the breeding season commences (Pinowski, 1965 b, 1966, 1967 a). Often house sparrows will build winter nests, but for tree sparrows the nest building in the autumn is closely connected only with the activities of the autumnal sexual activity.

Both species defend their nests against other sparrows, the defense being mainly intrasexual, i.e. males repel males and females repel females. Males will accept females other than their mates during this period (Summers-Smith, 1963; Pinowski, 1967 a), but polygyny is extremely rare for both species (Summers-Smith, 1954; Pinowski, personal observation).

In central Poland, which constitutes an optimal area for *P. montanus*, autumn- and winter-foraging flocks are usually large. However, when the snow cover reaches a depth of 15 to 20 cm, much of their potential food supply becomes unavailable and the flocks break up into much smaller social units that scatter widely in their search for food. On the edge of their range, or in less desirable biotopes, especially in mountainous areas subject to severe winters, flocks are never large (Pinowski, 1964, 1968). In contrast with the house sparrow, the European tree sparrow tends to forage in large flocks throughout the entire year, except for the breeding season, whereas the house sparrow will form large flock associations only in late summer and early autumn.

During the winter and breeding season, the house sparrow forages close to the area of the breeding colony, mainly in and around livestock pens, in villages, and in cities where they disperse outward from the city core in the morning and return in late afternoon (Summers-Smith, 1954; Fallet, 1958 a).

Important in the social behavior of these two species is the high degree of synchronization of their activities throughout the year. Such synchronized behavior minimizes inter- and intraspecific conflict during the breeding season because each group establishes a mutually exclusive area, in which it carries out its life history events, and which is instrumental in providing for anti-predator mobbing behavior.

For the most part, both species are sedentary, some 80% never moving from their natal areas, but there is evidence for long-distance dispersal or perhaps migration in some populations through a few recoveries in Europe of house sparrows several hundred kilometers from their banding sites. Populations along the northern edge of the range of both the house sparrow and tree sparrow show some migratory movement (Krüger, 1944; Kalela, 1949; Summers-Smith, 1956; Gaginskaya, 1967; Ulfstrand, Roos, Alerstam & Österdahl, 1974; Noskov *et al.*, 1965; Noskov, Ziman & Rezhÿï, 1975; and others).

In central Europe the tree sparrow rarely disperses over 10 km, but in western and northern Europe and in Japan dispersal of several hundred kilometers in a southerly direction is known (Table 3.1).

Population densities

Data on the preferred breeding biotope and densities within various biotopes must be known before the dynamics of a species can be fully understood. There are differences in biotopes occupied by various races of both *P. montanus* and *P. domesticus*. A survey of many published reports covering many different areas of Europe and Asia provides a general picture of population densities across the continent where these two species co-exist (Figs. 3.1 and 3.2, Appendixes 3.1 and 3.2).

Table 3.1. *Dispersal records for banded sparrows*

Location	% banded young	Number of recoveries	Distance in kilometres					Authority
			0–10 (%)	11–50 (%)	51–100 (%)	101–300 (%)	Above 300 (%)	
			P. montanus					
Niigata, Japan	100	121	62.9	4.9	0	19.0	12.2	Kuroda (1966a)
Leningrad, USSR	100	75	93.3	1.3	1.3	1.3	1.3	Noskov & Gaginskaya (1969)
Belgium	100	49	89.8	4.1	0	4.1	2.0	Verheyen (1957)
Germany	?	30	77.0	6.5	6.5	10.0	0	Lambert (1936)
Dresden, GDR	100	34	91.2	8.8	0	0	0	Creutz (1949)
Czechoslovakia	100	52	96.2	3.8	0	0	0	Balát (1975)
Wolfsburg, GFR	100	22	100.0	0	0	0	0	Scherner (1972)
Dziekanów Leśny, Poland	100	1382	100.0	0	0	0	0	Pinowski (1965a, 1968)
			P. domesticus					
Oklahoma, USA	?	2031	99.9	0.05	0	0.05	0	North (1968)
Wisconsin, USA	?	57	100.0	0	0	0	0	Beimborn (1967)
Stuttgart, GFR	98	881	99.2	0.5	0	0	0.3	Preiser (1957)
Bonn, GFR	100	30	96.6	0	3.5	0	0	Rademacher (1951)
Bonn, GFR	96.4	39	97.5	0	0	2.5	0	Przygodda (1960)
Kiel, GFR	59	45	98.0	0	2.0	0	0	Speyer (1956)
Gdańsk, Poland	100	88	97.8	2.2	0	0	0	B. Pinowska (unpublished)
Dziekanów Leśny, Poland	100	194	100.0	0	0	0	0	J. Pinowski (unpublished)
Poland	?	50	96.0	2.0	0	2.0	0	*Acta Ornithologica*[a]

[a] Domaniewski (1933, 1934); Domaniewski & Kreczmer (1936, 1937); Rydzewski (1949 a, b); Szczepski (1951); Szczepski & Szczepska (1953).

House sparrow densities show considerable variability across Europe and into Central Asia (Fig. 3.1). In the western part of Europe the values are fairly low, ranging from 22 and 35 birds km^{-2} on small islands, approximately 200 in rural and urban areas to 619 in the London area. Values increase rather steadily across the continent until they reach highest average levels in their preferred biotope in southeastern Europe (*ca* 1890 birds km^{-2}). Thereafter, they decrease across the steppe regions of the USSR. In central Asia densities are again very high, owing to village populations of the migrant *P. domesticus bactrianus*. Population densities of *P. d. bactrianus* reported from cities in the Fergonskaya Valley of south-central Asia (40° N, 72° E) are 780 to 1670 birds km^{-2} (Sharipov, 1974). For the house sparrow an average density value is approximately 630 birds km^{-2} throughout the range we have just discussed (Appendix 3.1).

In North America the breeding bird survey conducted by the US Fish and Wildlife Service and the Canadian Wildlife Service (Robbins, 1973) indicates regions of relative density achieved in the past 100 years. The largest and densest populations occur in mid-continent agricultural areas which are quite similar to the Old-World preferred biotope. Data are inadequate for providing absolute density estimates in the New World to compare with Old-World values.

The house sparrow reaches its highest average densities (*ca* 1200 to 1300 birds km^{-2}) in association with human dwellings (categories 14, 17 and 18, Table 3.2), especially where there is livestock. For the most part, house sparrows are restricted to these associations, but range into the local countryside 1 to 2 km from their nesting centers. Where human settlements are sporadic, the distribution is coincident and densities are low as well. This occurs especially in the northern part of the species' range and in newly occupied areas (Merikallio, 1958; Summers-Smith, 1959; Haartman *et al.*, 1963–1972; Vierke, 1970; Robbins, 1973; Smith, 1973).

There is also pronounced geographic variation in the density of breeding *P. montanus montanus* populations (Fig. 3.2, Appendix 3.2). Average densities range from 29 birds km^{-2} in the west (England) to peaks of 2444 birds km^{-2} in central Europe and to lower levels in southeastern Europe at the base of the Caucasus. Densities greater than 3000 birds km^{-2} have been reported in forest and steppe areas of south-central Europe (Eliseeva, 1960, 1961; Korodi Gál, 1960). Westward in cooler oceanic provinces peak densities of 271 birds km^{-2} have been reported in Denmark in optimum biotope (Joensen, 1965). Low populations are known from Ireland and Scotland (Sharrock, 1974), southeastern Finland (Merikallio, 1958), the Baltic regions of Poland and the taiga of Asia (Baikal) (0.2 birds km^{-2}, Izmaïlov & Borovitskaya, 1967). In general, there is a south to north gradient, the northern areas having lower densities. From our survey we obtain an average density, weighted over all biotopes, of 422 birds km^{-2}

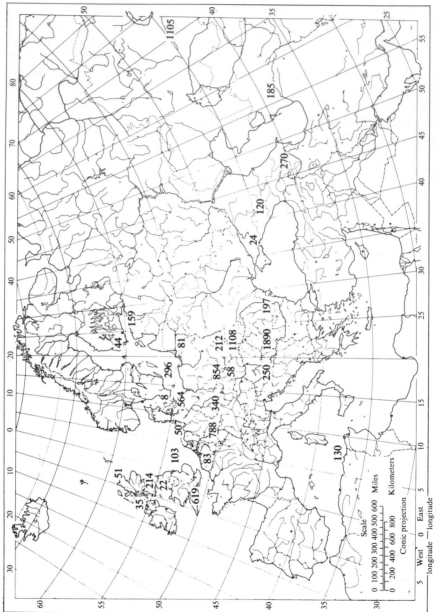

Fig. 3.1. Breeding density estimates of *P. domesticus* in Eurasia and North Africa. Values are birds per km².

58

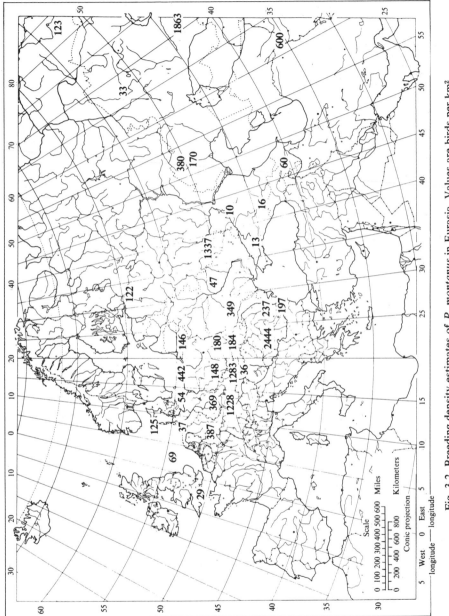

Fig. 3.2. Breeding density estimates of *P. montanus* in Eurasia. Values are birds per km².

59

for the Eurasian continent. Other races of the European tree sparrow show similar variability. The density of *P. montanus transcaucasicus* Buturlin in and near villages of USSR has been reported to be 60 birds km⁻² (Drozdov, 1965). *P. montanus dilutus* Richmond has very high breeding densities (2800 to 4000 birds km⁻²) in villages and cities of central Asia (Formozov, 1944; Sharipov, 1974). This latter race has much the same set of habits as the house sparrow of Europe. In eastern Asia *P. montanus saturatus* Stejneger prefers human settlements. Their densities vary from 242 to 625 birds km⁻² in Japan (Kuroda, 1973 a).

  P. hispaniolensis (Spanish sparrow) populations nest in dense colonies, often numbering 800 000 pairs, in shelterbelts within Kazakhstan, USSR. In one area of Kazakhstan an estimated 2 600 000 birds nested in one square kilometer (Gavrilov, É. I., 1974).

Biotope preferences and differences

Table 3.2 lists the most important breeding biotopes for *P. montanus montanus* and *P. domesticus domesticus*. From the literature values we have estimated breeding densities in each biotope. For the tree sparrow, these densities (*Y*) conform to an exponential regression curve on the biotope number (*X*) defined by the equation

$$Y = 1254\,e^{-0.276X} \quad (r = -0.978) \tag{3.1}$$

A similar curve for the house sparrow has the equation

$$Y = 85.5\,e^{0.134X} \quad (r = 0.803) \tag{3.2}$$

  The difference in biotope preferences and mutual exclusion between the two species is very evident. In areas where there is apparent overlap, house sparrows occupy building complexes and tree sparrows nest in trees and other places. Apparent niche overlap agrees, as far as the limited data warrant, with the theoretical concepts of Cody (1974).

  The interrelation between the two species is not uniform or static throughout their ranges. In western Europe, the house sparrow replaces the tree sparrow by monopolizing the available nesting places (Summers-Smith, 1963; Pinowski, 1967 a). In southwest Asia, on the other hand, the tree sparrow exerts dominance over the house sparrow in its occupancy of nesting areas inside villages and cities. In regions where *P. domesticus bactrianus* and *P. hispaniolensis* have co-existed in the past, introduction of shelterbelts has induced an increase of the Spanish sparrow and a decline in *P. d. bactrianus* (Gavrilov, É. I., 1974).

  Breeding densities vary differently in the house and tree sparrows with the size of the biotope available (Table 3.2). For the house sparrow,

Table 3.2. *Gradient of breeding populations of* P. montanus *and* P. domesticus *in different biotopes. The gradient has been set on the basis of decreasing density per unit area of* P. montanus *breeding populations*

Biotopes	Average biotope size (ha)	
	P. montanus	*P. domesticus*
1 Deciduous forests over 50 years old with nest boxes	18	—
2 Old parks in villages and open fields with trees	8	17
3 Old orchards	23	—
4 Young orchards with nest boxes	36	—
5 Pine forests, one to 50 years old with nest boxes	25	—
6 Deciduous forests over 100 years old	74	—
7 Riparian areas and cemeteries	20	22
8 Old parks in larger towns	22	24
9 Small allotments and gardens in towns	17	19
10 Pine forests 50 to 100 years old	27	—
11 Villages	45	—
12 Deciduous forests 50 to 100 years old	114	—
13 Deciduous forests one to 50 years old	24	—
14 Suburban areas with one family houses	—	65
15 Pine forests one to 50 years old	16	—
16 Commercial and shopping areas	105	41
17 Residential areas with apartment complexes	30	26
18 Family dwellings with livestock	—	146

density (Y – individuals km^{-2}) increases in relation to continuous tracts of suitable biotope available ($X =$ km^2) at an exponential rate

$$Y = 13.7X^{0.88} \quad (r = 0.657) \tag{3.3}$$

whereas the density of the tree sparrows declines by

$$Y = 2021X^{-0.836} \quad (r = -0.536) \tag{3.4}$$

The tree sparrow prefers small tracts of deciduous forests or forest edges. As the size of the forest block increases, the extent of forest edges decreases; hence the decline in its densities. Since the tree sparrow lives in more nearly natural biotopes which are subject to erratic random changes, its population densities are less predictable from area to area than those of the house sparrow that occupies the more stabilized agricultural and urban environments (Pinowski, 1967a).

Breeding season

The start of the breeding season, counted from the date when the first egg is laid, varies considerably with latitude and altitude, weather conditions, principally temperature, and the age composition of the population. A better indicator for the onset of egg-laying is the average date when the first egg is laid for the population, not the day when the first egg *per se* was laid. Unfortunately, such data are very scarce. Our analysis of materials, covering 14 years at Dziekanów Leśny, Poland, yielded a significant correlation ($P < 0.01$) between the date the first egg was laid and the average for the population. For *P. domesticus* the relationship is

$$Y = 21.82 + 0.50X \quad (r = 0.67) \tag{3.5}$$

and for *P. montanus*

$$Y = 22.91 + 0.39X \quad (r = 0.70) \tag{3.6}$$

where $Y =$ average date when the first egg was laid in the first clutches in the population and $X =$ date of first egg in first clutch (date counted in days from 1 April).

The mean time for initiating egg-laying varies linearly with latitude for both the house sparrow and tree sparrow, although the evidence for the latter shows a poorer fit than that for the house sparrow (Appendixes 3.3 and 3.4). For the house sparrow the equation is

$$Y = 2.48 + 1.94X \quad (r = 0.882) \tag{3.7}$$

and for the tree sparrow

$$Y = 74.2 + 0.77X \quad (r = 0.341) \tag{3.8}$$

where $Y =$ calendar days from 1 January and $X =$ latitude in decimal degrees. At the equator (Singapore, 1° N) tree sparrow clutches are started by the end of January (Ward & Poh, 1968). In the north (Great Budworth, England, 53° 30' N) the tree sparrow does not begin to lay until four months later (Boyd, 1932). In Baroda, India (22° 18' N) the house sparrow starts laying eggs in the second week of February when the average temperature in the daytime rises to about 22 °C (Naik & Mistry, 1973). At Muonio, Finland (67° 30' N), when spring comes early, house sparrows start laying eggs early in May but more often in mid-May (Montell, 1917).

House sparrows have been reported in the mountains as high as 4572 m (Kendeigh, 1973 c, d). Studies on the house and tree sparrows which were carried out during the same years in Poland and Romania in the Carpathian Mountains and also in Annaberg, Harz Mountains, GDR, indicate significant delays in the onset of egg-laying with increasing altitude (Table 3.3). The onset of egg-laying at the higher altitudes is comparable to that at higher latitudes.

Adult female house sparrows, two years or older, lay eggs before one-year-old females in their first reproductive season (Summers-Smith, 1963; Seel, 1968 a; Dawson, 1972). This also appears to be true in the tree

Table 3.3. *The effect of altitude on the onset of egg-laying and on the duration (number of days) of the breeding season (Mackowicz et al., 1970; Ion, 1973; Pinowski & Wieloch, 1973; Schlegel, in preparation)*

Year	Kraków, Poland (321 m)		Nowy Targ, Poland (593 m)		Zakopane-Kuźnice, Poland (1023 m)		Iasi, Romania (639 m)		Cimpulung Moldovenesc, Romania (639 m)		Annaberg-Buchholz GDR (600 m)	
	Date	Breeding season (days)	Date	Breeding season (days)	Date	Breeding season (days)	Date	Breeding season (days)	Date	Breeding season (days)	Date	Breeding season (days)
P. domesticus												
1967	—	—	—	—	—	—	2/4	132	16/4	104	—	—
1968	4/4	106	28/4	83	4/5	86	27/4	144	12/4	108	—	—
1969	28/4	61	3/5	—	—	—	—	—	—	—	—	—
1970	11/4	—	18/4	—	—	—	—	—	—	—	—	—
1971	8/4	96	4/5	57	1/5	108	—	—	—	—	—	—
1972	30/3	127	—	—	15/4	101	—	—	—	—	—	—
P. montanus												
1967	—	—	—	—	—	—	17/4	115	27/4	94	—	—
1968	6/4	121	27/4	99	—	—	12/4	120	18/4	97	26/4	—
1969	21/4	105	3/5	118	—	—	—	—	—	—	6/5	—
1970	20/4	—	23/4	—	—	—	—	—	—	—	1/5	—
1971	15/4	116	1/5	108	—	—	—	—	—	—	18/4	—
1972	10/4	—	23/4	100	—	—	—	—	—	—	—	—

63

sparrow (de Bethune, 1961; Seel, 1968 a), but there are fewer quantitative data available.

If the duration of the breeding season is defined as the time interval between the date of the first egg in the initial clutch and the date at which the last nestling leaves the last brood, the amount of variability from year to year at a locality may be as great as the mean duration at latitudes 25° apart. The variability would doubtless be considerably less if based on mean dates of initiating egg-laying of the population as a whole and the mean dates when the last young of all clutches become fledged, but such data are not available.

In the wet, sub-tropical climate of regions around Baroda, India (22° 18′ N), the breeding season of the house sparrow lasts from the middle of February to the middle of October (about 245 days) with only a short summer break. Sporadic broods may be recorded even in December (Naik & Mistry, 1973). At Mexico City (19° N) in the western hemisphere, house sparrows are reported to reproduce throughout the year, but in many southern regions the breeding season is interrupted by hot spells between April and August (Wagner, 1959). At Muonio, Finland (67° 30′ N), by contrast, the breeding season normally lasts only 70 days, except during exceptionally warm springs when egg-laying may begin during early instead of mid-May (Montell, 1917). At Ivalo, Finland (68° 40′ N), the season may be only 50 days long, from mid-May to early July, as on 5 July 1972 Pinowska and Pinowski could find only one nest with nestlings although there were several flocks of juveniles in the vicinity.

With the tree sparrow, the breeding season ranges from 150 days at Singapore along the equator (Ward & Poh, 1968) to 131 days at Bujor Targ, Romania (45° 51′ N) (Ion, 1973) to 61 days near Gdańsk, Poland (54° N) (Pinowska, unpublished).

In the southern hemisphere, the breeding season of the house sparrow lasts from October to January (about 90 days) in Zanzibar, Africa (16° S) (Moreau, 1950) and from September to January (about 120 days) in New Zealand (40° S) (Dawson, 1973).

When house sparrows build their nests inside buildings or other heated places, the breeding season begins earlier and lasts longer, even into the autumn and winter (Young, 1962; Haartman et al., 1963–1972; Novotný, 1970; Will, 1973). Usually also, the duration of the breeding season is longer at low altitudes than at high ones except when temperature inversion occurs, as was noted during two different years between Nowy Targ and Zakopane-Kuźnice, Poland (Table 3.3).

The duration of the breeding season depends upon the weather, food supply and age composition of the population. Photoperiod and perhaps protein levels (Jones & Ward, 1976) control the development of the gonads, but the onset of egg-laying is then largely determined by temperature, especially during the three or four days of most rapid growth of the ovum preceding laying (Kendeigh 1941; Il'enko, 1958; Mackowicz, Pinowski & Wieloch, 1970). Naik & Mistry (1973) observed at Baroda, India, that clutches were started when the temperature rose 5 to 8 °C during periods of days with temperatures continuously above 15 °C.

Many authors believe that the termination of the breeding season is also determined by weather conditions, perhaps in relation to available food supply (Perrins, 1965; Newton, 1967; Lack, 1968). Moulting, however, follows closely after nesting, and moulting needs to be completed before the beginning of cold weather. The disadvantages of a late moult in young birds and the resulting delay or absence of autumn sexual behavior may well be the main selection pressure conditioning the end of the breeding season even while weather and food conditions are favorable (Pinowski, 1968). In central Poland, the tree sparrow terminated its breeding season between 14 August and 1 September during a 14-year study and the house sparrow a little earlier, from 3 to 31 August. The average number of days from the departure of the last nestling from its nest to a mean daily temperature below freezing is 90±9 days in the tree sparrow and 88±23 days for the house sparrow (calculated from data in Pinowski, 1968; Mackowicz *et al.*, 1970; Novotný, 1970; Strawiński & Wieloch, 1972; Pinowski & Wieloch, 1973; Wieloch & Fryska, 1975). On the average, house sparrows in central Europe spend 80 days moulting (Zeidler, 1966; Dol'nik, 1967 *a*) and at Kangasala, Finland 52–64 days (P. Rassi, personal communication). Nesting must, therefore, terminate in time to permit moulting to be completed before weather conditions become stressful.

Number of broods

The exact number of broods per year raised by female sparrows can only be obtained from banded individuals, and such data are few (Summers-Smith, 1963; Naik, 1974; P. Rassi, personal communication). Approximations may be obtained by assuming that each pair remains at one nest over the entire breeding season; we have numerous examples on the number of broods per nest.

At Baroda, India, the average number of clutches laid per nest by house sparrows is 4.3 (Naik & Mistry, 1973; Naik, 1974). In middle latitudes, the average number is two or three, but there may be as many as four or five during exceptionally warm years or when the nests are in heated buildings (Seel, 1968 *a*; Will, 1969, 1973; Novotný, 1970; Wieloch & Fryska, 1975; Anderson, 1977). In central Poland, the average number per nest site varied over 14 years between 2.0 and 2.7 (Pinowska & Pinowski, 1977). Northward, the average number decreases to one per year (Keleïnikov, 1953), but when the summer is exceptionally warm, two or three

65

M. I. Dyer, J. Pinowski & B. Pinowska

Table 3.4. *Average number of broods per nest during a breeding season for the house sparrow and tree sparrow*

Location	Latitude	Average number of broods	Authority
	P. domesticus		
Baroda, India	22° 18' N	4.3	Naik (1974)
Lahore, Pakistan	31° 30' N	2.1	Mirza (1973)
Oxford, England	51° 46' N	2.1	Seel (1968 a)
Slezké Rudoltice, Czechoslovakia	50° 13' N	2.6	Novotný (1970)
Kraków, Poland	50° 04' N	2.0	Pinowski & Wieloch (1973)
Dziekanów Leśny, Poland	52° 20' N	2.1	Pinowska & Pinowski (1977)
Gdańsk, Poland	54° 20' N	1.7	Strawiński & Wieloch (1972)
Plainview, Texas, USA	34° 10' N	3.9	Mitchell, Hayes, Holden & Hughes (1973)
Plainview, Texas, USA	34° 10' N	3.0	Mitchell *et al.* (1973)
Portage des Sioux, Missouri, USA	38° 55' N	2.55	Anderson (1973)
McLeansboro, Illinois, USA	38° 07' N	2.2	Will (1969)
Coldspring, Wisconsin, USA	42° 50' N	1.5	North (1973 a)
Kangasala, Finland	61° 27' N	1.6	Rassi (unpublished)
Oulu, Finland	65° 00' N	2.0	Alatalo (1975)
	P. montanus		
Portage des Sioux, Missouri, USA	38° 55' N	2.64	Anderson (1973)
Sapporo, Japan	43° 0' N	2.0	Abé (1969)
Bzenec, Czechoslovakia	49° 00' N	1.97	Balát (1971)
Sokolnice, Czechoslovakia	49° 20' N	1.9	Balát (1971)
Dresden, GDR	51° 03' N	1.6	Creutz (1949)
Oxford, England	51° 46' N	1.6	Seel (1968 a)
Dziekanów Leśny, Poland	52° 20' N	2.4	Pinowski (1968) Mackowicz *et al.* (1970)
Wolfsburg, GFR	52° 25' N	1.79[a] 1.27[b]	Scherner (1972)

[a] Data for 1968.
[b] Data for 1969.

broods may be raised, even above the Arctic Circle (Montell, 1917, Table 3.4).

The average number of broods raised per nest of tree sparrows in central Poland varied from 1.9 to 2.7 over a period of 14 years or about the same as for the house sparrow (Pinowski, 1968; Mackowicz *et al.*, 1970; Pinowski & Wieloch, 1973; Pinowski, unpublished data). The averages for different localities listed in Table 3.4 are slightly lower than those for the house sparrow. Occasionally birds may raise four broods (de Bethune, 1961; Anderson, 1975; Wieloch & Fryska, 1975).

Clutch size

Clutch size in the house sparrow normally varies from two to eight eggs (Summers-Smith, 1963), but in the sub-species *P. domesticus bactrianus* clutches of 10 have been recorded (Gavrilov, É. I., 1974). Clutches tend to increase in size from south to north in both the eastern and western hemispheres (Fig. 3.3) but little can be made of this fact at this time. In studying geographic variations in clutch size, one must allow for both seasonal and possible yearly fluctuations. Exceptionally small clutches occur in England (Summers-Smith, 1963; Seel, 1968 b). Likewise, smaller than average clutches (3.8) have been reported in New Zealand (Dawson, 1973). There is no clear gradient in clutch size according to longitude across continental Europe, as has been reported in some other species (Lack, 1954; Klomp, 1970), although in the eastern parts of the ranges of both *P. domesticus domesticus* and *P. d. bactrianus* large clutches have been recorded (Rustamov, 1958; Gavrilov, É. I., 1974).

Clutch size in the tree sparrow commonly varies from three to eight eggs (Pinowski, 1968), although Gavrilov (1974) has found clutches of nine eggs in Kazakhstan, USSR. Average clutch sizes vary from 3.7 in Singapore (P. Ward, personal communication) to 5.6 eggs in Japan (Abé, 1970). Near Peking, China, clutches average five eggs (Chia, Bei, Chen & Cheng, 1963), as they do also in the one locality in the central USA where the species occurs (Anderson, 1973).

First and second clutches are generally larger than third or later ones for both *P. domesticus* and *P. montanus* (Pinowski, unpublished data; Balát, 1971; Eliseeva, 1961). However, when clutch size is compared with time of year there is an interaction with date so that the distribution of clutch size *per se* throughout the breeding season assumes a parabolic relationship. The curve for *P. domesticus* is

$$Y = 4.589 + 0.128X - 0.012X^2 \qquad (3.9)$$

and for *P. montanus* it is

$$Y = 3.85 + 0.384X - 0.027X^2 \qquad (3.10)$$

where Y = clutch size and X = date (10-day periods beginning 1 April). Nonetheless, there are variations on this pattern which are dictated by seasonal environments. At Baroda, India, where first clutches are usually begun in February, they are large during the spring monsoon period, decline in size during the months of July and August, and increase to the yearly maximum as breeding accelerates again in September and October (Naik & Mistry, 1973). Seasonal variations in clutch size have been reported to be correlated with photoperiod (Seel, 1968 b), food availability (Lack, 1954), and those temperatures at which the female has

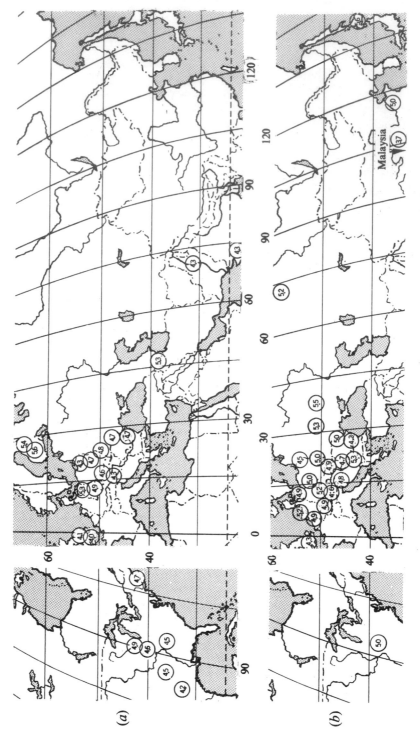

Fig. 3.3. Geographic variation in clutch size for the house sparrow (*a*) and tree sparrow (*b*).

Table 3.5. *Annual variation in clutch size in various places for P. domesticus*

Location	Years											Authority
	1963	1964	1965	1966	1967	1968	1969	1970	1971	1972	1973	
McLeansboro Illinois, USA	—	—	—	4.47	4.55	4.30	—	—	—	—	—	Will (1969)
Portage des Sioux, Missouri, USA	—	—	—	—	—	4.14±0.7[a]	4.60±0.6	4.60±0.9	4.68±0.9	4.62±0.9	4.68±0.9	Anderson (1977)
Slezké Rudoltice, Czechoslovakia	4.36±0.6	4.42±0.8	4.23±0.7	—	—	—	—	—	—	—	—	Novotný (1970)
Wieniec, Poland	—	—	—	—	—	—	—	—	4.49±0.9	4.50±0.9	4.40	Pinowska (unpublished)
Gdańsk, Poland	—	—	—	—	—	—	4.00±1.0	4.80±0.2	4.60±0.2	4.50	—	Strawiński & Wieloch (1972); Wieloch & Fryska (1975)
Dziekanów Leśny, Poland	4.82±0.13	4.72±0.1	4.53±0.1	4.94±0.1	4.61±0.1	4.73±0.1	4.84±0.1	4.70±0.2	4.64±0.2	4.53	4.67	Pinowska & Pinowski (1977)
Rzepin, Poland	—	—	—	—	4.50±0.7	4.60±0.7	—	—	—	—	—	Mackowicz et al. (1970)
Turew, Poland	—	—	—	—	—	—	—	—	4.59±1.0	4.70	4.70	Wieloch & Fryska (1975)
Kraków, Poland	—	—	—	—	4.80±1.2	4.80±0.9	4.50±0.2	4.40±0.8	4.32	4.49	4.60	Pinowski & Wieloch (1973)
Nowy Targ, Poland	—	—	—	—	5.00±2.2	4.20±1.4	4.20±1.3	4.80±0.1	—	—	—	Pinowski & Wieloch (1973)

[a] Part of year only.

69

Table 3.6. *Mean egg-loss rates* (%)

Area	P. domesticus	P. montanus
Continental Europe	29.8[ab]	28.3[b]
British Isles	16.0[ab]	9.6[b]
North America	38.7[b]	24.0
Far East	—	12.3
Indian subcontinent[c]	28.0	—
New Zealand	23.5	—

[a] Significantly different, $P < 0.05$.
[b] Significantly different, $P < 0.01$.
[c] Tropical and temperature regions of India and Pakistan.

her maximum productive energy (Chapter 5); doubtless all three factors interact.

When there are year-to-year changes in the average clutch size for the tree sparrow across Europe, our analyses indicate such changes tend to occur simultaneously in many widely separated areas, this occurrence having a frequency of 69% ($P < 0.01$). This information indicates the extent to which tree sparrows are sensitive to general climatic conditions during development of their clutches. On the other hand, we find no such regularity in fluctuations of house sparrow clutches among the different study areas that have been reported, nor is there such regularity in the variations of average annual clutch size within a single area (Table 3.5).

Egg loss

Measurements of the mean rates of all egg loss for *P. domesticus* and *P. montanus* show two trends: (1) loss rates in continental Europe for both species are higher than those recorded in the British Isles, and (2) those for the house sparrow of North America are higher than those of Europe or the British Isles. House sparrow egg loss rates in Europe are comparable to those on the Indian subcontinent and New Zealand (Table 3.6, Appendixes 3.5 and 3.6). It is of interest to compare egg mortality rates between species where data are available. In England, *P. domesticus* rates are higher than those of *P. montanus* ($P < 0.01$); however, for continental Europe there are no differences between the two species ($P > 0.1$).

For the Spanish sparrow inhabiting large colonies with open nests situated in tree crowns, Gavrilov (1962) recorded only 5% egg loss near Alma-Ata, USSR (43° 15' N). But for another species, the cape sparrow *P. melanurus* (Müller) of South Africa, Siegfried (1973) recorded an egg loss rate of 36%.

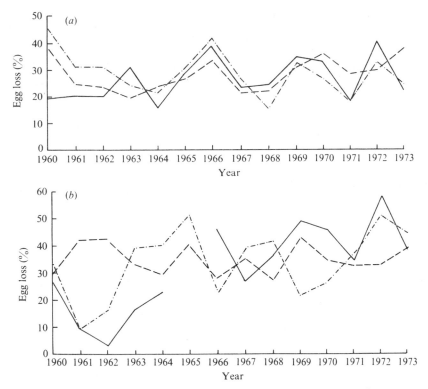

Fig. 3.4. Egg loss in the house sparrow (*a*) and European tree sparrow (*b*) over 14 years at Dziekanów Leśny, Poland: —— first brood; – – – second brood; – · – third brood (Pinowska & Pinowski, 1977).

The rates for all egg losses vary from year to year for *P. domesticus* and *P. montanus* (Fig. 3.4). Average egg loss rates are slightly higher for the house sparrow and the coefficient of variation is higher in nearly all instances (Table 3.7). Egg loss is usually higher in later broods when compared to the first brood (Creutz, 1949; Blagosklonov, 1950; Eliseeva, 1961; Pinowski, 1968; Novotný, 1970; Naik & Mistry, 1973), although Will (1969) noted in a three-year study in southern Illinois, USA, that eggs laid in mid-season broods were most subject to loss.

Will (1969) reported the lowest egg loss rates for the house sparrow in clutches of four, even though the most common clutch size was five. In England Seel (1968*b*) did not find any differences in egg loss relative to clutch size with either the house sparrow or tree sparrow. In Poland the highest egg loss rates for the tree sparrow were in clutches of three and four eggs and somewhat lower in the largest clutches of seven eggs (Pinowski, 1968). Clutches of five eggs, which were most frequent throughout the season, suffered fewest losses.

71

Table 3.7. *Average egg loss rates* ($\overline{X}$) *and coefficient of variation* (CV) (%) *for 14 years at Dziekanów Leśny, Poland* (*Pinowska & Pinowski, 1977*). *Original data in Appendixes 3.5 and 3.6*

	Clutch			All clutches
	1	2	3	
P. domesticus				
$\overline{X}$	24.4	34.6	33.3	31.6
CV	56.2	18.2	37.1	35.9
P. montanus				
$\overline{X}$	26.4	28.4	31.2	27.7
CV	30.8	22.2	33.5	20.1

Table 3.8. *Percentages of infertile eggs in relation to total eggs laid*

Area	*P. domesticus*	*P. montanus*	Authority
Missouri, USA	2.0	2.3	Anderson (1973)
Czechoslovakia	2.0	—	Novotný (1970)
Czechoslovakia	5.5	7.5	Balát (1971, 1974)
England	8.0	4.0	Seel (1968 *b*)
England	15.6	—	Dawson (1972)
Poland	—	3.5–7.2	Pinowski (1968)
GFR	—	14.0	Scherner (1972)
USSR	—	6.8–15.0	Koval' & Samarskiï (1972)
GDR	—	27.8	Creutz (1949)
Belgium	—	29.0	de Bethune (1961)

For the house sparrow, few infertile eggs are laid (Table 3.8). For the tree sparrow the number of infertile eggs laid is much higher, ranging from about 2% to nearly 30% (Table 3.8). The highest percentage of infertile eggs has been recorded for final broods in both species (Pinowski, 1968; Novotný, 1970). For the house sparrow there is an exponentially increasing rate of infertility with subsequent broods. Over a three-year period Novotný's data show the following relationship:

$$Y = 0.759\,e^{0.442\,X} \quad (r = 0.698) \tag{3.11}$$

where Y = percentage of infertile eggs and X = brood number. In addition, the last eggs laid in all clutches by the tree sparrow show a high proportion of infertility (Eliseeva, 1961). Infertile eggs on the average weighed less than those that ultimately hatched (Dawson, 1972).

In Alma-Ata, USSR, 4.9% of the eggs laid by the Spanish sparrow are infertile (Gavrilov, 1962). The highest incidence of infertile eggs for this species occurred in clutches of three eggs.

For the house sparrow a small proportion of the embryos dies during development; the percentage seems to be fairly constant for many areas: 3.7%, Oxford (Seel, 1968 b); 2.1%, Brno, Czechoslovakia (Balát, 1974); and 2.4%, Portage des Sioux, USA (Anderson, 1973). The same data for the tree sparrow are: 1.5% (Seel, 1968 b); 1.4% (Balát, 1972 b); and 2.2% (Anderson, 1973). For the most part, death usually occurs in the early stages of development (Anderson, 1977). In addition, a small percentage of house sparrow and tree sparrow eggs are broken by incubating birds (Pinowski, 1968; Novotný, 1970).

Studies made on mortality rates during hatching show both the house sparrow and the tree sparrow with 1% losses (Seel, 1968 b). In India a high mortality rate was recorded during hatching, varying from 5 to 35% during different months of the breeding season (Naik & Mistry, 1973). A relatively high rate of mortality (12 to 30%) was recorded during incubation and hatching by Novotný (1970). Dawson (1972) reported rates of 9% in England.

In central Europe the percentage of tree sparrow eggs lost decreases (hatching success, Y, increases) significantly for the first broods as environmental temperatures (X) increase giving a linear relationship of

$$Y = 64.05 + 1.84X \quad (r = 0.645) \tag{3.12}$$

By the time second and third broods are formed, this relationship no longer exists (Pinowski, 1968). Will (1969), however, reported high egg loss rates during warm summer periods when house sparrow nest temperatures reached 44 °C. His explanation was that the embryos in various stages of development died because of overheating. The adverse effects of high temperatures have also been reported in Mexico (Wagner, 1959).

Other causes of death are from diseases, such as salmonellosis (Pinowska, Chyliński & Gondek, 1976), and from predation by birds, e.g. *Dendrocopus major* L. (Balát, 1971), and small mammals (*Sciurus vulgaris* L., *Martes martes* L. (Balát, 1971); *Apodemus silvaticus* (L.) (Eliseeva, 1961); and *Eliomys quercinus* L. (de Bethune, 1961)).

So far we have discussed the causes of egg losses in nests in which at least one or more eggs hatched. But most often, overall mortality rates are influenced dramatically by loss of the entire clutch due to various causes. The percentage of eggs lost when entire clutches were destroyed in studies on the house sparrow ranges from none in Poland (Mackowicz et al., 1970) to 59% in Wisconsin, USA (North, 1973 a). For the tree sparrow the percentage of eggs destroyed has a much smaller range (2.4 to 13.8) in Poland (Mackowicz et al., 1970) (Table 3.9).

The degree to which whole clutches are destroyed depends upon the type of nest substrate and placement in terms of susceptibility to seizure by competitors and predators. Anderson (1973) showed that 31% loss rates

73

M. I. Dyer, J. Pinowski & B. Pinowska

Table 3.9. *Per cent egg loss when entire clutches are destroyed*

Location	P. domesticus	P. montanus	Authority
Portage des Sioux, USA	18.9	10.6	Anderson (1973)
McLeansboro, Illinois, USA	18.4	—	Will (1969)
Coldspring, Wisconsin, USA	59.0	—	North (1973 a)
Brno, Czechoslovakia	9.0	—	Balát (1974)
Sokolnice, Bzenec, Czechoslovakia	—	2.4	Balát (1971)
Nowy Targ, Poland	0.0	4.0	Mackowicz et al. (1970)
Kraków, Poland	12.7	13.8	Mackowicz et al. (1970)
Rzepin, Poland	2.0	2.4	Mackowicz et al. (1970)
Dziekanów Leśny, Poland	12.8	11.4	Mackowicz et al. (1970)
Wolfsburg, GFR	—	25.8	Scherner (1972)

Table 3.10. *Mean nestling loss rates (%) (Appendixes 3.5 and 3.6) for nests both partly successful and unsuccessful*

Area	P. domesticus	P. montanus
Continental Europe	25.9	20.5
England	53.6	37.1
North America	41.1	27.0
Far East	—	13.0
Indian subcontinent	59.0	—
New Zealand	23.5	—

in protected nest boxes were lower than losses in other locations (38%) ($P < 0.01$). Balát (1973) reported high loss rates when house sparrows nested in abandoned nests of *Delichon urbica* L. and *Hirundo rustica* L. that were subject to failure with the additional weight of the new nest. Loss rates (nearly 50%) for Spanish sparrow nests are high because nests are placed high in exposed areas of tree crowns (Gavrilov, 1962).

Nestling mortality

The highest average mortality rates of house sparrow nestlings occur in the Indian subcontinent and the lowest in New Zealand (Table 3.10). All of the differences between areas are statistically significant (no comparisons were possible for New Zealand, however).

The mortality rates of the tree sparrow nestlings are highest in England and lowest in the Far East (Table 3.10). The difference in mortality between England and continental Europe is significant ($P < 0.01$), but the data are too few to test for differences with North America. Nestling

74

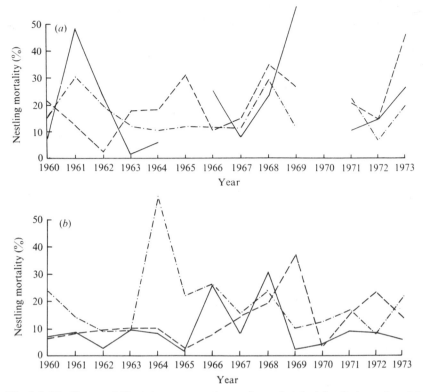

Fig. 3.5. Nestling mortality rates as percentages of eggs hatched for *P. domesticus* (a) and *P. montanus* (b) over 14 years at Dziekanów Leśny, Poland: —— first brood; – – – second brood; –·– third brood (Pinowska & Pinowski, 1977).

mortality rates are lower everywhere ($P < 0.01$) for the tree sparrow compared to the house sparrow.

Fluctuation in mortality rates from one year to another accounts for much of the variability in the data presented (see also Appendices 3.5 and 3.6) and is well shown in the study conducted in Poland (Fig. 3.5; Pinowska & Pinowski, 1977). On a year-to-year basis *P. domesticus* and *P. montanus* show very little agreement. Average mortality rates are lower for *P. montanus*; however, this species shows the greatest amount of variability, this pattern holding true for all three broods (Table 3.11). In contrast with this study and others of shorter duration in Poland, reports of nestling losses in the house sparrow in Czechoslovakia (Novotný, 1970) and in the tree sparrow in USSR (Eliseeva, 1961) indicate that there is very little or no annual fluctuation.

For the house sparrow per cent nestling mortality (Y) can be fitted to significant parabolic curves in relation to brood size (X):

M. I. Dyer, J. Pinowski & B. Pinowska

Table 3.11. *Average nestling death rates* ($\bar{X}$) *and coefficient of variation* (CV) (%) *for 14 years at Dziekanów Leśny, Poland* (*Pinowska & Pinowski, 1977*)

	Broods			
	1	2	3	All broods
P. domesticus				
$\bar{X}$	21.1	21.1	16.6	19.5
CV	80.7	55.2	45.7	63.4
P. montanus				
$\bar{X}$	9.6	13.1	19.6	14.1
CV	88.5	69.7	65.5	77.4

$$\text{England:} \quad Y = 23.2 + 14.4X - 1.64X^2 \tag{3.13}$$

$$\text{USA:} \quad Y = 12.8 + 19.4X - 2.52X^2 \tag{3.14}$$

These curves predict that the highest mortality rates for the house sparrow occur in broods with four and five nestlings in England (Seel, 1970) and North America (Will, 1969). Encke's (1965) work in the GDR shows greatest mortality in the larger broods, but there are only records for one year at each of two localities.

For *P. montanus* the picture is somewhat different, from data in the GFR (Scherner, 1972) a parabolic relationship is expressed by the equation

$$Y = 9.56 + 18.2X - 2.67X^2 \tag{3.15}$$

but Pinowski's (1968) data from Dziekanów Leśny, Poland, indicate that mortality rates are highest in broods having one nestling and decrease exponentially as the broods become larger:

$$Y = 30.27X^{-0.669} \quad (r = -0.920) \tag{3.16}$$

The study by Seel (1970) shows highest rates in broods of three with a larger amount of scatter for broods of two and four. Generally, mortality rates decrease as brood size increases (Fig. 3.6).

Per cent nestling mortality (Y) decreases linearly with age (X = age classes 1, 2, 3; for 5-day periods of nestling life) for both sparrows. For *P. domesticus* the predicted equation is

$$Y = 26.8 - 7.8X \quad (r = -0.848) \tag{3.17}$$

And for *P. montanus* it is

$$Y = 15.2 - 4.3X \quad (r = -0.793) \tag{3.18}$$

There are several points of interest in this relationship. Firstly, there appears to be a south–north gradient in that at the northern border of each

76

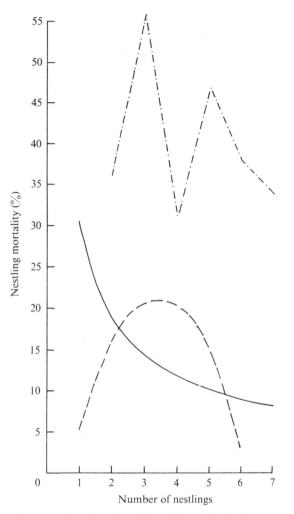

Fig. 3.6. Mortality rates of nestlings for different clutch size for the tree sparrow: —— Poland (Pinowski, 1968); – – – GFR (Scherner, 1972); – · – England (Seel, 1970).

species' range, the mortality rates for the youngest age class (zero to five days) are 51% and 58% higher than the mean values for the house sparrow and the tree sparrow, respectively. Secondly, the slopes and intercept values suggest that mortality rates are greater for *P. domesticus* than for *P. montanus*. Thirdly, mortality rates are approximately the same in both species in latter phases of the nestling developmental period. We have not been able to test these statements statistically because the data available are not sufficient, but the trends are strong.

There are various causes of mortality in nestlings. From the seventh day and increasingly so up to and including the twelfth day, nestlings run a risk of falling out of the nest; at this age they are still unable to live independently of their parents (Novotný, 1963, 1970). This happens most frequently with nests built in trees (Gavrilov, 1962) and with house sparrows that build on insecure structural features (Mackrodt, 1967–1968).

Long periods of rainy weather with drops in temperature increase death rates (Eliseeva, 1961; Pinowski, 1968; Mitchell *et al.*, 1973). A linear correlation exists in the tree sparrow between per cent nestling survival (Y) and daily mean temperature (X) (J. Pinowski, unpublished data).

$$Y = 68.54 + 1.18X \quad (r = 0.555) \tag{3.19}$$

Entire broods may be lost when temperatures drop below freezing (P. Rassi, personal communication). Very high temperatures may also cause mortality (Wagner, 1959; Will, 1969; Dawson, 1972; Keil, 1973).

House sparrow nestlings succumb to nest infestations of fly larvae (T. Tomek, personal communication) and to salmonellosis (Pinowska *et al.*, 1976) and tuberculosis (Balát, 1972 *a*), which they catch from their parents and which often causes mortality.

Destruction of nestlings by predators occurs more frequently with the tree sparrow that nests in tree holes or nest boxes, often far from human settlements, than with the house sparrow, the close associate of man. Tree sparrow nestlings are lost to the same predators that destroy eggs. House sparrow nestlings are sometimes killed by mammals, e.g. *Rattus norvegicus* (Berkenhout) (Mitchell & Hayes, 1973), *Dryomys nitedula* Pallas (Jordania, 1970) and birds, e.g. *Turdus merula* L. (Stohn, 1971). The wryneck *Jynx torquilla* L. often appropriates tree sparrow nests for its own eggs after which the tree sparrow nestlings die (Sokołowski, 1929; Keil, 1957; Mikhel'son, 1958; Eliseeva, 1960; Pinowski, 1967 *a*; and others). Capturing the adult birds on the nest often causes them to desert. Desertion is more frequent in the tree sparrow during the period of egg-laying (68%) than during incubation (34%) or when nestlings are present (7%). Of 68 cases where house sparrows were captured on the nest, 36% of them deserted (Pinowski, Pinowska & Truszkowski, 1973).

Survival of fledglings and juveniles

After leaving the nest, fledgling and juvenile birds must develop flying ability, learn to feed by themselves, and recognize potential predators, undergo moulting and sustain increasingly severe winter conditions. The fledglings are under the care of their parents for about 14 days before they become independent juvenile birds (Summers-Smith, 1963). Only about 65% of young house sparrows survive to one month of age and 19% to

12 months. The survival curve (Summers-Smith, 1963) of house sparrow fledglings and juveniles follows the equation

$$Y = 102.9X^{-0.688} \quad (r = -0.990) \tag{3.20}$$

where Y = percentage of juveniles surviving and X = month after fledging. Since there is a progressive decline in survivors with time, a smaller percentage of the juveniles that fledged early in the year enter into the breeding the next year than those that fledged late in the season, assuming that all young of the year attain sexual maturity at the same time.

The survival curve for juvenile European tree sparrows in Poland is based on a two-year study by Pinowski (1968) and is expressed by the equation

$$Y = 118.8X^{-0.911} \quad (r = -0.980) \tag{3.21}$$

The slope of the curve is steeper and only 12% survive for 12 months. Pinowski estimated that 15 to 20% survive to the first breeding season (May), 3 to 4% to the second, and 1 to 2% to the third. A differential survival rate occurs between the broods; that of the first brood is approximately 35% lower than the second and third broods, and that of the second brood is lower than the third.

Survival of adults

The survival rates of adults are largely based on recovery of banded (ringed) birds. From such data, statistics on survival or mortality may be estimated (Lack, 1951; Farner, 1955; Quick, 1963). Until recently a major difficulty commonly encountered was the incomplete recovery of banded birds, which gives rates that are over-estimated for mortality and underestimated for survival. Haldane (1955) developed mortality/survivorship ratios to avoid this difficulty and recently, Seber (1970), Robson & Youngs (1971), and Anderson, Kimball & Fiehrer (1974) have worked out a procedure using birds of unknown age. This latter method is a stochastic extension using maximum likelihood procedures of the general theory by Jolly (1965). Recently, this method has been expanded to allow usage of several different age classes, including marked young of the year. From recovery data, life tables can also be developed showing mortality, survival and life expectancy for each age class (Eberhardt, 1971, 1972) and this we have done for data collected by various authors.

Estimates obtained from dynamic life table methods (Hickey, 1952; Quick, 1963) define steep exponential decay curves of Deevey's (1947) Type III survivorship. Averages for all available data give curves having the following equations for *P. domesticus*:

$$Y = 2952.7e^{-1.03X} \quad (r = -0.979) \tag{3.22}$$

and for *P. montanus*:

$$Y = 4607.7\,e^{-1.483X} \quad (r = -0.943) \tag{3.23}$$

where Y = number of birds surviving and X = age in years.

There is considerable variability among populations, mostly due to truncation of recovery periods four or five years after banding. It is interesting to note that for both England and North America, where there are sufficient data to make comparisons, females have higher survival rates (the Y-intercept and slope, $-b$, is smaller) than males (Table 3.12). During the breeding season female mortality is often higher, but outside the breeding season it is usually accepted that male mortality is higher (Summers-Smith, 1963). It would seem on the average that males are subject to greater mortality than females, perhaps because of their visibility and vulnerability owing to more prominent displays and behavior.

The life table data for *P. montanus* gives a very low survivability level, mainly because of the low numbers of recaptures and the fact that few were recaptured after the third and fourth years. From the slopes expressed in Table 3.12 we can see no apparent differences between reporting sites, although that possibility is not ruled out.

Survival rates for the house sparrow, obtained using these different estimators, vary in England from 45 to 62.9% per year (Table 3.12). Lack's (1951) method of calculation gives the lowest values, most of which are lower than those reported by Summers-Smith (1963). These values assume that the amount of mortality becomes stabilized after the birds become one year old (Botkin & Miller, 1974). The mean age of the adult population appears to be about 1.7 years.

The information on house sparrows for the Netherlands (Table 3.12) indicates a much lower survival rate than in England, but again, this conclusion is based on a relatively small sample and a paucity of recoveries after three years. However, the data from the USA provide average survival rates that are only slightly higher than in the Netherlands and well below those in England. One cannot be sure whether England presents a more favorable environment for this species, whether the higher survival rates are due to a better and more adequate set of data, or whether the lower survival rates in other areas are real and are compensated for by the higher fledging success (lower nestling mortalities) noted in preceding sections. Survival rates for the tree sparrow are mostly below even the lowest values for the house sparrow in the USA and especially in Poland (Table 3.12).

There are few data about mortality levels during various months of the

Table 3.12. *Survival rates (% per year) of all age classes based on estimates obtained from several methods*

Population	Sex	N	Author	Lack (1951) s_x	Haldane (1955) s_x	Seber (1970) s_x	Life table[a] survival rate	Authority
						P. domesticus		
England	♂	95	60.0	45.0±3.7	53.0±1.6	—	$Y = 4590.3\ e^{-1.156X}\ (r = -0.983)$	Summers-Smith (1963)
England	♀	106	64.0	51.0±3.4	62.0±1.2	—	$Y = 3744.0\ e^{-0.997X}\ (r = -0.985)$	Summers-Smith (1963)
England	♂, ♀	201	62.0	48.0±2.5	58.0±1.0	—	$Y = 4449.6\ e^{-1.702X}\ (r = -0.986)$	Summers-Smith (1963)
England	♂, ♀	513	—	53.4±1.5	60.0±1.2	—	$Y = 1798.4\ e^{-0.607X}\ (r = -0.999)$	A. S. Cheke and British Trust for Ornithology (unpublished)
England	♂, ♀	3316	—	—	—	62.9[b]	—	A. S. Cheke and British Trust for Ornithology (unpublished)
Netherlands	♂, ♀	56	—	31.7	—	—	$Y = 2272.0\ e^{-0.913X}\ (r = -0.985)$	Van Balen & Speek (unpublished)
USA	♂, ♀	278	41.5	31.5±2.3	41.0±4.1	—	$Y = 2781.8\ e^{-0.948X}\ (r = -0.787)$	Anderson (1977)
North America	♂	81	—	32.5	—	—	$Y = 3852.3\ e^{-1.2517X}\ (r = -0.997)$	Dyer (unpublished data) US Fish and Wildlife Service
North America	♀	79	—	37.3	—	—	$Y = 1537.3\ e^{-0.7258X}\ (r = -0.988)$	Dyer (unpublished data) US Fish and Wildlife Service
North America	♂, ♀	310	—	34.0	—	—	$Y = 2287.8\ e^{-0.963X}\ (r = -0.990)$	Dyer (unpublished data) US Fish and Wildlife Service
All							$Y = 2952.7\ e^{-1.03X}\ (r = -0.979)$	
						P. montanus		
USA	♂, ♀	91	33.4	28.3±4.0	33.0±5.9	—	$Y = 5385.2\ e^{-1.511X}\ (r = -0.995)$	Anderson (1977)
Poland	♂, ♀	282	—	17.3±2.0	17.0±12.1	—	$Y = 11803.8\ e^{-2.255X}\ (r = -0.987)$	Pinowski (unpublished)
Czechoslovakia	♂, ♀	92	—	33.3	—	—	$Y = 4501.5\ e^{-1.328X}\ (r = -0.993)$	Balát (1975)
All							$Y = 4607.7\ e^{-1.483X}\ (r = -0.943)$	

[a] Data are computed from life table estimates based on 1000 birds alive on 1 January of their first year of life.
[b] Analysis made on adults only.

year for birds. Those reports that do address this problem for seed-eating birds show that the winter is the period of greatest stress and many species may be limited by winter resource supplies (Pulliam & Enders, 1971; Fretwell, 1972 a; Wiens, 1974 b). However, data available from house sparrows clearly indicate exactly the opposite view (Summers-Smith, 1963: 141). Lessened winter mortality likely has an important meaning for the life cycle of the house sparrow, especially in England where winters are often mild, there is close association with man and thus there are apparently sufficient food resources. This picture is not seen elsewhere for house sparrow populations (Fallet, 1958 a) and is not the normal situation for the tree sparrow in Europe (Pinowski, 1968), these latter conditions satisfying situations noted above for passerines that are not strict commensals of man.

Life expectancy

Life expectancy was calculated from the life tables after the manner of Deevey (1947), Hickey (1952) and Quick (1963), rather than the more recent procedure reported by Botkin & Miller (1974). The procedure followed permits separate determination of life expectancy for each age class.

In England, the data obtained by Summers-Smith (1963) show that life expanctancy declines linearly with age (Table 3.13). Data for the years 1954 to 1973 (Cheke, unpublished) show, however, a definite parabolic decrease. Data from the Netherlands (Van Balen & Speek, unpublished) and from Missouri, USA (Anderson, 1977) show increases in the second year of life and sharp decreases thereafter. Our analysis of unpublished data from the US Fish and Wildlife Bird-Banding Laboratory for all of the USA shows that life expectancy for the female increases during both the second and third years before it declines, but not for the male. Life expectancy is higher for the female at all ages, in both data from the USA and England, with reduction to 0.5 year not attained until the sixth year, while in the male it is reached in the fourth and fifth years, respectively. We suspect that the data from the Netherlands and perhaps Missouri are incomplete for the older years. Some birds in England are known to live 10 years (Summers-Smith, 1963).

Life expectancy curves for the tree sparrow decrease linearly and values are low at all ages compared with the house sparrow. This suggests that population turnover rates are higher in the tree than in the house sparrow.

The number of young produced by a female in a given breeding season is a direct function of the physiological status of the bird (Jones & Ward, 1976; Pinowska, 1976), phenological events in the nest and available food for the young. The principal parameters measured to determine production are numbers of eggs laid, egg losses and mortality in nestlings. For instance, in spite of the very large number of eggs laid by *P. domesticus* in India, success is very low because of high rates of egg loss and nestling mortality. Other conditions are shown in Appendix 3.7. For *P. montanus* the highest production of young is reported generally for regions with highest egg production (Appendix 3.8).

In a stable population, production of fledglings (number of female fledglings per female adult) per year ($\bar{m}$) must be such that the number of young surviving (s_0) equals the number of adults dying ($1-s$). Since the average production per female and adult survival (s) have been estimated for a few populations (Table 3.14), the survival rate of young can be calculated by the Henny & Wight (1969, 1972) equation, $\bar{m} = (1-s)/s_0$. The range of values thus obtained for production and survival of young (*ca* 36%) and adult (*ca* 63%) birds in England (Table 3.14) appears intuitively reasonable, but there appear to be some minor discrepancies.

In order to obtain juvenile survival rate estimates necessary to maintain a stable population, we calculate levels of 23 to 26%, higher than those measured by Summers-Smith (1963) or predicted by our exponential decay curves discussed earlier. We suspect that the production estimates are somewhat low for we realize the difficulty an investigator has in obtaining representative values. Using the production values of Anderson (1977), his estimates of adult survival plus the survival estimates for *P. domesticus* in North America, we obtain estimates of 21% for juvenile survival, a value close to that which Summers-Smith (1963) reported for a marked population in England. We have tried to predict the adult survival rate for a stable population of *P. montanus* in Poland, where sufficient data are lacking, but there are estimates of production and juvenile survival from Pinowski's (1968) work. By substituting and solving for *s* we obtain a value of 57.8% for a stable population; again a value intuitively reasonable.

Henny, Overton & Wight (1970) reported an approach derived from their matrix methods for analyzing life history data to show whether populations are increasing or decreasing. In their equation $\mu = m_1 s_0 - 1 + s_1$ one can estimate μ, the annual rate of change in the population, given that production (m_1) and survival values (s_0 and s_1) are known for the various age classes (see Henny *et al.*, 1970 for explanation). According to this procedure, estimates which have been derived previously for both the house sparrow in England and the tree sparrow in Poland indicate de-

Table 3.13. *Comparison of life expectancy models for Passer. Y = life expectancy, X = age*

Population	Sex	N	Model	r	Mean age (years) life table	Seber (1970)	Authority
P. domesticus							
England	♂	95	$Y = 1.258 - 0.197X$	-0.979	1.82	—	Summers-Smith (1963)
England	♀	106	$Y = 1.435 - 0.197X$	-0.982	2.03	—	Summers-Smith (1963)
England	♂, ♀	201	$Y = 1.533 - 0.173X$	-0.981	1.93	—	Summers-Smith (1963)
England	♂, ♀	513	$Y = 1.406 + 0.191X - 0.027X^2$	—	—	1.67 ± 0.17	Cheke (unpublished)
Netherlands	♂, ♀	56	$Y = 1.297 - 0.233X$	-0.805	1.47	—	Van Balen & Speek (unpublished)
North America	♂	81	$Y = 0.988 - 0.146X$	-0.946	1.48	—	Dyer (unpublished) US Fish and Wildlife Service
North America	♀	79	$Y = 1.082 + 0.417X - 0.105X^2$	—	1.59	—	Dyer (unpublished) US Fish and Wildlife Service
North America	♂, ♀	310	$Y = 1.0 + 0.176X - 0.049X^2$	—	1.52	—	Dyer (unpublished) US Fish and Wildlife Service
Missouri, USA	♂, ♀	278	$Y = 0.929 + 0.139X - 0.047X^2$	—	1.39	—	Anderson (1977)
P. montanus							
Missouri, USA	♂, ♀	91	$Y = 1.017 - 0.126X$	-0.990	1.39	—	Anderson (1977)
Poland	♂, ♀	282	$Y = 0.798 - 0.105X$	-0.964	—	—	Pinowski (unpublished)
Czechoslovakia	♂, ♀	52	$Y = 1.15 - 0.16X$	-0.992	—	—	Balát (1975)

Table 3.14. *Population parameters concerned with rates of production of fledglings, post-fledging and juvenile mortality, and age independent adult mortality to maintain a stable population for species of granivorous birds (see also Appendixes 3.5 and 3.6)*

Population	Production female fledglings per adult female ($\bar{m}$)	Estimated post-fledging and juvenile survival[a] (s_0)	Adult female survival (s)	Authority
		P. domesticus		
India	2.35	—	—	Naik (1974)
Missouri, USA	2.29	20.9	51.0	Anderson (1977)
Poland	1.95	—	—	Pinowska & Pinowski (1977)
Czechoslovakia	3.35	—	—	Novotný (1970)
England	1.61	26.1	58.0	Summers-Smith (1963)
England	1.61	23.2	58.6	Cheke (unpublished)
Finland	2.06	—	—	Rassi (unpublished)
		P. montanus		
Japan	4.53	—	—	Abé (1969)
Czechoslovakia	2.59	—	—	Balát (1971)
GDR	1.70	—	—	Creutz (1949)
Poland	3.52	12.0	57.8[b]	Pinowski (1968)
GFR	2.72	—	—	Scherner (1972)
England	2.32	—	—	Seel (1968 b)

[a] Solving for s_0 by: $\bar{m} = 1 - s/s_0$, Henny & Wight (1969).
[b] Solving for s by above equation.

creasing population status (Table 3.15). However, the estimate, -15.6% per year, for the house sparrow may be within the limits of error for this method of analysis, but the larger negative value for the tree sparrow, mainly due to the low survival estimate for adults (Tables 3.12 and 3.15), may indicate that these central European populations are actually declining. Such a hypothesis is worth special note in order to determine if the population is actually in decline, but if not, to determine what can be done to bring the model and the estimates into correspondence with the real situation.

A third, more robust method of analysis, a matrix model approach devised by Lebreton (1973), showed that house sparrow populations in Poland were relatively stable. The annual rate of increase fluctuated around a value of 1.0 (11-year average = 0.9826), until 1968, after which a marked decline occurred (Table 3.16). Pinowski believes that the fluctuations are realistic, especially when analysing the severity of winter conditions in 1968–1970. Since 1970 house sparrow numbers have been obviously higher during the breeding period, mainly because of increased

85

M. I. Dyer, J. Pinowski & B. Pinowska

Table 3.15. *Estimates of population annual rate of change for house sparrows of England and the tree sparrow of Poland. The equation $\mu = m_1 s_0 - 1 + s_1$ from Henny et al. (1970) is used for the calculation*

Species	Location	m_1	$s_0{}^a$	s	Estimated proportion annual rate of change (μ)
P. domesticus	England	1.61	0.164	0.58	−0.156
P. montanus	Poland	3.52	0.124	0.17	−0.394

[a] Estimates from author's own work (see earlier section).

potential for breeding in newly developed housing tracts (J. Pinowski, personal observation).

The family Icteridae

Social structure

Much is known about the individual and territorial behavior of icterids, particularly the red-winged blackbird *Agelaius phoeniceus* (Allen, 1914; Nero, 1956 a, b; Orians & Christman, 1968). But little is known about their flock movements and/or feeding habits during the period of the year they spend in large social units, a time that constitutes the majority of their life span.

The breeding season commences in April and May and lasts until late June or early July in normal years. Redwings traditionally nest in cattail (*Typha*) marshes (Allen, 1914), but in this century have expanded dramatically into upland vegetation (Graber & Graber, 1963), mostly in old-field complexes. Grackles *Quisculus quiscula* nest mainly in deciduous and coniferous tree plantations, but there has been some spread into marshes (Wiens, 1965; M. I. Dyer, personal observation). Cowbirds *Molothrus ater* are social parasites and lay their eggs in the nests of many different species, particularly certain New-World warblers (*Parulidae*) and fringillid sparrows (Young, 1963 a). After the breeding season, adults and young alike gather in mixed sex and species flocks and start forming flock concentrations along traditional bird migration pathways.

These three blackbird species are migratory; however, their migrations usually are performed somewhat leisurely. The birds leave their breeding grounds in August and arrive on their wintering grounds in October and November (Meanley, 1971). They spend the winter in concentrated roosting areas in southern USA; many of the roosts are immense. The birds

Table 3.16. *Annual rates of increase (number in individuals) for the house sparrow population at Dziekanów Leśny, Poland (Pinowska & Pinowski, 1977) as determined from the Matrix model developed by Lebreton (1973)*

Year	Annual rate of increase
1960	0.9986
1961	0.9488
1962	1.0506
1963	0.9947
1964	1.0765
1965	0.9554
1966	0.9823
1967	1.0137
1968	1.0202
1969	0.9135
1970	0.8544
11-year average:	0.9826

range outward from their roosting centers at dawn in mixed flocks to forage during the day, much after the patterns discussed by Hamilton, Gilbert, Heppner & Planck (1967) and Wiens & Dyer (1975 a), then return to roost during the night.

Populations appear to stay together throughout much of the year (Dyer, 1964). Unpublished data obtained in Ontario, Canada and Ohio, USA, provides additional evidence that once flocks are formed during July, August and September, they remain coherent, at least until they start their southward migration. The record from one trapping site where several hundred birds were marked and recaptured shows that as long as a flock unit lived in the area of the trapping site, there was a relatively high probability for recapture, and furthermore, a large percentage of the retrapped birds were captured simultaneously on the first day. Also, there seemed to be mutual exclusion of flocks, or at least sub-populations. While one population unit was in the trap vicinity, no others were being captured, but after the flock presumably left the area, a new wave of birds moved in. This was repeated at least four times. The persistence of association between individual birds is further shown by two cases where grackles and cowbirds banded from the same flocks in 1966 and 1967 in southwestern Ontario and four cases involving grackles and redwings from central Ohio were found together in an Ohio roost two to five and a half years later (1970) where they had been killed in an experimental lethal-control program (Mitchell, Dyer & Coon, unpublished report, US Fish and Wildlife Service).

Table 3.17. *Production values for red-winged blackbird in North America*

Authority	Biotope[a]	Sex ratio female:male	Clutch size	Hatching success (% eggs)	Fledging success (% nestlings)	Number produced per female[b]
Smith (1943)	m	1.6–2.8:1.0	5.04[c]	72.0	82.4	3.0
Orians (1961a)	m	2.8–3.7:1.0	3.7	—	—	2.61
Meanley & Webb (1963)	m	1.9:1.0	3.3	—	—	4.2
Case & Hewitt (1963)	m	2.2:1.0	3.5	—	—	2.49 (0.87)
		1.9:1.0	3.56	—	—	2.68 (0.72)
Goddard & Board (1967)	m	1.6:1.0	3.4	40.1	59.6	0.81 (0.64)
Collier (1968)	—	2.7:1.0	—	—	—	—
Laux (1970)	—	3.3:1.0	3.61	62.3	66.2	1.49
Holm (1973)						
1965	m	—	3.58	58.0	41.8	0.87
1966	m	3.02:1.0	3.44	63.6	49.7	1.09
1967	m	2.71:1.0	3.58	53.6	41.8	0.83
Dyer (unpublished)						
1965	m	—	2.80	67.5	63.5	1.20
1966	m	—	2.78	82.6	62.6	1.59
1967	m	—	3.01	68.0	41.4	0.70
Dolbeer (1976)	u	—	3.45	—	—	2.89 (0.96)
Williams (1940)	m	—	3.19	72.0	67.0	— (1.57)
Robertson (1972)	m	—	3.27	67.0	65.0	2.71 (1.44)
	u	—	3.19	57.0	46.0	2.49 (0.85)

Reference	Habitat[a]	Sex ratio				
Wood (1928)	—	—	3.17	72.6	66.0	2.19 (0.95)
Wood (1938)	m	—	3.25	84.6	63.6	2.1 (1.75)
Perkins (1928)	m	—	3.46	88.0	64.4	2.0 (1.96)
Young (1963b)						
1959	m	—	3.07	53.8	53.0	0.88 (0.55)
1960	m	—	3.22	44.6	40.0	0.58 (0.39)
Beer & Tibbits (1950)	m	—	3.18	79.4	65.9	1.87 (1.44)
Brenner (1966)	m	2:1	3.24	68.5	88.0	2.67 (1.69)
	m	2.1:1	2.79	56.3	83.0	2.14 (0.83)
	m	1:1	3.69	68.0	48.0	1.58 (1.0)
	m	0.4:1	2.43	78.9	80.0	2.40 (1.50)
	m	0.9:1	3.24	69.0	50.0	1.43 (1.05)
Snelling (1968)	—	—	3.43	59.5	50.1	1.02
Wiens (1965)	—	—	3.45	69.3	60.9	1.46
$\bar{X}$	—	2.08:1	3.28	66.25	60.00	1.86 (1.12)
S.D.		±0.855	±0.30	±11.79	±13.77	±0.871 (±0.455)

[a] m = marsh habitat, u = upland or old-field habitat.
[b] Per successful nest in open numbers, per known nesting attempt in parentheses.
[c] Not included in computation of mean.

Red-winged blackbird

Red-winged blackbird populations occur in two major biotopes: cattail marshes or wetlands and dry upland areas, notably old-field complexes or cultivated hayfields. Such upland biotopes have become more widely dispersed through eastern North America with the conversion of the original grassland and forests into agricultural land. The breeding stock of the two ecotypes may be more or less distinct (Dyer, 1964; Stone, 1973; Dyer & Abramsky, in preparation), except along transition zones between the two. The situation is complex, however, since the two biotopes may co-exist in a mosaic where their separation is indistinct (Stone, 1973). The species deserves further watching with respect to the development of the two ecotypes (Hesse & Lustick, 1972) and how this may affect their natality and mortality patterns.

In contrast with the house and tree sparrows where the sex ratio of the adults is approximately 50:50 and they are monogamous in their mating behavior, the sex ratio of adult redwings on the breeding grounds favors the females, and the males are polygynous (Orians, 1961 *a*). Females breed when they are one year old, but first-year males have only limited success setting up and keeping territories, commonly not breeding until the second year.

The number of each sex is apparently equal in the early nestling stage (Williams, 1940), but possibly more females than males fledge (Dyer, unpublished data). The number of non-breeding males is difficult to determine, but on their breeding grounds females outnumber males on the average about two to one (Table 3.17).

The red-winged blackbird is multi-brooded and the later clutch sizes may be smaller (Collins, 1968), but the average clutch size, grouping all clutches together, is 3.28 (Table 3.17). The species may lay four eggs per clutch and occasionally five, but an average clutch size of 5.04 reported by Smith (1943) is clearly too high.

Approximately two-thirds of the eggs hatched (66%) in nests from which at least one nestling was produced and of these, a slightly lower percentage (60%) of the young fledged (Table 3.17). The difference in percentages is not, however, statistically significant. Probably predation is the most important mortality factor for both eggs and nestlings.

The production rate in nests that produce young is nearly two per female (1.86, Table 3.17), but if all nests are included it drops to nearly one (1.12). There is difficulty in assessing production because it is not known precisely how many nests are built per female per year (Dolbeer, 1976). We assume that the value per successful nest is close to the true population value.

Adult survival rates were computed using the same variety of pro-

90

Table 3.18. *Survival rates (% per year) of adult red-winged blackbirds based on several methods of computation*

Population	Sex	N	Author's	Lack (1951)	Haldane (1955)	Seber (1970)[d] I	Seber (1970)[d] II	Life table[a] survival rate	Authority
Massachusetts, USA	♂, ♀	177	51.1	49.1	—	—	—	$Y = 2024.8\ e^{-0.643X}\ (r = -0.998)$	Fankhauser (1967)
Massachusetts, USA	♂, ♀	695	53.4	53.7	—	—	—	—	Fankhauser (1967)
North America	♂	1646	42.1	41.8	—	—	—	—	Fankhauser (1971*a*)
North America	♀	705	43.3	45.1	—	—	—	—	Fankhauser (1971*a*)
North America	♂	123	42.6	41.1	—	—	—	—	Fankhauser (1971*a*)
North America	♀	70	59.1	52.4	—	—	—	—	Fankhauser (1971*a*)
Ontario, Canada	♂	801	—	40.9[b]	53.0[b]	—	—	$Y = 10617.2\ e^{-1.666X}\ (r = -0.960)$	Dyer (unpublished, 1965–1972)
Ontario, Canada	♂	92	—	48.3[c]	57.0[c]	—	—	$Y = 4483.4\ e^{-1.087X}\ (r = -0.928)$	Dyer (unpublished, 1965–1972)
Ontario, Canada	♂, ♀	57	—	30.5[b]	39.0	—	—	$Y = 4425.9\ e^{-1.357X}\ (r = -0.992)$	Dyer (unpublished, 1965–1972)
Ontario, Canada	♂, ♀	864	—	40.2[b]	51.0[b]	—	—	$Y = 10617.2\ e^{-1.666X}\ (r = -0.921)$	Dyer (unpublished, 1965–1972)
Ontario, Canada	♂, ♀	96	—	47.8[b]	55.0[b]	—	—	$Y = 4522.0\ e^{-1.109X}\ (r = -0.926)$	Dyer (unpublished, 1965–1972)
Ontario, Canada	♂	33	—	—	—	65.1	55.7	—	Dyer (unpublished, 1965–1972)
Ontario, Canada	♀	70	—	—	—	56.9	39.3	—	Dyer (unpublished, 1965–1972)
Ontario, Canada	♂, ♀	100	—	—	—	56.7	—	—	Dyer (unpublished, 1965–1972)
Michigan, USA	♂	93	—	—	—	62.1	—	—	Laux (1970)
Michigan, USA	♂	31	62.0[b]	59.2	—	—	—	—	Laux (1970)

[a] Data are computed from life table estimates based on 1000 birds alive on 1 January of their first year of life.

[b] Calculations of recaptures, known age birds.

[c] Calculations of recoveries, known age birds.

[d] Calculations on one year old or older birds, two methods are used (Anderson *et al.*, 1974).

cedures as for the house and tree sparrows (pp. 79–82) and conform to the Type III curve of Deevey (1947) with an equation of

$$Y = 2024.8\,e^{-0.643\,X} \quad (r = -0.998) \tag{3.24}$$

where Y = number of birds surviving and X = age in years. Curves for localities other than Massachusetts, as defined by equations in Table 3.18, show differences, mainly because recoveries do not exist past five years. The slopes are much steeper, and even though there is a good fit to the data, these values must be viewed with caution because they give estimates of mortality higher than that which is likely to exist. For Massachusetts, data giving 51% survival were obtained by shooting and the figure of 53% was obtained from data collected by other methods of recovery (Fankhauser, 1967); the difference between the two is not significant. In the two sets of data for the whole of North America, survival rates of males are lower than those of females. This difference between sexes is contradicted in Dyer's (unpublished) data for Ontario and Ohio where females have the lower rate, especially from the Seber (1970) estimate upon which considerable reliance can be placed. Where we have combined recoveries for males and females we feel the value 56.7% is quite realistic for this population.

Life expectancy values for the redwing are parabolic in time, according to Fankhauser's (1967) data (much the same as some of the *Passer* curves) and decrease linearly according to Dyer's data from the Lake Erie basin (Table 3.19). Life expectancy values appear to differ very little when one compares data obtained from recoveries (those obtained from dead birds reported *ca* 200 to 2000 km distant) to recaptures obtained in the area where the birds were marked. Since Fankhauser's (1967) data contain more recoveries of older age classes, the curve obtained from those data are more representative of life expectancy values. Estimates of mean age of redwing populations, given in Table 3.19, range from 0.94 to 2.14, indicating a relatively low average age in the population because of high mortality rates, especially for the younger age classes.

The production rate averages 0.93 ± 0.46 female fledglings per female adult; and if adult female survival is taken as 57%, then the fledgling and juvenile survival rate becomes 46% according to the Henny *et al.* (1970) equation. Such a high survival rate for young birds is unlikely. If we assume a production rate of the mean plus one standard deviation $(0.93 + 0.87 = 1.80)$, the survival rate estimate for the young of the year is 0.24, a more realistic value, and could then represent a population that is more or less stable. Otherwise, with the parameters obtained in this synthesis we would have to conclude that the population is in decline.

Table 3.19. *Comparison of life expectancy estimations for icterids. Y = life expectancy, X = age in years. Mean age is determined from life table estimates and from Seber's (1970) methods*

Population	Sex	N	Model	r	Life table	Mean age			Authority
						Life table	Seber	Seber	
Agelaius phoeniceus									
East Coast, USA		325	$Y = 1.649+0.077X-0.021X^2$	—	2.14	—	—	—	Fankhauser (1967)
Lake Erie Basin recoveries	♂, ♀	96	$Y = 1.632-0.24X$	−0.986	1.92	—	—	—	Dyer (unpublished)
Lake Erie Basin recoveries and recaptures	♂, ♀	864	$Y = 1.123-0.16X$	−0.943	1.67	—	—	—	Dyer (unpublished)
Lake Erie Basin recaptures and recoveries of adults	♂, ♀	128	—	—	—	1.22 / 0.72	0.94 / 0.29	1.17 / 0.42	Dyer (unpublished)
Quiscalus quiscula									
Lake Erie Basin recoveries	♂, ♀	37	$Y = 1.559-0.198X$	−0.985	1.70	—	—	—	Dyer (unpublished)
Lake Erie Basin recaptures	♂, ♀	418	$Y = 1.103-0.117X$	−0.822	1.38	—	—	—	
Lake Erie Basin recaptures and recoveries of adults		449	—	—	—	1.27 / 0.42	1.76 / 0.52	1.74 / 0.41	Dyer (unpublished)
Molothrus ater									
Lake Erie Basin recaptures	♂, ♀	72	$Y = 1.121+0.175X-0.046X^2$	—	1.72	—	—	—	Dyer (unpublished)
Lake Erie Basin		308	—	—	—	1.27 / 0.40	1.27 / —	1.28 / 0.38	

Common grackle

Even though the common grackle shares many of the same habits as the redwing – flocking in large numbers, frequenting agricultural and urban ecosystems, and sharing some of the same nesting areas (Wiens, 1965) – it does have markedly different breeding behavior and population structure. For instance, unlike the redwing, it is not polygynous and depends more upon tree groves for its breeding colonies. Erskine (1971 a) has developed a model of breeding biotope overlap among six icterid species that includes estimates of the degree to which grackles and redwings share breeding areas. Production values are available from several papers, although there are not as many reports available for the grackle as for the redwing.

Clutch size has been intensively investigated (Table 3.20), the average for North America being approximately 4.45 eggs per active nest. Average hatching success is 63.3% from five studies, fledging success is 59%, and our estimate of production of young per female is 1.87. It is unfortunate that there are no more studies from which production can be estimated, especially from areas where grackles are most common (Erskine, 1971 a: 364).

As with the other species studied in this volume, the common grackle survival curve is Type III (Deevey, 1947). The grackle survival curve obtained from life table computations of males and females in the Lake Erie basin shows a very steep exponential regression:

$$Y = 7165.6 e^{-1.1635 X} \quad (r = -0.989) \tag{3.25}$$

where Y = number of birds surviving and X = age in years. Even though the fit of the data to the exponential model is good, grackle recovery rates, in years subsequent to banding, were very low, and the life table representation must be viewed with caution.

In Fankhauser's (1971 b) rather large samples for the eastern USA, the estimates of survival percentages are somewhat lower than percentages based on recoveries of other species (Table 3.21). This is also true for Dyer's (unpublished) data from Ontario and Ohio. Different methods of calculating survival rates give varying results, but we consider that Seber's (1970) method I gives the best estimates. Thus for the central continental population of the USA and Canada a survival rate of 60.7% per year for the adult population is probable, a level somewhat higher than that noted for redwings and *Passer*.

Life expectancy values for the common grackle decrease linearly with age for the Lake Erie basin population of the USA and Canada. Here again caution must be employed. This species is known to be fairly long-lived (Fankhauser, 1971 b), and the life expectancy curve is also likely to be

Table 3.20. *Production values for common grackle in North America*

Authority	Location	Clutch size	Hatching success (% eggs)	Fledging success (% nestlings)	Production number per female
Maxwell & Putnam (1972)	Ohio	4.21	48.8	66.7	2.6
Peterson & Young (1950)	Wisconsin	4.65	72.6	64.6	2.18
Wiens (1965)	Wisconsin	4.40	—	—	—
Willson, John, Lederer & Muzos (1971)	Illinois	4.72	—	—	—
Jones (1969)	Kentucky	4.70	—	—	—
Johnston (1964)	Kansas	4.70	—	—	—
Erskine (1971 *a, b*)	Canada	4.23	—	—	—
Long & Long (1968)	Illinois	4.21	—	—	—
Snelling (1968)	Wisconsin	3.98	66.3	34.7	0.915
Wiens (1963)	Wisconsin	4.42	80.1	65.8	2.33
Dyer (unpublished)	Canada	4.33	48.7	63.2	1.33
$\bar{X}$	—	4.45	63.3	59.0	1.87
S.D.	—	±0.25	±14.15	±13.65	±0.71

parabolic as has been noted for the redwing. The mean age of the adult populations is estimated to be between 1.38 and 1.70 years (Table 3.19).

Since we have no estimates of survival rates of juveniles, or young birds of the year, it is not possible to estimate population status as we have provided for *Passer*. However, using the Henny *et al.* (1970) methods, we estimate that juvenile survival needs to be 21% to maintain a stable population, a value which is quite in line with those reported earlier in this chapter.

Brown-headed cowbird

Although not usually classed as agricultural pests, cowbirds occupy the same agricultural ecosystems and consume many of the same food resources as the red-winged blackbird and common grackle. In recent years there has been concern regarding the role of the cowbird in causing a population decrease of the Kirtland's warbler *Dendroica kirtlandii* (Baird), in Michigan (Mayfield, 1975), and increasingly heavy management programs have been directed against the cowbird (Shake & Mattsson, 1975).

Production values for brown-headed cowbirds have been summarized by Young (1963 *a*) and Wiens (1963), with other reports by McGeen & McGeen (1968) and McGeen (1972). The number of eggs laid per host varies from 1.49 to 1.38 (Table 3.22). The percentage that hatch is 38.1% and 61.7% of those nestlings fledge. The number ultimately produced per host nest is estimated to be 0.35 to 0.32, a function of the relatively low

95

Table 3.21. *Survival rates (% per year) of common grackles in eastern North America*

Authority	Sex	Population	N	Lack (1951)	Haldane (1955)	Life table survival rates	r	Seber (1970)[c] I	II	Other (by author)
Fankhauser (1971 b)	♂	Not known	1751	45.4	—	—	—	—	—	45.0
Fankhauser (1971 b)	♀	Not known	1521	44.3	—	—	—	—	—	44.0
Fankhauser (1971 b)	♂	Not known	897	50.3	—	—	—	—	—	49.0
Fankhauser (1971 b)	♀	Not known	611	55.1	—	—	—	—	—	53.3
Dyer (unpublished)	♂, ♀	Ontario, Canada	418	31.0[a]	39.0	$Y = 7165.6\ e^{-1.635\,X}$	−0.989	—	—	—
Dyer (unpublished)	♂, ♀	Ontario, Canada	37	46.6[b]	54.0[b]	$Y = 2797.8\ e^{-0.902\,X}$	−0.996	—	—	—
Dyer (unpublished)	♂	Ontario, Canada	179	—	—	—	—	59.8	52.2	—
Dyer (unpublished)	♀	Ontario, Canada	157	—	—	—	—	59.3	56.2	—
Dyer (unpublished)	♂, ♀	Ontario, Canada	336	—	—	—	—	60.7	55.7	—

[a] Estimates obtained from recaptures.
[b] Estimates obtained from recoveries of dead birds.
[c] Estimates obtained from one-year-old or older birds.

Table 3.22. *Production values for brown-headed cowbirds in North America*

Authority	Eggs per nest	Hatching success (% eggs)	Fledging success (% nestlings)	Number produced per host nest	Number produced per female cowbird
Young (1963 *a*)	1.49	38.1	61.7	0.35	6[a]
Wiens (1963)	1.38	—	23.0[b]	0.32	—

[a] Estimated.
[b] Percentage fledged is of eggs, not percentage of nestlings as presented elsewhere in this volume.

number being hatched. Once through the critical egg phase, survival of nestlings is fairly high. Young estimated that six young would be produced per year per adult female cowbird. The cowbird has become an essentially indeterminate layer (McGeen & McGeen, 1968) and apparently compensates for the low productive rate per host nest by 'flooding' the hosts' nests wherever suitable conditions can be found, anywhere from five or six to 30 eggs a season can be laid (Young, 1963 *a*; McGeen, 1972; Payne, 1965, 1973).

A Type III survival curve also clearly describes the conditions for this species (Y = number of individuals surviving, X = age in years):

$$Y = 2296.4\,e^{-0.849 X} \quad (r = -0.996) \tag{3.26}$$

Even though the data are somewhat truncated, the curve projects reasonably well to older age classes, i.e. using the exponential decay curve the last members of any given cohort do not die until the tenth year. Other estimators show considerable differences, but we consider the best survival estimates to be those provided by the Seber (1970) method (Table 3.23) which indicate that survival is 55.6% per year for adults.

The only data that give reasonable predictions about life expectancy come from the Ontario–Ohio recapture data. Life expectancy is best fitted by a parabolic curve (Table 3.19), the highest value being reached by two-year-old birds.

To obtain estimates of juvenile survival rates we have used the Henny *et al.* (1970) equations; we have assumed a production level of three females per year per adult female (Young, 1963 *a*) and an adult survival rate of 55.6%. We then estimate juvenile survival to be 14.8% within a stable population. This value is much lower than we have estimated elsewhere for granivorous birds and may be conservative. If so, then the combination of high production potential and what is really average

Table 3.23. *Survival rates (% per year) of brown-headed cowbirds in eastern North America*

Population	Sex	N	Other	Lack (1951)	Haldane (1955)	Seber (1970)[c]		Life table	r	Authority
						I	II			
Not known	♂	2336	36.4[a]	36.3	—	—	—	—	—	Fankhauser (1967)
Not known	♀	1165	31.0[a]	30.4	—	—	—	—	—	Fankhauser (1967)
Not known	♂	378	48.5[b]	50.0	—	—	—	$Y = 2296.4\, e^{-0.849\,X}$	−0.991	Fankhauser (1967)
Not known	♀	142	40.4[b]	38.3	—	—	—	—	—	Fankhauser (1967)
Ontario, Canada	♂, ♀	72	—	41.0[a]	45.0[a]	—	—	—	—	Dyer (unpublished)
Ontario, Canada	♂, ♀	36	—	49.3[b]	53.0[b]	—	—	—	—	Dyer (unpublished)
Ontario, Canada	♂	188	—	—	—	58.5	—	—	—	Dyer (unpublished)
Ontario, Canada	♀	28	—	—	—	—	38.2	—	—	Dyer (unpublished)
Ontario, Canada	♂, ♀	216	—	—	—	55.6	—	—	—	Dyer (unpublished)

[a] Estimates obtained from recaptures.
[b] Estimates obtained from recoveries of dead birds.
[c] Estimates obtained from one-year-old or older birds.

survival conditions would show that the population is expanding, a fact that is asserted by Mayfield (1975) and others. It is noteworthy that these data were taken during a period of apparent population increase and also during a time in which intensive control efforts have been carried out.

As with the other two icterid species, survival rates calculated from returns are lower than those based on recoveries (Table 3.23). All the data with this species show a higher survival rate for males than for females, with a range of 36 to 59% in the males and 30 to 38% in the females. These rates lie well below those for the other icterids. The average age for adults in the population is estimated to be 1.72 from life table estimates and 1.28 from Seber's (1970) method.

Discussion and summary

Estimates of density for the house sparrow exist for many places in Europe, a center of investigation on this species, and much is now being assembled in North America about its distribution and relative density through the breeding bird census (Robbins, 1973). But little is known about densities of the house sparrow elsewhere in the world, nor of the densities of the tree sparrow across Asia. Summers-Smith (1963) predicted that the house sparrow would eventually reach every inhabitable place, despite efforts to prevent this in many areas, such as has occurred in Australia. In addition to this continued expansion of its range (Chapter 2, this volume), there are indications that the house sparrow may be increasing its density in Europe, mainly owing to continued expansion of housing throughout the continent, a development which, for the most part, will favor increases in the numbers of colonies. Not only is there an increase in potential habitat, but with concomitant increases in farm livestock numbers in Europe, there is also an expanded potential for food throughout the year. In cities, feeding of birds is thought to be important, especially during the winter months. In the USA, within the past 20 years, the increased popularity of horse riding has created a situation much like that which existed 50 to 70 years ago during the heavy use of draft horses, thus there could easily be an increase in house sparrows in the future.

The establishment of gardens, orchards and parks in Europe is beneficial to the tree sparrow, but this is countered by the use of herbicides and pesticides and the development of monocultures. Thus, it is difficult to predict what the final outcome will be for various populations of this species.

The distribution of house sparrows and tree sparrows when they inhabit the same regions is of special interest. While we do not know of definitive studies reporting descriptions of niche separation or overlap, the conditions discussed for equations 3.1 and 3.2 strongly suggest mutual exclusion. This potential is closely related to the observation that as total biotope size increases, population density of the house sparrow increases, whereas just the opposite occurs for the tree sparrow (equations 3.3, 3.4). With changes in land-use characteristics throughout the world, this condition will likely have the most dramatic effect on population sizes for these two species.

The density and thus the population size of the tree sparrow is more subject to fluctuation than is that of the house sparrow. *P. montanus* lives in a less predictable environment than *P. domesticus* in its association with man; thus the tree sparrow is subject to greater environmental fluctuations during its lifetime. Such exposure to density-independent controlling variables, such as severe winters with a deep snow cover, limits the size of the tree sparrow populations to low levels, especially in central Europe (Pinowski, 1968). In Asia the role of the tree sparrow in the community changes dramatically to that of the house sparrow of Europe and there are major increases observed in population levels (Sudilovskaya, 1954; Rustamov, 1958; Gavrilov, É. I., 1974).

Perhaps the most salient feature to examine in studies of population dynamics is that of the nestling survival rates (Ricklefs, 1969 *a, b*). In a comprehensive review, Ricklefs pointed out that tree-hole nests suffer far less mortality, both in terms of individual and whole nest losses, than do open nests. *Passer* is representative of tree-hole nesting species and the three icterid species are typical of open-nest species.

The results we provide unfortunately do not include sufficient parameters to compare all mortality values used by Ricklefs, but some comparisons can be made. The egg and nestling survival data for *P. domesticus* and *P. montanus* from Tables 3.6 and 3.10 compare favorably with other tree-hole nesters. Also, the survival rates are equal to or much higher in all cases than those reported for the three open-nesting icterid species (Tables 3.17, 3.20 and 3.22; Ricklefs, 1969 *a*). For a species such as *Passer* there is relatively little predation pressure, in contrast with open-nesting birds, and thus the nesting stage is largely governed by mortality causes due to uncertain food supplies available to the parents. The icterids must cope with both stresses in order to survive. As the estimated adult mortality rates are approximately the same, it appears that there are major differences in the juvenile survival period. It is possible that icterids, once they leave the nest, have a relatively constant mortality rate, but that *Passer* may not, the first year levels being rather high. If such is the case, it is quite obvious that various species, even though broadly classed as granivores, are subject to quite different selective forces. In order to survive, therefore, these granivorous species must evolve different strategies to cope with these diverse selective forces.

Ricklefs (1969 *a*) concluded that starvation owing to uncertain food supplies governs the mortality rates during the nestling phase for several species. This is probably true for the granivores that we are considering here, although for the open-nesting species (icterids) weather could play a major part, since Francis *et al.* (unpublished) have shown that up to 40% of growth variance of nestlings is governed by micro-climatic changes. If the micro-climatic stress is sufficiently severe, death of young can easily follow the slow-down of growth noted by these authors. Hole-nesting species do not experience this degree of micro-climatic effect. Since they do not suffer the degree of predation that open-nesting species do, it is then likely that food is the main limiting factor regulating the survival of young in the nest.

From the information presented concerning egg and nestling losses and mortality rates of Old and New World house sparrow populations, it appears that there are major differences in these life history processes. In nearly every case, mortality levels seem to be higher in North America than in Europe and we hypothesize that despite its century-old existence in the New World, this species may still be adjusting its production and mortality processes in response to selection pressures. Thus, it is probable that mortality functions regulate the size of the population in any given area, a condition noted for most passerine species where there are sufficient data to analyze the situation (Murton, 1972). Such a hypothesis seems likely in view of the fact that birth rate, i.e. clutch size potential, varies little from year to year. If there is regulation of the population from the standpoint of more or less young being raised, it comes from the addition of an extra brood during the breeding season, or perhaps from lack of development of a brood, rather than in changes of clutch size *per se*.

Thus, depending upon the conditions in any given area, number of clutches, average clutch size, egg loss and nestling mortality all combine to give the number of young raised by a female in a breeding season. Such is true, of course, for any avian species, but the important thing to note from this synthesis is that rarely do these events interact in the same way from year to year, or from area to area. It is these variations of differing selective forces that dictate numbers of young fledged during the breeding season each year, which in turn affect recruitment rates and cohort size. These numbers vary considerably, ranging from 2.6 to 7.7 fledglings per female for the house sparrow and 3.4 to 9.0 for the tree sparrow (Appendixes 3.7 and 3.8).

It has been thought that blackbird densities have risen dramatically in North America during this century. While there is some information to indicate that this is so for some areas (Graber & Graber, 1963), there are no data to substantiate an increase on a continental basis and it is probably true they have always been quite numerous. It is difficult, if not impossible, to follow and monitor very large numbers of birds throughout the year.

For this reason any information about their population status must be obtained by indirect methods. There have been attempts to monitor and count blackbird roosting concentrations throughout the southern USA wintering grounds (Webb & Royall, 1970); however, at best these counts become simply indices. The best time for getting good counts, or even indices, is on the breeding ground. Two types of surveys exist, one from the North American breeding bird census (Robbins & Van Velzen, 1974) and the other from an intensive survey in the Lake Erie basin (Dyer, Siniff, Curtis & Webb, 1973). In this latter survey, data were presented that show how the redwing breeding population fluctuated in various areas and biotopes during a three-year census period. There was a statistically significant change during each of the three years. Data added to that survey indicate that on the fourth year it was no longer possible to demonstrate statistical variation throughout the region. Thus it would seem that at least three to five years' data must be accumulated before significant changes or trends in population size can be described. To describe controlling variables, even longer periods will be required. However, it is necessary to look at the resource area in which granivorous birds reside. In the same region of the redwing survey noted above, a five-year census, taken in an Ohio county where intensive agriculture is developed, showed that there was little or no variation in the number of territorial males (Dyer, 1970). The conclusion reached was that in areas where large amounts of nutrients or energy are put into agricultural crop production with an emphasis on cereal grain harvest, a certain and constant amount 'leaks out' through the trophic structure to produce or support a constant amount of biomass in the avian population. Thus, the redwings of the intensively cultivated areas of the middle portion of North America very likely are the beneficiaries of present agricultural methods. This constant input of fertilizers and energy source apparently prevents natural fluctuation of the populations, in contrast with what happens in nearby regions where populations undergo fluctuation. In this respect, we urge that careful attention be given to better development of monitoring programs, not only for the granivorous species with which we have dealt in this volume but also for other granivores as well.

Many of the populations about which we have given synopses have been subjected to lethal control operations within recent years. Continuation of such control operations consequently could have pronounced effects on the overall natality–mortality processes described in this chapter that could result in changes of population sizes.

In the section *Reproductive biology* in Chapter 8, this volume, Wiens & Johnston look at *r* and *K* selective strategies for granivorous birds. They note that birds are very low in fecundity in contrast with other organisms. Yet from the information we present here we hypothesize that the 'less favorable', less predictable environment in which *P. montanus* lives may tend to make it more of an *r* strategist than *P. domesticus*, especially if we take into account the continuum of conditions theorized by Pianka (1974). Several of the tree and house sparrow's production and survivorship parameters fit the pattern predicted by Pianka when the two species are compared. For the case of the red-winged blackbird, we cannot demonstrate clear differences between production parameters of birds nesting in upland and wetland biotopes, but when one considers the overall environment of the two conditions, the upland biotope is more severe and less predictable overall than the wetland biotope. Robertson (1972, 1973) concluded that the wetland biotope was 'optimal' for red-wings, but Wiens & Dyer (1975 a) criticized his conclusions on the basis that since most of the *total* population of mid-North American production took place in the uplands, it is not possible to determine qualitatively which biotope is optimal. Robertson's position has also been questioned by Dolbeer (1976) who showed that upland areas produce as many fledglings per territory as wetland areas. The situation is still worthy of further study. For instance, Dyer & Abramsky (in preparation) have shown that growth rates of nestlings are significantly higher in upland biotopes than in wetland, and that fledging weights, as estimated by the asymptote of Dyer & Abramsky's logistic model, are lower for upland birds. They suggest that there is an apparent selection pressure to produce birds with these growth characteristics in uplands, and perhaps a different set of selective pressures to produce nestlings which grow more slowly and fledge at higher weights in wetlands. It is thus of considerable interest to us that such individual processes, when put together in a population sense, fit the *r* and *K* continuum suggested by Pianka (1974) and that this continuum can be demonstrated *intraspecifically*. If these hypotheses are borne out, it will be important in the future to regard redwing populations, as well as other granivorous species, in the context of MacArthur & Wilson's (1967) *r* and *K* selection theory. In this respect, many of the natality and mortality processes presented for the few granivores reviewed here ought to be intensively re-investigated in order to test this hypothesis.

In the GDR and GFR, attempts were made to rid villages of house sparrows by poisoning, but to date they have not succeeded. The after effects showed that within a short period of time invaders from nearby colonies filled in the 'hole' left in the community when the resident birds were killed (Mansfeld, 1950; Gersdorf, 1951; Preiser, 1957). In China, European tree sparrows were eradicated from large regions by a concerted effort carried out for several days (Summers-Smith, 1963). Unconfirmed reports state that in some areas *P. montanus* has been re-introduced from populations in Pakistan. In the USA there has been a growing sentiment against large overwintering flocks of blackbirds in the south and recently

103

the first major program to eradicate entire roosts developed, although such action had been going on for about a decade on an experimental basis (LeFebvre & Seubert, 1970). Treatment was not widespread until 1975, at which time approximately 4.5 million redwings, grackles, cowbirds and starlings were killed. At the present time there is no way to determine whether the 4.5 million birds, approximately 1% of the continental population, represents replacement or additive mortality. However, reports by Fretwell (1972a) and others lead us to consider that the wintertime mortality is the highest throughout the year for each of these four species and the treatment mortality is very likely to be additive.

In the southern republics of the USSR, massive poisoning programs have been carried out for about 20 years against populations of *P. hispaniolensis* and *P. domesticus bactrianus* (Shtegman, 1956; Golovanova & Zusmanovich, 1961; Gavrilov, 1965; Golovanova, 1966). As with the icterids, nothing is known at this time about the effect on these populations. The same type of actions have been taken against the Spanish sparrow and house sparrow in North Africa, again without reports of effects on the population size (Bortoli, 1973), but it is certainly of heuristic interest to monitor these actions to determine what happens in the future.

The role of flocking in various birds has long been of interest (Darling, 1952; Emlen, 1952) and it is especially important to these few granivorous species considered here. While we are not able to contribute to the reasons for flocking behavior, it is quite obvious that it is important and that it may enhance the survival potential of individual members when within flocks. Much of the information within this chapter, particularly juvenile and adult survival, represents events carried out during the non-breeding season, the majority of the life history of birds. Thus alterations in flock size ought to have demonstrable effects on the population, especially if density-dependent processes can be demonstrated. One example of how flocks operate is well demonstrated by differences in European tree sparrow foraging habits during open periods of the winter and during periods of heavy snow cover. Under favorable conditions large flocks form, but with heavy snow the flocks disperse widely, thus allowing for better usage of available forage. After the snow disappears the flocks reform and then find it advantageous to use single patches of forage resources more thoroughly. Presumably this is the type of strategy many birds use, but it is difficult to obtain information on such conditions for those species which maintain large foraging flocks such as blackbirds, *Passer* and *Quelea*.

In summary, we feel that the important granivores of the world are currently well-adapted species with relatively stable population sizes over a long-term period even though some, like quelea, fluctuate widely about such a mean (P. Ward, personal communication). Even though there are observable fluctuations in specific locations, there is no reason to believe that *Passer* and icterid species will change their population status dramatically in the near future. With continued land-use changes now evident on several continents, and with an increased emphasis on growth of cereal grains, it is possible that populations could increase. However, changes in food supplies alone are not sufficient grounds on which to base predictions about the growth of these bird populations. Also, if these bird species become super-abundant, or particularly troublesome in isolated areas, we should expect management programs directed at regulating numbers. Such programs, reviewed in Chapter 7, this volume, often change population structure one way or another, and it is necessary to keep their effects in mind when interpreting time- and area-specific dynamics of granivorous bird populations in various locations around the world.

We thank many members of the scientific community for contributing information to this chapter. P. Rassi, Helsinki, Finland; Rauno Alatalo, Oulu, Finland; Anthony Cheke, Oxford, England; J. H. van Balen, Arnhem, the Netherlands; Robert Spencer, British Trust for Ornithology, Tring; and the US Fish and Wildlife Service Bird-Banding Laboratory all provided unpublished data. We thank L. Jaraczewski, Warsaw, Poland; Zvika Abramsky and David M. Swift, Natural Resource Ecology Laboratory, Colorado State University, Fort Collins, Colorado; Dr Ken Burnham, US Fish and Wildlife Service, Fort Collins, Colorado for technical help; and J. Pinowska-Borońska, Warsaw, Poland, for translating much of our literature cited in this paper. Dr S. C. Kendeigh, Dr J. A. Wiens, Dr Peter Ward, Dr T. Anderson, Dr R. F. Johnston, Dr V. R. Dol'nik, Dr K. Petrusewicz and Dr J. B. Cragg all gave helpful suggestions and critique to this chapter. To them we are very grateful.

The work has been supported in part by the USA National Science Foundation Grant GB-42700, by the Institute of Ecology, Polish Academy of Sciences, and the support services of the Natural Resource Ecology Laboratory, Colorado State University, through NSF Grants GB-41233X, BMS73-02027 AO2 and DEB73-02027 AO3. Travel funds have been supplied by the IBP, London.

4. Biomass and production rates

J. PINOWSKI & A. MYRCHA

The preceding chapter has dealt with population dynamics as it refers to individuals. Production rates, however, are based on biomasses and all energy equations (Chapter 5, this volume) are based on weight. We are concerned in this chapter, therefore, with the analysis of weights of eggs, nestlings and adults and how they vary with time and locality, to serve as a background for the chapters that follow. This analysis deals only with the house sparrow *Passer domesticus* (L.), tree sparrow *P. montanus* (L.), red-winged blackbird, *Agelaius phoeniceus* (L.), and common grackle, *Quiscalus quiscula* (L.), as suitable data are available only on these species of granivores.

The genus *Passer*

Eggs

Variation in successive clutches

Successive clutches are identified by the time of year in which they are laid and do not necessarily represent successive clutches of the same individual except in Romania and Hungary. In the most numerous data, from Czechoslovakia (Fig. 4.1: 3*a*, *b*, *c*), the lightest eggs of *P. domesticus* were laid in the second clutches, the differences in weight between the first and second clutches being statistically significant ($P < 0.05$) for 1963 and 1965. In India, *P. domesticus indicus* breeds from February to October with an interruption in the summer. The heaviest eggs occur in the first clutches after the summer break (Fig. 4.1: 2*a*, *b*). The only data for *P. montanus* are from the Netherlands (Fig. 4.1: 1*a*, *b*, *c*), and here the third clutches appear to be the lightest, but the differences between clutches are not statistically significant. Thus, there is no clear certainty that the weight of *Passer* eggs varies consistently between clutches, although this has been found in some other species (Nice, 1937; Kendeigh, 1941; Coulson, 1963; Gromadzki, 1966; Perrins, 1970; Pikula, 1971).

Annual variation

Variations in the mean weights of eggs from year to year in house and tree sparrows are not statistically significant (Table 4.1).

J. Pinowski & A. Myrcha

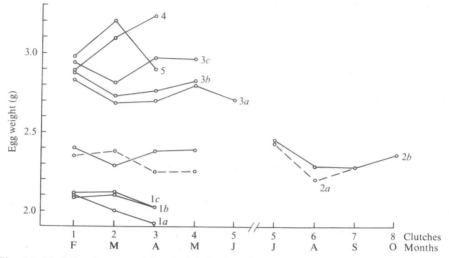

Fig. 4.1. Variation in egg weights during the breeding season; for India (2a, b) average weights in clutches of eggs are given by months; for other localities average weights in all clutches are given by successive clutches. *P. montanus*: (1a) Loenen, (1b) Oosterhout, (1c) Schuilenburg, the Netherlands (Van Balen, Brugge & Korf, unpublished). *P. domesticus*: (2a) 1972, (2b) 1971, Baroda, India (Naik & Mistry, in preparation); (3a) 1963, (3b) 1965, (3c) 1964, Slezké Rudoltice, Czechoslovakia (Novotný, 1970); (4) Iasi, Romania (Ion, 1973); (5) Báczsalmás, Hungary (J. Rékási, unpublished).

Egg weight and clutch size

In India the weight of house sparrow eggs in three-egg clutches is significantly greater ($P < 0.001$) than in other size clutches (Table 4.2). In the great tit *Parus major* L. and blue tit *P. caeruleus* L., heaviest eggs have also been found in medium-sized clutches (Winkel, 1970).

Egg weight and order laid

Successive eggs of the house sparrow tend generally to decrease in weight (Table 4.2), although in individual clutches this is sometimes not evident or even reversed. The decrease between successive eggs averages 2.14% in two-egg clutches, 2.08% in three-egg clutches, 2.13% in four-egg clutches, and 1.9% in five-egg clutches. The decrease in weight between successive eggs is statistically significant in all instances except in the two-egg clutch and between the third and fourth eggs in the five-egg clutch. This trend agrees with what has been found in gulls (Vermeer, 1969) but is opposite to the increase in weight that occurs in the house wren *Troglodytes aedon* Vieillot (Kendeigh, Kramer & Hamerstrom, 1956).

108

Table 4.1. *Variations in egg weights from year to year*

Year	Number of eggs	Weight (g)	Locality	Authority
		House sparrow		
1963	280	2.73±0.24	Czechoslovakia	Novotný (1970)
1964	481	2.94±0.28	Czechoslovakia	Novotný (1970)
1965	418	2.79±0.37	Czechoslovakia	Novotný (1970)
1968	142	2.83	GDR	Wendtland (unpublished)
1969	122	2.64	GDR	Wendtland (unpublished)
1970	28	3.05	GDR	Wendtland (unpublished)
1971	95	2.78	GDR	Wendtland (unpublished)
		Tree sparrow		
1957	30	2.24[a]	Poland	Busse (1962)
1958	110	2.23[a]	Poland	Busse (1962)
1959	137	2.14[a]	Poland	Busse (1962)
1960	174	2.09[a]	Poland	Busse (1962)
1961	243	2.17[a]	Poland	Busse (1962)

[a] Egg weights calculated from the size according to Dawson's (1972) formula: $W = Klb^2$, l = length, b = breadth, $K = 0.0005442$ g mm^{-3} density^{-1}.

Table 4.2. *Weights of house sparrow eggs* (g) *in clutches of different size, Naik & Mistry* (in preparation)

	First egg	Second egg	Third egg	Fourth egg	Fifth egg	Whole clutch
One-egg clutches						
$\bar{X}\pm$s.d.	2.30±0.17	—	—	—	—	2.30±0.17
c.i.	2.23–2.37	—	—	—	—	2.23–2.37
N	27	—	—	—	—	27
Two-egg clutches						
$\bar{X}\pm$s.d.	2.34±0.23	2.29±0.26	—	—	—	2.32±0.24
c.i.	2.27–2.43	2.21–2.37	—	—	—	2.26–2.36
N	44	44	—	—	—	88
Three-egg clutches						
$\bar{X}\pm$s.d.	2.43±0.29	2.39±0.24	2.33±0.26	—	—	2.38±0.27
c.i.	2.39–2.47	2.36–2.42	2.30–2.36	—	—	2.36–2.40
N	221	221	221	—	—	663
Four-egg clutches						
$\bar{X}\pm$s.d.	2.40±0.26	2.34±0.23	2.30±0.24	2.25±0.24	—	2.32±0.25
c.i.	2.37–2.43	2.31–2.37	2.27–2.33	2.22–2.28	—	—
N	315	315	315	315	—	1260
Five-egg clutches						
$\bar{X}\pm$s.d.	2.38±0.24	2.34±0.18	2.31±0.23	2.28±0.23	2.20±0.31	2.30±0.25
c.i.	2.33–2.43	2.30–2.38	2.26–2.36	2.23–2.33	2.13–2.27	—
N	74	74	74	74	74	370

$\bar{X}$ = mean; s.d. = standard deviation; c.i. = 95% confidence interval ($\bar{X}\pm2$ s.e.).

Table 4.3. *Geographical variations in egg weights (g) in the house sparrow*

Sub-species	Locality	Latitude, longitude	Number of eggs	Weight of eggs ($\bar{X}\pm$s.D.)	Authority
Asia					
indicus	Baroda, India	22° 18′ N, 73° 13′ E	2408	2.34±0.25	Naik & Mistry (in preparation)
hyrcanus	Talysz, USSR	39° N, 49° E	?	2.36	Mustafaev (1969)
griseogularis	Turkmenistan, USSR		25	2.65	Keleïnikov (1953)
bactrianus	Alma-Ata, USSR	43° 05′ N, 77° E	80	2.35±0.03	Gavrilov & Korelov (1968)
domesticus	Alma-Ata, USSR	43° 05′ N, 77° E	62	2.54±0.04	Gavrilov & Korelov (1968)
Europe					
domesticus	Bujor Targ, Romania	45° 51′ N, 27° 57′ E	150	3.08	Ion (1973)
domesticus	Romania		177	2.93	Cătuneanu & Theis (1965)
domesticus	Bácsalmás, Hungary	46° 10′ N, 19° 20′ E	202	3.04	J. Rékási (unpublished)
domesticus	Trnava, Czechoslovakia	48° 30′ N, 17° 50′ E	133	2.78	Matoušek (1956)
domesticus	Slezké Rudoltice, Czechoslovakia	50° 13′ N, 17° 43′ E	1099	2.85±0.04	Novotný (1970)
domesticus	Saxony, GFR		100	2.94	Niethammer (1937)
domesticus	England		868	2.79	Bumpus (1896) from Pearson (1902)
domesticus	Oxford, England	51° 46′ N, 1° 15′ W	362	2.80	Dawson (1972)
domesticus	South England		687	2.85	Pearson (1902)
domesticus	Jordenstorf, GDR	53° 50′ N, 12° 35′ E	387	2.77	A. Wendtland (unpublished)
domesticus	Belgium		308	2.94	Verheyen (1967)
domesticus	Great Britain		100	3.00	Witherby, Jourdain, Ticehurst & Tucker (1948)
domesticus	Kangasala, Finland	61° 27′ N, 24° 03′ E	54	2.82	P. Rassi (unpublished)
USA					
domesticus	Ithaca, New York	42° 45′ N, 76° 30′ W	54	2.94	Weaver (1943)
domesticus	Buffalo, New York	43° N, 79° W	21	2.62±0.15	Ar et al. (1974)
domesticus			868	2.73	Bumpus (1896) from Pearson (1902)
New Zealand					
domesticus	Christchurch	43° 42′ S, 172° 38′ E	217	2.88	Dawson (1964)

110

Table 4.4. *Geographical variations in egg weights (g) in the tree sparrow*

Sub-species	Locality	Latitude, longitude	Number of eggs	Weight of eggs	Authority
malaccensis	Jawa, Indonesia	8° S, 110° E	?	1.97	Hoogerwerf (1949)
malaccensis	Singapore	1° N, 104° E	57	2.00	Ward & Poh (1968)
saturatus	Taiwan	24° N, 121° E		2.13	Hartert (1903–1910)
saturatus	Peking, China	39° 55′ N, 116° 25′ E	158	2.01±0.002	Chia, Bei, Chen & Cheng (1963)
saturatus	Sapporo, Japan	43° 00′ N, 141° 30′ E	342	2.27±0.20	Abé (1969)
montanus	Romania		100	2.08	Dombrowski (1912)
montanus	Bujor Targ, Romania	45° 51′ N, 27° 57′ E	150	2.11	Ion (1973)
montanus	Romania		256	2.13	Cătuneanu & Theis (1965)
montanus	Bácsalmás, Hungary	46° 10′ N, 19° 20′ E	75	1.92	J. Rékási (unpublished)
montanus	Trnava, Czechoslovakia	48° 30′ N, 17° 50′ E	35	2.22	Matoušek (1956)
montanus	Warsaw, Poland	52° 10′ N, 21° E	694	2.15	Busse (1962)
montanus	Dziekanów Leśny, Poland	52° 20′ N, 20° 50′ E	68	2.13	Pinowski (1967b)
montanus	Saransk, USSR	54° N, 44° E	75	2.37	Maïkhruk (1974)

J. Pinowski & A. Myrcha

Geographical variations

The weight of house sparrow eggs is clearly less at Baroda, India, and at three localities in the USSR than elsewhere (Table 4.3). These localities in Asia occur at relatively low latitudes but include the smaller bodied sub-species, *P. domesticus indicus*, *P. d. hyrcanus*, and perhaps *P. d. griseogularis* and hence are not definitely related to latitude (Sudilovskaya, 1954; Gavrilov & Korelov, 1968). In some other species and even sub-species, smaller individuals are known to lay smaller eggs (Nice, 1937; Snow, 1960; Coulson, 1963). At all other localities in Europe, USA and New Zealand variations in egg weights are small and within the range of variations within a season.

Tree sparrow eggs average least in weight in Hungary and Indonesia and greatest in weight in Japan and at Saransk, USSR (Table 4.4). There is no correlation with latitude.

Loss during incubation

Egg weight decreases during incubation mainly because of evaporation of moisture through the shell. In a study done in New Zealand with the house sparrow (Dawson, 1964), this loss varied between 0.01 and 0.05 g egg^{-1} day^{-1}. It averaged 0.013 ± 0.0018 before incubation, 0.032 ± 0.0011 during incubation and 0.070 ± 0.0055 g egg^{-1} day^{-1} immediately before hatching. These differences are all statistically significant ($P < 0.05$). In the USA, Weaver (1943) found the daily weight loss during incubation to average 1.34% or 0.038 g egg^{-1} day^{-1}.

At one locality in the Netherlands, the eggs of the tree sparrow lost 0.0164 g egg^{-1} day^{-1} and at two other localities, 0.0229 g egg^{-1} day^{-1} (Van Balen, Brugge & Korf, unpublished). In the Ferganskaya Valley of the USSR, the eggs of the house sparrow, *P. d. griseogularis*, lost 0.71% of their original weight per day during the 13 days of incubation and the eggs of the tree sparrow, 0.86% per day over its 11-day incubation period (Kashkarov & Puzankova, 1974).

These egg losses during incubation in *Passer* correspond to those that occur in other species that lay eggs similar in size and weight, e.g. *Acanthis cannabina* (L.), *Sylvia communis* L., *Fringilla coelebs* L., *Prunella modularis* (L.), *Erithacus rubecula* (L.), *Emberiza citrinella* L. (Groebbels & Möbert, 1927, after Drent, 1970).

Young

The increase in weight of house and tree sparrow nestlings follows the logistic growth curve that is typical for passerines generally (Ricklefs, 1968). In order to compare the growth rate in successive broods and in

112

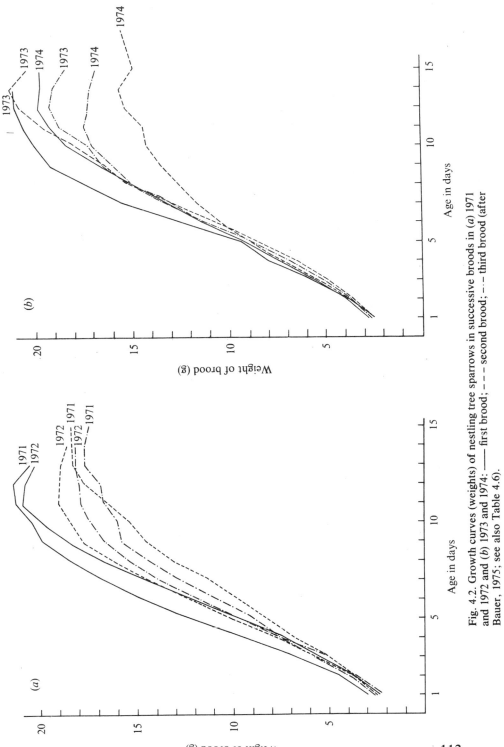

Fig. 4.2. Growth curves (weights) of nestling tree sparrows in successive broods in (a) 1971 and 1972 and (b) 1973 and 1974: —— first brood; – – second brood; – – third brood (after Bauer, 1975; see also Table 4.6).

113

Table 4.5. *Summary of growth parameters in the house sparrow*

Locality	Brood 1			Brood 2			Brood 3			Authority
	k	A	r^2	k	A	r^2	k	A	r^2	
Bujor Targ, Romania	0.4203	24.40	0.974	—	—	—	—	—	—	Ion (1973)
Dziekanów Leśny, Poland, 1968	0.5416	23.64	0.996	—	—	—	—	—	—	Pinowska & Pinowski (1977)
Dziekanów Leśny, Poland, 1973	0.5821	23.26	0.993	—	—	—	—	—	—	Pinowska & Pinowski (1977)
Slezké, Rudoltice, Czechoslovakia	0.5989	21.56	0.981	0.2624	25.04	0.975	—	—	—	Novotný (1970)
Oxford, England	0.5528	21.61	0.977	—	—	—	—	—	—	Dawson (1972)
Portage des Sioux, USA	0.7343	22.83	0.993	—	—	—	—	—	—	Anderson (1973)
Coldspring, USA	0.6270	25.50	0.974	—	—	—	—	—	—	North (1973a)

	Brood 1			Brood 2			Brood 3			Brood 4			Brood 5		
	k	A	r^2	k	A	r^2	k	A	r^2	k	A	r^2	k	A	r^2
Turew, Poland[a]	0.5109	24.97	0.992	0.4343	22.04	0.998	0.5001	25.09	0.986	0.4884	20.28	0.988	0.4806	24.76	0.990
Baroda, India[b]	0.5192	15.05	0.971	0.4990	14.73	0.980	0.4946	15.75	0.971	0.5180	15.88	0.994	0.5852	16.80	0.975
	Brood 6			Brood 7			Brood 8			Brood 9					
	0.6426	19.24	0.969	0.5851	18.63	0.990	0.5754	17.86	0.970	0.5238	18.05	0.992			

k = the rate of growth of the young throughout the nestling period ($g\ day^{-1}$), A = fledging weight of the individual, as estimated by the asymptote determined from the logistics curve, r^2 = the goodness of fit of the data by the non-linear least-squares method.

[a] 1–5, Wieloch & Fryska (1975).

[b] 1–9, Naik (unpublished).

Table 4.6. *Summary of growth parameters in the tree sparrow*

Locality	Brood 1			Brood 2			Brood 3			Authority
	k	A	r^2	k	A	r^2	k	A	r^2	
Peking, China	0.6942	18.17	0.985	—	—	—	—	—	—	Chia et al. (1963)
Sapporo, Japan	0.4201	16.86	0.998	0.4661	21.15	0.994	0.4442	16.93	0.993	Abé (1969)
Sokolnice, Czechoslovakia	0.5535	19.57	0.995	—	—	—	—	—	—	Balát (1971)
Bzenec, Czechoslovakia	0.5071	20.23	0.992	—	—	—	—	—	—	Balát (1971)
Breclav, Czechoslovakia (1971)	0.5298[a]	21.67	0.998	0.3547[a]	19.47	0.996	0.4769[a]	17.74	0.997	Bauer (1974)
Breclav, Czechoslovakia (1972)	0.4688	21.93	0.998	0.4836	19.70	0.997	0.4769	18.71	0.999	Bauer (1974)
Breclav, Czechoslovakia (1973)	0.4763	22.07	0.998	0.3964	22.30	0.997	0.4544	19.63	0.997	Bauer (1974)
Breclav, Czechoslovakia (1974)	0.4178	20.67	0.998	0.4582	15.81	0.976	0.5090	17.09	0.983	Bauer (1974)
Iasi, Romania	0.3991	22.30	0.989	—	—	—	—	—	—	Ion & Saracu (1971)
Portage des Sioux, USA	0.8416	17.03	0.982	—	—	—	—	—	—	Anderson (1973)
Turew, Poland (1969)	0.5327	22.22	0.985	—	—	—	—	—	—	Wieloch & Fryska (1975)
Turew, Poland (1973)	0.3943	24.43	0.999	—	—	—	—	—	—	Wieloch & Fryska (1975)
Dziekanów Leśny, Poland (1968)	0.5407	20.54	0.992	—	—	—	—	—	—	Pinowski (1968)
Dziekanów Leśny, Poland (1969)	0.4587	19.20	0.990	—	—	—	—	—	—	Mackowicz, Pinowski & Wieloch (1970)
Dziekanów Leśny, Poland (1973)	0.5759	20.39	0.976	—	—	—	—	—	—	Pinowski & Wieloch (1973)
Schuilenburg, Netherlands	0.5356	19.15	0.999	—	—	—	—	—	—	J. H. van Balen (unpublished)
Oosterhout, Netherlands	0.4839	22.28	0.998	—	—	—	—	—	—	J. H. van Balen (unpublished)
Loenen, Netherlands	0.4762	22.45	0.996	—	—	—	—	—	—	J. H. van Balen (unpublished)

k = the rate of growth of the young throughout the nestling period (g day^{-1}), A = fledging weight of the individual, as estimated by the asymptote determined from the logistics curve, r^2 = the goodness of fit of the data by the non-linear least-squares method.

[a] The differences are statistically significant between the growth rates (k) of broods 1 and 2 ($P < 0.05$) but not significant for broods 2 and 3. The differences of fledging weights (A) are significant only between broods 1 and 3 ($P < 0.01$).

. Pinowski & A. Myrcha

different localities, values of k (g day^{-1}) were calculated. Likewise, fledging weight was determined from the asymptote of the curves (Dyer & Abramsky, in preparation).

Successive broods

In both species of *Passer*, measured growth rates vary from year to year and between localities (Fig. 4.2, Tables 4.5, 4.6), but the differences are mostly not significant. The data for the house sparrow at Turew, Poland, show the highest growth rate for the first brood and the lowest for the second. First broods also have highest growth rates in Czechoslovakia (Novotný, 1970). At Baroda, India, where there may be nine broods raised during a year, highest growth rates occur after the summer break, correlating with the largest eggs which are laid at this time also (Naik, unpublished). In the tree sparrow, Bauer's (1974) data for Czechoslovakia show lower growth rates for the second brood than either the first or third in two of the four years.

The weight of the young at fledging, the number of days that they stay in the nest and their growth rates are interrelated. Fig. 4.2 shows clearly that broods with fast growth rates attain greatest weights before leaving the nest and fledge in the fewest number of days. The last broods of house sparrows often stay in the nest longer than any other brood (Novotný, 1970). In India, growth rates are fastest and weights at fledging are greatest for broods five to nine, after the summer break (Table 4.5).

Growth rates are influenced by temperature and the quantity and quality of food. Temperatures both above and below normal depress growth rates (Dawson, 1972; Novotný, 1973). The mean maximum weights attained by 44 tree sparrow nestlings in Japan in the spring at an average ambient temperature of 13.5 °C was 16.75 ± 1.13 g; that of 38 nestlings during the summer at 20.8 °C was 20.60 ± 2.93 g, a significant difference ($P < 0.01$). The mean duration of the nestling period in the spring was 17.5 ± 0.3 days; in the summer, 14.0 ± 0.2 days ($P < 0.05$). There was a higher proportion of plant material in their food during the spring than in the summer which may also have exerted an effect (Abé, 1969). In Czechoslovakia and Poland the diet of tree sparrow nestlings of first broods consists mainly of lepidopteran and coccinellid larvae, that of second broods that grow the slowest consists of orthopterans, while that of third broods is frequently garbage (Myrcha, Pinowski & Tomek, 1973; Bauer, 1974, 1975). Similar differences between brood diets have been noted by Berger (1959) and Grün (1964a).

Geographical variations

The principal geographical difference noticeable in the growth of nestlings (Tables 4.5, 4.6) is the significantly lower weight ($\bar{X} = 16.89$ g) attained by

116

Table 4.7. *Changes in specific calorific values (gcal g^{-1}) of nestlings during growth*

Age in days	Tree sparrow				House sparrow	
	(1)	(2)	(3)	(4)	(5)	(6)
0	—	—	—	—	847	768
1	—	—	913±76	697±43	789	825
2	1100	800	—	670±20	887	882
3	—	—	—	777±22	916	939
4	—	—	947±95	821±6	904	996
5	—	—	—	991±19	979	1053
6	—	—	—	1017±15	1150	1110
7	1250	930	—	990±15	1240	1167
8	—	—	1412±131	1250±33	1190	1224
9	—	—	—	1188±11	1230	1281
10	—	—	—	—	1507	1338
11	—	—	1927±149	1590±59	1486	1395
12	2420	1620	1916±107	1493±14	1505	1452
13	—	—	1940±195	1336±101	—	1509
14	—	—	2025±112	1633±136	1492	1566
15	—	—	—	—	1509	1623

(1) Bauer (1975), Breclav, Czechoslovakia; (2) Pinowski (1967 b), Powsin near Warsaw, Poland; (3) Van Balen *et al.* (unpublished data), the Netherlands; (4) Myrcha & Pinowski (1969), Dziekanów Leśny near Warsaw, Poland; (5) Myrcha *et al.* (1973), Dziekanów Leśny near Warsaw, Poland; (6) Blem (1975), Richmond, Virginia, USA.

the Indian population of house sparrows (Naik & Mistry, in preparation) than elsewhere, for instance, Poland ($\bar{X} = 23.43$ g). The lower weight of the young at fledging in India is correlated not with growth rate ($\bar{X} = 0.5492$ g day^{-1}), which is very similar to the growth rate elsewhere, but with the generally lower weight characteristic of adult birds in this area (see below).

Calorific production

The specific calorific value of nestling sparrows decreases during the first two or three days after hatching, as the unused egg yolk absorbed into the body becomes metabolized, and then increases throughout most of nest life (Table 4.7), as is characteristic of species generally (Ricklefs, 1967). The increase is caused primarily by a decrease in the proportion of water in the body and secondarily by an increase in lipids (Myrcha & Pinowski, 1969; Myrcha *et al.*, 1973). The equation for the increase in kcal g^{-1} of fresh tissue at various ages in days (d) for the house sparrow ($N = 62$) is:

$$\text{kcal g}^{-1} = 0.7918 + 0.0547d \quad (r = 0.935) \tag{4.1}$$

and for the tree sparrow ($N = 73$) it is:

$$\text{kcal g}^{-1} = 1.0885 + 0.0733d \quad (r = 0.896) \tag{4.2}$$

117

Blem (1975) gives the average rate of increase in the total calorific content of the body from day one to 14 as 2.85 ± 0.16 kcal day^{-1}. Myrcha *et al.* (1973) found the total increase per bird from hatching to the time of fledging in the house sparrow to be: first brood, 31.6 kcal; second brood, 31.5 kcal; third brood, 35.0 kcal, and in the tree sparrow: first brood, 27.7 kcal; second brood, 27.5 kcal; third brood, 25.7 kcal (differences statistically significant).

After leaving the nest, fledgling and juvenile house sparrows continue to increase in weight until about 80 to 100 days of age, when they reach the weight of adults (P. Rassi, unpublished data). In England, young birds increase in weight from June to August (O'Connor, 1973). There are no data available for the house sparrow, but for the tree sparrow the specific calorific value g^{-1} fresh weight continues to increase after they leave the nest until moulting is completed (Myrcha & Pinowski, 1970).

Total Production

To obtain total production, one must include not only the young that leave the nest but also the biomass production of eggs that do not hatch and hatched young that do not fledge. The nutrients and energy in these eggs and young, as well as in adult birds that die each year, are utilized by predators and decomposers and, hence, are part of the impact that birds make on the ecosystem.

There is variation from year to year and between localities in the amount of this total production, but overall it averages 142.7 ± 26.3 g per pair in the house sparrow and 143.8 ± 41.4 g per pair in the tree sparrow (Table 4.8). In terms of energy, the average production is 217.9 ± 45.7 kcal per pair in the house sparrow and 204.9 ± 59.8 kcal per pair in the tree sparrow.

When population densities are known (Chapter 3, this volume) and assuming that all birds breed, the biomass production per area per year of living adults may also be calculated. There are considerable variations in production between biotopes and localities and usually smaller variations from year to year in the same biotope (Table 4.9).

Adults

Adult birds gain in weight during the daytime as a result of feeding and lose weight at night when the energy stored in the body during the daytime, principally as fat, is metabolized. These processes are generally in balance, except in cool weather when there may be a change in weight because larger fat reserves are laid down on the body and during warm weather when the reverse occurs (Kendeigh, Kontogiannis, Masac & Roth, 1969; Pinowski & Myrcha, 1970).

118

Table 4.8. *The biomass (g) produced by one breeding pair per year.* (*For methodology of counting, see Mackowicz et al., 1970.*)

Locality	Year	Eggs lost	Nestlings lost	Young leaving the nest	Total production	Authority
House sparrow						
Poland						
Gdańsk	1969	3.7	21.2	67.2	92.1	(1)
	1970	2.5	11.3	111.2	125.0	(1)
Dziekanów	1967	8.6	34.4	138.4	181.5	(2)
Leśny	1968	6.0	24.5	139.1	169.6	(2)
	1969	6.1	19.5	96.4	122.0	(1)
	1970	7.2	14.4	98.4	120.0	(1)
Rzepin	1967	4.7	15.5	133.5	153.6	(2)
	1968	3.5	23.2	117.7	144.4	(2)
Kraków	1967	4.0	42.5	114.2	160.7	(2)
	1968	3.0	11.4	117.5	131.9	(2)
Nowy Targ	1967	1.1	22.0	157.0	180.1	(2)
	1968	2.2	32.1	97.8	132.1	(2)
		4.4±2.1	22.7±9.2	115.7±23.2	142.7±26.3	
Tree sparrow						
Gdańsk	1969	1.3	2.0	62.7	66.0	(1)
	1970	2.4	27.1	88.5	118.0	(1)
Dziekanów	1967	6.5	13.3	187.3	207.1	(2)
Leśny	1968	5.0	18.5	165.5	189.0	(2)
	1969	5.0	13.0	82.0	100.0	(1)
	1970	7.6	11.3	170.1	189.0	(1)
Rzepin	1968	5.0	21.6	96.4	123.0	(2)
Kraków – town	1967	2.5	68.0	37.3	107.8	(2)
Kraków – farm	1967	2.5	5.2	136.0	143.7	(2)
	1968	5.0	10.1	123.7	138.8	(2)
	1969	2.0	10.1	88.8	100.1	(1)
	1970	2.8	4.2	132.0	139.0	(1)
Czechoslovakia						
Sokolnice	1970	3.9	39.1	120.2	163.2	(3)
	1971	2.6	19.6	169.5	191.7	(3)
Bzenec	1970	5.9	48.2	158.0	212.1	(3)
	1971	7.0	17.9	176.6	201.5	(3)
		4.2±1.9	20.6±17.1	124.7±43.6	143.8±41.4	

(1) Pinowski & Wieloch (1973); (2) Mackowicz *et al.* (1970); (3) Balát & Toušková (1972).

In permanently resident species, such as *Passer*, there tends to be a progressive gain in weight from mid-summer to mid-winter due to increased fat storage on the body, and a loss in weight during the spring, which is correlated inversely with prevailing ambient temperatures (Baldwin & Kendeigh, 1938; O'Connor, 1973). Superimposed on these seasonal trends are fluctuations that sometimes differ between sexes. During the moulting period, i.e. from August to October, there is an

119

Table 4.9. *Biomass production per unit area* ($kg\ km^{-2}\ year^{-1}$)

Locality	Year	Biomass production	Authority
	House sparrow		
Poland			
Gdańsk – village	1973	2.5	Wieloch & Fryska (1975)
Rzepin – village	1967	35.6	Mackowicz et al. (1970)
	1968	30.0	Mackowicz et al. (1970)
Dziekanów Leśny – village	1970	3.5	Pinowski & Wieloch (1973)
Kraków – farm	1970	21.3	Pinowski & Wieloch (1973)
Kraków – town	1970	29.0	Pinowski & Wieloch (1973)
	Tree sparrow		
Gdańsk – village	1973	0.8	Wieloch & Fryska (1975)
Dziekanów Leśny – village	1961	20.8	Pinowski (1967 b)
	1962	14.6	Pinowski (1967 b)
	1970	0.7	Pinowski & Wieloch (1973)
Rzepin – village	1968	4.6	Mackowicz et al. (1970)
Kraków – farm	1970	45.4	Pinowski & Wieloch (1973)
Kraków – town	1970	3.4	Pinowski & Wieloch (1973)
GFR			
Wolfsburg – suburbs, village	1968	11.4	Scherner (1972)
	1969	6.9	Scherner (1972)
Czechoslovakia			
Breclav – alluvial forest	1971	1.0	Bauer (1975)
	1972	1.2	Bauer (1975)
	1973	1.2	Bauer (1975)
	1974	1.1	Bauer (1975)

increase in water content within the body and an increase in weight of plumage (Korelus, 1947; Dol'nik, 1967a; Myrcha & Pinowski, 1970). A decrease in male body weight recorded in October in three of the four localities (Fig. 4.3) has been attributed to increased sexual displays and growth of reproductive organs – but this decrease did not occur in females in two of three localities (Fig. 4.4). During the spring, the seasonal decrease in weight resulting from the decrease in fat reserves is more or less offset in the female by the increase in size of the ovary and oviduct as breeding commences. The body composition, and hence the calorific value of its tissues, varies with the addition and utilization of lipids, but an average value of 1.567 kcal g^{-1} (Blem, 1975) would be a useful approximation.

In North America, house sparrows progressively increase in weight from south to north (Johnston & Selander, 1964; Blem, 1973a), due in large part to greater fat reserves on the body. In Europe, one study of birds in an oceanic climate (England, Norway, Denmark) indicated the reverse trend (Johnston & Selander, 1973a), but other studies further inland in continental climates showed an increase in weight from south to north

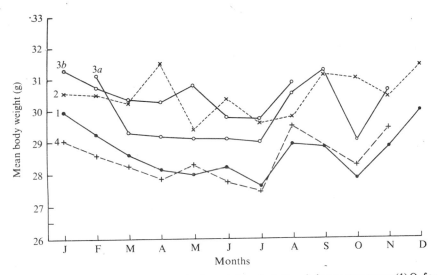

Fig. 4.3. Monthly variations in mean body weights of adult male house sparrows: (1) Oxford, England (O'Connor, 1973); (2) Czechoslovakia (Folk & Novotný, 1970); (3*a*) 1956, (3*b*) 1957, Moskva, USSR (Il'enko, 1962); (4) Austin, Texas, USA (Selander & Johnston, 1967).

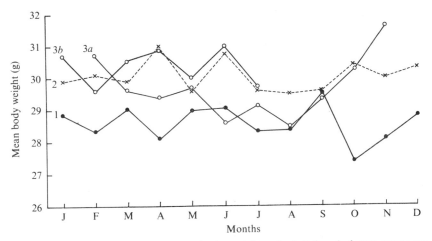

Fig. 4.4. Monthly variations in mean body weights of adult female house sparrows: (1) Oxford, England (O'Connor, 1973); (2) Czechoslovakia (Folk & Novotný, 1970); (3*a*) 1956, (3*b*) 1957, Moskva, USSR (Il'enko, 1962).

121

J. Pinowski & A. Myrcha

in both house and tree sparrows (Danilov, Nekrasov, Dobrinskii & Kopein, 1969; Nordmeyer, Oelke & Plagemann, 1973). At Baroda, India (22° 18′ N), adult house sparrows weigh only 20.5 g, much lighter than populations elsewhere (Naik & Mistry, in preparation). The regressions of weight (W = grams) on latitude (north) (°L), for the house sparrow are:

$$\text{males } (N = 37): \quad W = 16.15 + 0.289°L \quad (r = 0.664, \ P < 0.01) \quad (4.3)$$
$$\text{females } (N = 37): \quad W = 16.77 + 0.262°L \quad (r = 0.699, \ P < 0.01) \quad (4.4)[1]$$

Red-winged blackbird *Agelaius phoeniceus*

Eggs

Egg weight of the red-winged blackbird is clearly less in tropical than in temperate regions; 73 eggs from Costa Rica weighed an average of 3.6 g (Orians, 1973) but 246 eggs from Wisconsin, USA, averaged 4.3 g (Beer, 1965). Orians (1973) wrote: 'eggs were about 0.5 g lighter in Costa Rica than in central Washington. However, since Costa Rican redwings are smaller than Washington redwings, the eggs are about the same proportion of adult weight.'

Young

The growth rate (k = g day^{-1}) follows the logistic growth curve (Dyer & Abramsky, in preparation). There is no demonstrable difference in k between males and females, but there is a major difference between the asymptotes for presumed fledging weights. For ten sets of redwing data from wetlands and seven sets from upland biotopes, both growth rates and asymptotic fledging weights in these two biotopes are significantly different. In upland areas, nestlings grow faster and fledge at lower weights.

The growth rates of nestlings in Pennsylvania, that hatched early in the breeding season did not differ statistically from those that hatched late (Brenner, 1964).

Values for the weight-specific energy content of nestlings (wet weight) increase during development from about 600 to 1300 gcal g^{-1} (Ricklefs, 1967), which is less than in *Passer* (see Table 4.7).

The production of biomass per female per year includes both the number of young raised to maturity, calculated from Table 3.17, and the eggs and young that were lost (Table 4.10). The weight of eggs was taken to be 4.3 g

[1] Data for both regressions obtained from the following papers: Danilov *et al.* (1969); Geiler (1959); Grimm (1954); Ion (1973); Löhrl & Böhringer (1957); Niethammer (1953); Nordmeyer *et al.* (1973); O'Connor (1973); Scherner (1974); Rékási (unpublished); Naik (unpublished).

Table 4.10. *Production of biomass (g female⁻¹ year⁻¹) for red-winged blackbirds*

Locality	Eggs lost	Nest-lings lost	Young leaving the nest	Total produc-tion	Authority
Cook County, Illinois	6.0	16.2	40.2	62.4	Smith (1943)
Michigan	5.8	19.2	61.1	86.2	Laux (1970)
Spokane County, Washington	6.4	30.6	32.3	69.4	Holm (1973)
Spokane County, Washington	5.4	27.8	40.5	73.7	Holm (1973)
Spokane County, Washington	7.3	29.1	30.8	67.3	Holm (1973)
Ohio	3.9	17.2	44.6	65.7	M. I. Dyer (unpublished)
Ohio	2.3	23.7	59.1	85.1	M. I. Dyer (unpublished)
Ohio	3.4	25.1	26.0	54.5	M. I. Dyer (unpublished)
New Haven, Connecticut	8.2	36.7	95.9	140.8	Robertson (1972)
New Haven, Connecticut	17.5	73.9	88.1	179.5	Robertson (1972)
Harrisburg, Pennsylvania	5.3	28.3	81.4	115.1	Wood (1928)
Harrisburg, Pennsylvania	2.6	30.4	78.1	111.0	Wood (1938)
Indianapolis, Indiana	1.8	27.8	74.4	104.0	Perkins (1928)
Stoddard, Wisconsin	6.1	19.7	36.1	61.9	Young (1963 b)
Stoddard, Wisconsin	7.7	22.0	23.8	53.5	Young (1963 b)
Madison, Wisconsin	3.1	24.3	69.5	96.9	Beer & Tibbits (1950)
Centre County, Pennsylvania	6.0	9.1	99.3	114.3	Brenner (1966)
Centre County, Pennsylvania	8.5	10.9	79.6	99.0	Brenner (1966)
Centre County, Pennsylvania	6.6	43.2	58.7	108.6	Brenner (1966)
Centre County, Pennsylvania	3.4	15.2	89.2	107.8	Brenner (1966)
Centre County, Pennsylvania	5.5	36.2	53.2	94.8	Brenner (1966)
Madison, Wisconsin	5.9	25.5	37.9	69.4	Snelling (1968)
Madison, Wisconsin	4.5	23.5	59.9	87.9	Wiens (1965)

($\bar{X}$ = 91.7±29.2)

(Beer, 1965), of nestlings zero to four days old, 10.0 g, and of older nestlings 30.4 g (Holcomb & Twiest, 1968). Of the total mortality of nestlings, 25% was calculated to have occurred within the first five days and 75% thereafter (Young, 1963 b). The number of nestlings lost each period was multiplied by their corresponding average weights. The number leaving the nest was multiplied by the average weight of the young for the given state (states as given in Table 4.10) or nearest state (Dyer & Abramsky, in preparation). The biomass produced per female per year averaged 91.7±29.2 g, with the greatest production occurring near New Haven, Connecticut, related to the high number of broods raised per female, and the smallest production in areas where there were considerable egg losses.

The biomass production per unit area was based on the density of singing males, assuming a ratio of two nesting females per male. Censuses of singing males have been made in Ohio and Michigan in the USA and

Table 4.11. *Biomass production* (g km^{-2} year^{-1}) *by populations of red-winged blackbirds. Density of males given in Table III of Dyer et al.* (*1973*)

Year	North-central Ohio	Northwestern Ohio	Michigan	Ontario
1968	5180	4184	4013	1708
1969	4824	4440	2021	2362
1970	3857	4426	3444	2476

Table 4.12. *Biomass production* (g km^{-2} year^{-1}) *by populations of red-winged blackbirds in Wood County, Ohio. Density of males given in Table IV of Dyer et al.* (*1973*)

Year	Upland biotopes			Wetland biotopes		Cultivated land	
	Hay	Pasture	Fallow	Marsh	Ditch	Tilled	Grain
1964	13520	854	4127	33017	24905	0	285
1965	12097	10531	4412	26328	17647	0	854
1966	11954	2419	4269	29317	32163	142.3	569
1967	10674	3985	6546	31594	27466	0	854
1968	14658	5123	5408	31309	26186	0	142

in Ontario, Canada, by Dyer, Siniff, Curtis & Webb (1973). Greatest biomass production was recorded in Ohio (Table 4.11) and in wetland biotopes (Table 4.12).

Adults

Adult male red-winged blackbirds of resident populations in central California are usually heaviest during late winter. Weights become significantly lower by the end of the breeding season in June and July and then increase slightly during late summer. The annual cycle for females is similar, except that females are heaviest during the egg-laying period. Body weight decreases during incubation and feeding of young (Payne, 1969). The slope of the regressions of body weight with time during the breeding season was found to be similar for males and females in Centre County, Pennsylvania (Brenner, 1968).

There is a latitudinal increase in weight from Oklahoma, USA, in a northwest direction to northern Alberta, Canada, in both males and females (Power, 1970).

Common grackle *Quiscalus quiscula*

The average weight of eggs of the common grackle, calculated from the weights of 256 eggs collected in Alberta, Ontario and Quebec, Canada and in Ohio and Pennsylvania, USA, is 6.3 g (Maxwell, 1970). The nestling growth curve is known only for east-central Illinois, USA (Willson, John, Lederer & Muzos, 1971). These data, together with Table 3.20, were used for calculating biomass production per breeding female (Table 4.13). Winds and rain contribute to a large extent to egg losses and to high mortality of nestlings (Wiens, 1965; Willson *et al.*, 1971; Maxwell & Putnam, 1972). The lack of data on nesting densities of this species prevents calculation of production per unit area.

Summary

The biomass of egg production in *Passer* varies between clutches, as does the number and weight of the eggs. Successive eggs in a clutch generally decrease in size, with eggs in three-egg clutches averaging a greater weight than those in clutches of other sizes. There is little significant variation from year to year or geographically, except in India where the eggs are lighter. A decrease in egg weight occurs during incubation because of the evaporation of moisture through the shell.

Growth rates of house sparrow nestlings are generally fastest in first broods. In India they are the fastest after the summer interruption of breeding; this is correlated with the largest eggs which are laid at this time also. Broods with fast growth rates attain greater weight before leaving the nest and fledge in the fewest number of days. Growth rates are influenced by ambient temperature and the quantity and quality of food. The weight of young at fledging in India are lower than elsewhere. Adult weight is not attained until 80 to 100 days after leaving the nest.

Except for the first couple of days, the specific calorific value of fresh tissue increases throughout nest life, that of the adult averaging 1.567 kcal g^{-1}. The total increase per bird from hatching to the time of fledging varies slightly between broods but averages 32.7 kcal for the house sparrow and 27.0 kcal for the tree sparrow.

The biomass of living adults produced per pair averages 142.7 ± 26.3 in the house sparrow and 143.8 ± 41.4 g in the tree sparrow. Production per female red-winged blackbird is much less, only 91.7 ± 29.2 g. This is the result of the smaller number of eggs laid, considerable egg loss and high mortality in nestlings. On the other hand, fledging red-winged blackbirds are nearly twice as heavy as the fledglings of *Passer*. The biomass produced per female common grackle (164.6 ± 36.28 g), which as a rule has only one brood in a breeding season but which is of even greater weight, exceeds

J. Pinowski & A. Myrcha

Table 4.13. *Production of biomass* ($g\ pair^{-1}\ year^{-1}$) *by common grackles*

Locality	Eggs lost	Nest-lings lost	Young leaving the nest	Total produc-tion	Authority
Ohio, USA	13.6	25.4	95.3	134.3	Maxwell & Putnam (1972)
Wisconsin, USA	8.0	43.7	152.8	204.5	Peterson & Young (1950)
Wisconsin, USA	8.4	63.2	64.1	135.7	Snelling (1968)
Wisconsin, USA	5.5	44.5	163.3	213.3	Wiens (1965)
Ontario, Canada	13.9	28.3	93.2	135.4	M. I. Dyer (unpublished)

$(\bar{X} = 164.6 \pm 36.2)$

that produced by the red-winged blackbird or by *Passer* which have several broods per year. The biomass per square kilometer varies widely with population density and between localities and biotopes.

Adult birds gain weight during the daytime, lose weight at night, vary in weight inversely with ambient temperature, and are generally heavier in the winter than in the summer.

Many individuals contributed information to this chapter, for which we are grateful. J. H. van Balen, T. Brugge, and B. Korf of Arnhem, the Netherlands; P. Rassi of Helsinki, Finland; J. Rékási of Báczsalmás, Hungary; and A. Wendtland of Jordenstorf, GDR, provided unpublished data. L. Jaraczewski and B. Pinowska of Warsaw, Poland, gave technical help. M. I. Dyer and Z. Abramsky calculated growth rates for *Passer*, and S. C. Kendeigh, J. A. Wiens, T. R. Anderson and M. I. Dyer, all of the USA, gave helpful suggestions and criticisms.

126

5. Avian energetics

S. C. KENDEIGH, V. R. DOL'NIK & V. M. GAVRILOV

When the population and biomass and how they fluctuate are known, then the impact of a species on its ecosystem is determined by how it affects the habitat, its competitive and social interrelations with other species, and its position and influence in the food web of the community. Birds have a minimal effect on the physical characteristics of the habitat, and we are not in this volume greatly concerned with their niche and behavioral interactions with other species. We will be evaluating their importance primarily from the amount of food that they consume. First, we must learn the food requirements of the individual. This information can then be integrated with population dynamics, biomass and productivity to determine the total effects.

Direct determination of the transfer of nutrients and energy through the avian component of an ecosystem may be obtained by measuring the nutrient and energy content of the foods ingested and the return of unused excreta to the ecosystem. The kinds of foods eaten and their relative proportions in the diet must be determined through stomach content analysis, observation, or other methods. The exact amount of food eaten is difficult to measure and there are no satisfactory direct methods of so doing with free-living birds, therefore, indirect methods must be employed. One method involves measurement of food consumption of birds, confined to small cages, under different environmental conditions. Values obtained for caged birds under controlled conditions may then be projected to free-living birds under natural conditions. This method, although involving many variables and including several assumptions, is practical for supplying reliable information.

In this chapter, we will be concerned with measuring the energy requirements of birds under a variety of controlled conditions. Analysis will be made of the various components in their energetics, including basal, standard and existence metabolism and the requirements for migration, reproduction and moulting. The ultimate objective is the calculation of the daily energy budget of birds under natural conditions and the translation of this budget into food consumed.

Natural conditions involve a number of parameters, but temperature is basically most important and at the same time most convenient, as the exposure of birds to particular temperatures can be readily controlled in the laboratory and fluctuations of temperature regularly measured under natural conditions. Photoperiod is also easily controlled in the laboratory and measured in the field, and its modifying influence on the interrelation of energy requirements and temperature will be considered. Likewise, the modifying effects of variations in wind velocity, insolation, radiation and other environmental parameters will be considered as far as the available information on their effects is known.

The house sparrow *Passer domesticus* (L.) will be given most attention as more is known concerning its energy relations than for any other wild species of bird. However, there are considerable data also on other species, especially granivorous birds, which will be drawn on freely, and presumably all birds, whatever their food habits, both passerines and non-passerines, have basically the same relations as does the house sparrow, so our findings can be given wide application. These relations will be quantified on the basis of the species' weights.

Our analysis will be ecologically oriented throughout. We will be less concerned with physiological mechanisms at the cellular or regulatory levels. Actually, our topic is one of physiological ecology, concerned with the whole organism and how it functions in its environment.

Although a polynomial regression may sometimes best describe the relationship of metabolic rate to temperature in individual instances, the form which it takes varies from one species to another. However, there is invariably a linear regression also, and since this linear regression describes the relation between metabolic rate and temperature in an adequate manner, it has been adopted in this paper. In the generalized equation $M = a - bt$, M is kcal bird^{-1} day^{-1}, a is the value of M at 0 °C, b is the temperature coefficient, kcal bird^{-1} day^{-1} °C^{-1} and T is ambient temperature, °C.

Metabolic rates vary between species as a function of weight (W = grams) and according to the allometric equation $M = aW^b \pm$ S.E.E. This equation gives a linear regression when expressed logarithmically: log M = log $a + b$ log $W \pm$ S.E.E. The Y-intercept, a, represents the value of M when $W = 1$, S.E.E. ($S_{y/x}$) is the standard error of the estimate, M, and shows the limits of confidence (± 2 S.E.E.) best at the mean value of W. Regression equations are derived by the least-squares method and are only given when they are statistically significant. Data obtained under different situations are often pooled when equations for the situations are not significantly different statistically. A confidence level of at least 5% is required, although a confidence level of better than 1% was obtained in nearly all instances. Differences between log regressions are evaluated both as to slope and elevation, elevation being the level of the values at the Y-intercept and slope the rate of change per unit weight. Correlation coefficients (r) are not given for the individual equations as they are high ($\geqslant 0.94$) except for some of the equations dealing with temperature coefficients (equation 5.9, $r = 0.87$, equations 5.17 to 5.21, $r = 0.69$–0.80), the lower critical temperature (equations 5.10 and 5.11, $r = 0.82$, equation

128

5.12, $r = 0.71$) and the existence metabolic rate for Emberizidae (equation 5.24, $r = 0.80$). All equations were obtained with the University of Illinois digital computer using SOUPAC programs for the regressions and Zar's (1974) programs for the evaluations.

Basal metabolism (BM)

Basal metabolism (BM) in the strictest sense is the rate of energy utilization by animal organs and tissues at complete rest, unstimulated by the digestion and assimilation of food or by low temperature. In higher animals, true BM cannot be measured because of the energy output of respiratory, cardiac and other muscular and glandular activities necessary for the existence of life. Because of this, Krogh (1916) suggested the term 'standard metabolism' in place of basal metabolism. However, retention of the term 'basal metabolism' in a less strict sense is useful for the rate of energy utilization of fasting, inactive animals in the zone of thermo-neutrality or at least at high, relatively unstimulating ambient temperature.

BM is commonly measured by the rate at which oxygen is consumed or carbon dioxide liberated after a short period of fasting so that food assimilation is not underway and the oxidation of body fat, as indicated by a low respiratory quotient, is the principal source of energy. Likewise, measurement values are utilized only during periods when observation or recorders indicate little or no overt activity. An understanding of how BM varies under different conditions is of importance, as basal metabolism is the foundation on which energy requirements for all other activities are superimposed. The BM also furnishes a reference base for use of coefficients to estimate the energy cost of other activities.

Sex

Although differences in BM between males and females are sometimes noted, their occurrences are erratic and inconsistent except when there are pronounced differences in weight. Generally, in species with sexes equal in weight, measurements of male and female are combined.

Daily rhythm

Measurements of BM of diurnally active birds taken at night generally give lower values than when made during the daytime even when the birds are maintained in the dark (Aschoff & Pohl, 1970*a*, *b*). Daytime values in our data (Appendix 5.1) range from 30 to 40% higher than night-time values in small birds (5–50 g) and 10 to 25% higher in large birds ($\geqslant$ 500 g). Higher

129

values of BM during the daytime are probably due to the greater muscle tone in birds that are normally active at this time (Dontcheff, Kayser & Reiss, 1935), and small birds are unable to relax in darkness during the daytime as much as large birds. King (1974) differentiates between the minimal BM obtained during the resting (sleeping) period and what he calls a 'fasting metabolic rate' which is the value obtained during the daytime or normal period of activity. Recognition of a minimal BM characteristic of each species has physiological significance and is best for comparative purposes, but to restrict the term to this minimal level runs into semantic difficulties and confusion because BM also varies seasonally and between populations adapted to different climatic regions.

A zone of thermo-neutrality (p. 134) is commonly demonstrable during the night, but some investigators have not found it in certain species during the daytime: e.g. house sparrow (Kendeigh, 1944; Hudson & Kimzey, 1966), evening grosbreak *Coccothraustes vespertinus* (Cooper) (West & Hart, 1966), horned lark *Eremophila alpestris* (L.) (Trost, 1972).

In nocturnally active birds, the daily rhythm is reversed (Gatehouse & Markham, 1970).

Seasonal changes

BM in the house sparrow has been shown (p. 178) to increase from September over winter and to be low during the spring and summer (Miller, 1939). An increase during the winter has also been demonstrated in the pigeon (Dontcheff & Kayser, 1934), but in the brambling *Fringilla montifringilla* L. (Pohl, 1971), and common redpoll *Acanthis flammea* (L.) (West, 1972a), BM is lowest in the spring and autumn.

In our data (equations 5.1 to 5.5), log regressions of BM on weight indicate that for passerines BM is higher in the winter than in the summer (Fig. 5.1). The slopes of the regression lines for BM are not different, however. The percentage difference between winter and summer BM is greater in small birds (11–19%) than in large birds (3%). The higher BM during the winter is correlated with greater endocrine and enzyme activity at the cellular level at that time (Kendeigh & Wallin, 1966; Barnett, 1970; Rising & Hudson, 1974). Non-passerines did not exhibit this seasonal difference.

Relation to species weight and taxonomy

A logarithmic relation of BM to weight has been known for many years. Lasiewski & Dawson (1967) were the first, however, to point out that this relation was quantitatively different between passerine and non-passerine species. The latest published compilation of data on the relation of BM

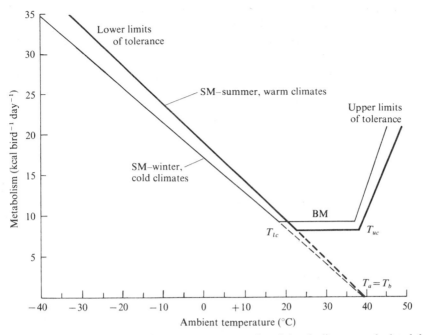

Fig. 5.1. Model showing adjustments to seasonal and local climates, calculated for a passerine bird weighing 25 g from equations 5.1, 5.2, 5.10, 5.11, 5.13 and 5.14 with upper critical temperatures and limits of tolerance based on studies of the house sparrow (Kendeigh, 1944; Hudson & Kimzey, 1966): T_{lc} = lower critical temperatures and t_{uc} = upper critical temperatures marking the limits of the zone of thermo-neutrality; T_a = ambient temperature, T_b = body temperature.

to body weight in different species is that of Dawson & Hudson (1970), but no recognition was given to variations in the data due to daily rhythms, seasonal acclimatization, or local adaptations. Aschoff & Pohl (1970b) give regression equations for day and night separately, but they are uncorrected for seasonal or local variations.

Our data on BM during the daytime are inadequate for separate analysis, but log regressions for summer and winter nights are significantly different in the Passeriformes, although not in the non-passerines.

Passerines

$$\text{Summer, night } (N = 41): \quad M = 0.8906 \, W^{0.6884} \pm 1.091 \qquad (5.1)$$

$$\text{Winter, night } (N = 35): \quad M = 1.110 \, W^{0.6577} \pm 1.116 \qquad (5.2)$$

Non-passerines

$$\text{Summer, night } (N = 30): \quad M = 0.5675 \, W^{0.7282} \pm 1.168 \qquad (5.3)$$

$$\text{Winter, night } (N = 12): \quad M = 0.6515 \, W^{0.7010} \pm 1.131 \qquad (5.4)$$

$$\text{Summer, winter, day, night } (N = 77): \quad M = 0.5224 \, W^{0.7347} \pm 1.237 \quad (5.5)$$

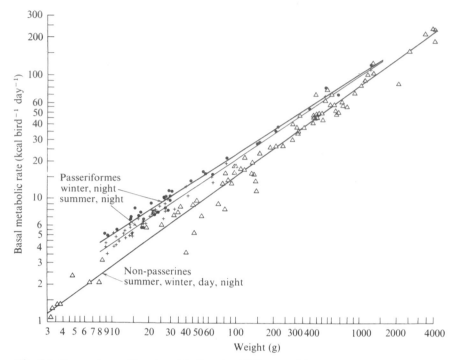

Fig. 5.2. Regressions of basal metabolic rates (BM) on weight.

The regressions for passerine species have higher elevations than those for non-passerines, and the one for winter has a different slope (Fig. 5.2). The passerines include relatively more small and tropical species and the non-passerines more large species. The regression for summer night in passerines (equation 5.1) and the pooled regression (equation 5.5) for non-passerines are not significantly different from ones given by Kendeigh (1969b), but the one for winter night in passerines (equation 5.2) has a higher elevation. The values obtained from these equations also agree fairly closely with those obtained by Aschoff & Pohl (1970b) and Lasiewski & Dawson (1967) but are probably more reliable as they are segregated for inherent variables of both time of day and season.

Zar (1968) was unable to find any differences in the BM between families of Passeriformes, but he found some differences between orders of non-passerines. Ligon (1968) also gives a separate equation for Strigiformes. We were able to obtain equations for Fringillidae, Corvidae and Muscicapidae among the Passeriformes and for the Anseriformes among the non-passerines, but none of them gave a value for the BM differing significantly from those provided by the generalized equations.

It is of considerable theoretical interest that segregating the data ac-

132

cording to time of day and season lowers the coefficient of weight towards $W^{0.67}$. The average of the coefficients in equations 5.1 and 5.2 is 0.6731, but if all our data for Passeriformes were combined ($N = 159$) it would be 0.7089, and for non-passerines (equation 5.5), 0.7347. These latter values are more nearly comparable to those of Lasiewski & Dawson (1967), 0.724 in Passeriformes and 0.723 in non-passerines, and Aschoff & Pohl (1970*b*), 0.704 and 0.726 in Passeriformes and 0.729 and 0.734 in non-passerines. Variation of the BM with $W^{0.67}$ is in conformity with Sarrus & Rameaux's *surface law* as developed by Rubner (1883) and Richet (1885).

The relation between BM and weight may be a compromise between two objectives: to provide power for the locomotor activity (especially flight) of birds of different size and to minimize heat loss. The required power for flight has been correlated with velocity and body weight as $W^{0.83}$ (Dol'nik, 1969), and heat transfer from the body to the surroundings as $W^{0.50}$ (pp. 134–6). If variations in the BM on weight were adapted only for activity purposes ($W^{0.83}$), there would be too great heat loss in large birds. On the other hand, adaptation of the BM to weight only for temperature regulation ($W^{0.50}$) would bring a power deficiency for flight in large birds. Natural selection has developed a relationship that is intermediate. The average value of $W^{0.83}$ and $W^{0.50}$ is in fact $W^{0.67}$.

Adaptation to local climate

Hudson & Kimzey (1966) demonstrated that BM averaged about 24% higher in three northern USA populations of the house sparrow than in a Texas population, correlated with differences in prevailing ambient temperatures. Desert and tropical species tend to have lower metabolic rates than related races or species from more mesic climates (Saxena, 1957; Yarbrough, 1971; Trost, 1972; Dawson & Bennett, 1973; Hinds & Calder, 1973) while arctic or nothern species tend to have higher rates (Winkel, 1951; West, 1972*b*; Kendeigh & Blem, 1974).

The data in Appendix 5.1 were segregated into species of 'northern' and 'southern' distribution and their regressions compared. 'Northern' species were those whose distribution during the breeding season lay mainly north of 40° N in North America and bounded the North and Baltic Seas in Europe. 'Southern' species were those that occurred mainly south of 40° N in North America, bounded the Mediterranean Sea, or were tropical. Species whose distribution was chiefly at the middle latitudes were not considered. The regression equations are not given as none of the differences between 'northern' and 'southern' species is statistically significant. However, the BMs for 'northern' birds in the summer at all weight levels are higher by 3 to 17% than are the BMs for 'southern' birds.

These latitudinal differences may be compared with the seasonal differences already noted (p. 130) and may be similarly related to differences in enzyme and hormonal actions. Obviously, any prediction of BM from generalized equations needs correction for adaptations to local climate if greater precision is required.

Temperature regulation (standard metabolism, SM)

In the zone of thermo-neutrality, body temperature is regulated primarily by changing the effectiveness of the body insulation, so that the rate of heat production (BM) is maintained at a constant level. With a drop in ambient temperature, however, there comes a point where insulation approaches its maximum capacity, below which body temperature cannot be maintained without an increase in heat production. This point is called the *lower critical temperature*. The lower critical temperature may be taken as the intersection of a flat line representing BM in the zone of thermo-neutrality and a rising line representing the regression of standard metabolism (SM) at decreasing ambient temperatures (Fig. 5.1). The regression of SM on temperature is often adjusted so that SM is zero at an air temperature equal to body temperature (39–40 °C).

The change from insulation or physical to metabolic heat regulation of body temperature is probably not, however, as abrupt as this would indicate. Likewise, a zone of thermo-neutrality is not always demonstrable especially during the daytime (see above). With continued rise in ambient temperature, there is another point, where insulation is reduced to a minimum. Above this point the heat produced by the body cannot be dissipated fast enough; it accumulates, there is a rise in body temperature, and consequently an increase in metabolic rate. This point is called the *upper critical temperature*. In order to avoid confusion we will use *standard metabolic rate* only in reference to energy requirements of fasting, resting birds exposed to temperatures below the zone of thermo-neutrality.

Temperature coefficients

The *temperature coefficient* (*b*) is that rate at which the SM changes per degree change in ambient temperature. Since the level of BM varies with time of day and season, the lower critical temperature and temperature coefficient may also be expected to vary.

On seven comparisons in five species (Appendix 5.1), the coefficient is higher during the daytime in four instances and lower in three. In the house sparrow (Kendeigh, 1969 a), the coefficient is lower in the daytime, meaning that the species is less responsive to a change in temperature then than at night. Seasonally, 24 out of 31 species have lower coefficients during the winter.

Significant regressions of temperature coefficients on weight, segregated with respect to season, are the following:

Passerines

Summer, night ($N = 41$): $b = 0.0846\ W^{0.5315} \pm 1.161$ (5.6)

Winter, night ($N = 35$): $b = 0.0728\ W^{0.5427} \pm 1.136$ (5.7)

Non-passerines

Summer, night ($N = 30$): $b = 0.0645\ W^{0.5734} \pm 1.109$ (5.8)

Winter, night ($N = 12$): $b = 0.0457\ W^{0.5886} \pm 1.333$ (5.9)

All the log regressions have the same slope, but the winter regressions have a lower elevation. These equations for summer and winter are sensitive not only to seasonal acclimatization (better insulation by plumage and fat) but also to seasonal changes of body weight. In general, winter birds are heavier than summer birds. Shilov (1968) has discussed seasonal changes in bird metabolism and plumage in considerable detail.

The regressions for passerines have a higher elevation than those of non-passerines, so that the temperature coefficients are significantly higher in birds of the same weight. This difference is obscured in data not segregated with respect to time of day and season (Calder & King, 1974).

Our values for the regression of temperature coefficients on weight for passerines, 0.5315 and 0.5427 kcal bird^{-1} day^{-1} °C^{-1}, and for non-passerines, 0.5734 and 0.5886, are somewhat higher than the composite regression of 0.492 obtained in a similar manner by Lasiewski, Weathers & Berstein (1967) for all birds. They are appreciably higher than the direct measurement of the heat conductance coefficient through the plumage of dead birds of 0.52 obtained by Herreid & Kessel (1967). In the living bird the rate of heat loss is faster than in the dead bird because of the circulation of blood to the skin and respiration. Nevertheless, all of these values are comparable.

The heat transfer coefficient through the plumage depends on three parameters: (1) thermal conductivity of the feather structure, (2) heat transfer rate over the air spaces within the plumage (radiation, convection), and (3) the size (volume) of these air spaces (Shilov, 1968). There is no reason to believe (at least there is no evidence available) that the rate of heat transfer along feather structures or across trapped air spaces varies between species, but the volume of air spaces would vary with the thickness of the plumage and the number of feathers involved, perhaps in ratio of one to the other.

The thickness of the plumage on a bird is, in turn, proportional to the

S. C. Kendeigh, V. R. Dol'nik & V. M. Gavrilov

ratio of the weight of the plumage to the surface area of the bird. The weight of the plumage, i.e. of the contour feathers only, varies with the bird's weight approximately as $W^{1.0}$ (Kendeigh, 1970). Since surface area varies as $W^{0.67}$, then the thickness of the plumage must vary as $W^{1.0}/W^{0.67}$ or $W^{0.3333}$. The number of feathers per bird varies approximately as $W^{0.1667}$. The size or volume of air spaces would then vary between species as $W^{0.3333}/W^{0.1667}$ or $W^{0.1666}$. Since this is the only parameter of the three listed in the above paragraph that varies with weight, the thermal conductivity of the plumage must change as $W^{0.1666}$. Using measured data on feather weights and number of feathers in species of different weight (Kendeigh, 1970), the empirical ratio $(W^{0.9591}/W^{0.6667})/W^{0.1779}$, is $W^{0.1145}$.

According to Fourier's law, heat flow from a flat plate varies as $k = h\,S/I(T_b - T_s)$, where k is the rate of heat transfer through unit thickness across unit area for a unit difference in temperature, h is thermal conductivity, S is the surface area, I is the thickness of the insulation, T_b is core temperature, and T_s is surface temperature. The heat transfer or temperature coefficients in birds (k), therefore, should vary with weight as

$$W^{0.1666} \times W^{0.6667}/W^{0.3333} \quad \text{or} \quad W^{0.5000}$$

or empirically as

$$W^{0.1145} \times W^{0.6667}/W^{0.2924} \quad \text{or} \quad W^{4888}.$$

The bird more nearly approaches a sphere than a plate, as called for in Fourier's law, especially when quiet with the feathers fluffed, but the relationship to weight would be essentially the same. Kleiber (1972) has similarly indicated that the rate of heat transfer is nearly proportional to the square root of body weight, ($W^{0.5}$), although he arrived at this coefficient in a different manner.

Lower critical temperature

The lower critical temperature (T_{lc}) varies with weight (Kendeigh, 1969a; Calder & King, 1974). With passerine species there is a seasonal variation in that it averages about 4 °C lower in winter than in summer, but with non-passerine species there are no significant differences between seasons or times of day.

Passerines

Summer, night ($N = 43$): $T_{lc} = 40.73\ W^{-0.1844} \pm 1.133$ (5.10)

Winter, night ($N = 35$): $T_{lc} = 39.81\ W^{-0.2497} \pm 1.260$ (5.11)

Non-passerines

Summer, winter, day, night ($N = 74$): $T_{lc} = 47.17\ W^{-0.1809} \pm 1.382$ (5.12)

The similarity in the exponents of W means that the lower critical temperature declines with increasing weight at the same rate in all taxa at all seasons. The lower levels of the a factor (Y-intercept) for passerine species in the winter compared with summer and their lower levels compared with non-passerine species is correlated with their higher BM (Calder & King, 1974).

Since the rate of heat transfer through the plumage has been shown to vary as $W^{0.50}$ and BM as $W^{0.67}$, then the lower critical temperature where SM = BM should vary with weight as $W^{0.50}/W^{0.67}$ or $W^{-0.17}$, or $T_b - T_{lc}$ as $W^{0.17}$. The exponents of W in equations 5.10 to 5.12 are a little high but approach the value of $W^{0.17}$. Kleiber (1972) gives this value as $W^{0.25}$ but he used $W^{0.75}$ for the relation of BM to weight which does not fit our data as well as $W^{0.67}$.

The width of the thermo-neutral zone is the number of degrees between the lower and upper critical temperatures. Shorter widths are more common among species of lesser weight, but the data are too few and heterogenous as to time of day and season to warrant serious statistical analysis.

One may attempt, however, a theoretical analysis of how the width of the thermo-neutral zone would vary with weight. At the upper critical temperature the feathers are flattened against the body to permit maximum heat transfer from the skin to the surroundings. The volume of air spaces between the feathers is reduced to the minimum. The rate of heat transfer through the feather structure is weight independent, so the heat transfer coefficient should vary with the weight of the bird only, as does the thickness of the plumage. As shown above, the slope of plumage thickness varies approximately as $W^{0.33}$. At the lower critical temperature, the plumage is fluffed to or near its maximum, and with abundant air trapped in spaces between the feathers, the heat transfer coefficient would vary as $W^{0.17}$. The width of the thermo-neutral zone should therefore vary as the ratio between these two coefficients or $W^{0.17}$. Empirically, the coefficient would be $W^{0.2924}/W^{0.1145}$ or $W^{0.1779}$. A modifying factor in this relationship is that evaporative cooling begins to increase appreciably in the zone of thermo-neutrality and may well affect the position of the upper critical temperature.

Standard metabolism at 0 °C

Since both the lower critical temperature and temperature coefficient vary with weight, the regression of SM on weight at 0 °C is likely to be different from that in the zone of thermo-neutrality. Log regression analysis shows that no differences occur between day and night in slope or elevation in either passerines or non-passerines; however, night-time equations are listed for better comparison with the BM. Likewise, there are no differences in slope between regressions for summer and winter, but elevations differ.

Passerines

$$\text{Summer, night } (N = 41): M = 3.457 \, W^{0.5277} \pm 1.133 \qquad (5.13)$$

$$\text{Winter, night } (N = 35): M = 3.092 \, W^{0.5313} \pm 1.098 \qquad (5.14)$$

Non-passerines

$$\text{Summer, night } (N = 30): M = 2.624 \, W^{0.5705} \pm 1.104 \qquad (5.15)$$

$$\text{Winter, night } (N = 12): M = 1.810 \, W^{0.5944} \pm 1.217 \qquad (5.16)$$

SM at 0 °C is lower in winter than in summer because of better plumage and fat insulation and as an effect of increased body weight resulting from the deposition of the less metabolically active fat. The higher exponents of W and lower Y-intercepts in the equations for non-passerine species are probably real, but an exact comparison with passerine species may be modified by the inclusion of more birds in captivity (zoos) among the non-passerines. Captive birds tend to be more obese, with plumage in poorer condition than in wild birds. The elevations of the four equations are not significantly different from those of earlier equations given by Kendeigh (1969*b*), but the slopes of the regressions, exponents of W, for passerine species are different.

Since the slopes of the regressions of SM on ambient temperature below the lower critical temperature depend on the rate of heat flow from the bird to the surroundings, and the rate of transfer varies as $W^{0.50}$, then the regression of SM at 0 °C on weight should also vary as $W^{0.50}$. This is approximately true for Passeriformes (0.5277 and 0.5313) and only slightly off for non-passerines (0.5705 and 0.5944). The change in the exponent of BM on weight (0.67) to that of SM on weight (0.50) probably comes at the lower critical temperature where insulation is at or near its maximum and rate of heat production becomes directly related to the rate of heat loss from the bird.

Small birds increase their metabolism from the lower critical temperature to 0 °C to a greater extent than do large birds. This increase is greater in the summer than in the winter and in non-passerines than in passerines (Kendeigh, 1969*b*).

Adaptation to local climate

The comparative analysis of metabolic parameters of species with 'northern' and 'southern' distribution, described above (p. 133), which showed an increased BM in cooler climates, also shows smaller temperature coefficients, lower lower critical temperatures of the zone of thermo-neutrality, and a lower SM at 0 °C in species adapted to cooler latitudes. This is supported by an analysis of heat transfer coefficients made by Drent & Stonehouse (1971). These adaptations to local climate, which may be genetically inherited, are similar to seasonal changes in metabolic adjustments that are definitely only somatic.

The house sparrow increases in weight at northern latitudes (Blem, 1974), and heat conductance, as measured on carcasses, decreases in a significant linear manner. Although 'northern' house sparrows have a greater total weight of plumage, the difference is not significant when calculated in terms of unit surface area. Dol'nik (1967a, b) found that 16 species not only possessed progressively greater fat reserves at more northern latitudes during the winter, but they also lost weight overnight at significantly lower rates. Probably both more plumage and more fat provide 'northern' birds with better insulation, and the better insulation is largely responsible for the differences noted.

Taxonomic differences

The consistent differences in metabolic rates between passerine and non-passerine species have been noted. Within the Passeriformes, no differences were found between the regression lines for SM at 0 °C for the families Fringillidae (summer, $N = 8$; winter, $N = 14$), Corvidae (winter, $N = 9$), or Muscicapidae (summer, $N = 13$) and the generalized ones for the order as a whole. Likewise, there was no significant difference between the regression line for Anseriformes (winter, $N = 7$) and the generalized ones for non-passerines. Differential evolution of metabolic rates (SM, BM) has apparently not developed very far at these lesser taxonomic levels.

Application

By using equations 5.1 to 5.5 for BM in the zone of thermo-neutrality, equations 5.10 to 5.12 for determining the lower critical temperature, and equations 5.13 to 5.16 for SM at 0 °C, it is possible to calculate an equation for SM on ambient temperature for any species whose weight is known to show temperature regulation requirements. Because of the variation between species, however, it is best to use measured rates, when these are available, rather than calculated ones. Since SM and BM are measured on fasting birds, early morning weights of wild birds, before they have replenished their fat reserves lost the preceding night, are most suitable. The actual cost of temperature regulation at any ambient temperature is the difference between the SM at that temperature and the BM.

Existence metabolism (EM)

Existence metabolism (EM) is the rate at which energy is used by caged birds maintaining a constant weight (± 1–2%) over a period of days when the birds are not undergoing reproduction, moulting, migratory unrest, growth, or fat deposition. The birds have a surplus of food and water always available, and there is a limited amount of activity permitted within the confines of the cage. EM is an integration of BM, temperature regulation, the heat increment of feeding and the energy expended in cage locomotor activity. It is determined by measuring the amount of food consumed (gross energy intake, GEI) and subtracting the amount eliminated or unassimilated from the intestine and kidneys (excretory energy). The difference is the amount metabolized (metabolized energy, ME). Since these measurements extend over a period of days, EM further integrates the accumulation of energy during the daytime (feeding period) and its loss from the body at night (non-feeding period).

Effect of photoperiod and season

In those species with data available at both short and long photoperiods (Appendix 5.2); the regression lines of EM on weight at both 0 and 30 °C in both passerine ($N = 34$) and non-passerine species ($N = 18$) have higher elevations under 15-hour photoperiods than under 10-hour photoperiods. In passerine species the difference in elevations or Y-intercepts is statistically significant, although in non-passerine species it is not. In neither group are the slopes of the regression lines at the two photoperiods significantly different.

Although the total food metabolized on a long photoperiod is greater than on a short one, it is not proportionately (50%) greater. For instance, according to the regression equations, the additional food metabolized by a 25-g passerine bird on a 15-hour photoperiod at 0 °C divided by the number of additional hours for feeding (five) averages only 21% of the average hourly rate of food metabolized on a 10-hour photoperiod. At 30 °C, it averages only 15%.

In general, feeding is most rapid in early morning, replacing most of the energy lost from the body over the preceding night. Sometimes there is increased feeding again in the evening, but once the preceding night's loss of energy is replaced, food consumption thereafter is sufficient just to take care of the current activity. The longer days are of no great advantage energy-wise at medium to high ambient temperatures except that the hourly rate of food assimilation is more leisurely, averaging only about 72% of the rate on the shorter photoperiods for the 25-g bird cited above. This is true to approximately the same extent in both passerine

140

and non-passerine species. Long photoperiods have a significant advantage, however, at very low ambient temperatures as they provide a longer period for replacing energy losses of the preceding night, and the nights when feeding is not done are shorter. Long photoperiods may also be used to full capacity during the migratory and breeding periods when many high-energy activities, e.g. fat deposition, reproduction, moulting, are involved.

For greater precision, the EM should therefore be analyzed separately for 'winter' (short photoperiod) and 'summer' (long photoperiod) months. The transition from 'summer' to 'winter' may best be taken at the mid-point or the end of the annual autumn moult and the transition from 'winter' to 'summer' at the beginning of general egg-laying in the spring. For the house sparrow at mid-latitudes we find it convenient to have 'summer' extend from May through October and 'winter' from November through April. For tropical and sub-tropical resident species, where the seasonal change in photoperiod is small, perhaps the equations for the 15+ photoperiod had best be used throughout the year as being more representative of heat-acclimatized birds.

Temperature coefficients

The linear regression of EM extends from the lowest ambient temperature that the birds can tolerate to, at least, the upper critical temperature of the zone of thermo-neutrality of the BM. There is generally no zone of thermo-neutrality in EM (Fig. 5.4) or it is only poorly defined. The sensitivity of metabolic rate to temperature is shown by the slope of these regressions or the temperature coefficients. The equations below show how the temperature coefficients ($b =$ kcal bird^{-1} day^{-1} °C^{-1}) vary with the weight of the species.

10± hour photoperiod, winter

Passeriformes ($N = 71$): $b = 0.1571 \, W^{0.2427} \pm 1.265$ \hfill (5.17)

Non-passerines ($N = 40$): $b = 0.1753 \, W^{0.3265} \pm 1.570$ \hfill (5.18)

Anseriformes ($N = 9$): $b = 0.3304 \, W^{0.2902} \pm 1.342$ \hfill (5.19)

15± hour photoperiod, summer

Passeriformes ($N = 70$): $b = 0.1909 \, W^{0.2162} \pm 1.162$ \hfill (5.20)

Non-passerines ($N = 70$): $b = 0.2761 \, W^{0.2818} \pm 1.497$ \hfill (5.21)

Psittaciformes ($N = 5$): $b = 0.2026 \, W^{0.3831} \pm 1.128$ \hfill (5.22)

The coefficients appear to vary with weight at a lower rate in Passeriformes than in non-passerines, but the differences are not statistically significant. The elevations of the regression lines are, however, signifi-

141

cantly different, so that the actual temperature coefficients are lower in passerines at both 10± and 15+ hour photoperiods. Of the six families of Passeriformes in which regressions were obtained (Muscicapidae, Emberizidae, Fringillidae, Estrildidae, Ploceidae, Corvidae), none showed significant variation from the general regressions of the Passeriformes under either 10± or 15+ hour photoperiods. Regression equations were obtained on five orders of non-passerines (Anseriformes, Galliformes, Gruiformes, Psittaciformes, Charadriiformes). Of these, the Anseriformes have higher coefficients at 10± hour photoperiods and the Psittaciformes have mostly higher coefficients at 15+ hour photoperiods than other non-passerine orders – all significant at the 5% level. Although the slopes of the regressions also appear to be different, the samples are too small and diverse to give them statistical significance.

It thus appears that non-passerines are more responsive to a change in ambient temperature than are the Passeriformes and that the Anseriformes and Psittaciformes are more responsive than other non-passerines. In terms of specific weight, heavier birds in all taxa are less responsive than lighter birds.

The slopes of regression lines for the temperature coefficients of EM are considerably flatter than those of the SM, indicating that the level of EM is less affected by a change in temperature than is SM. The average exponent or coefficient of weight for passerines and non-passerines combined is 0.2668 or only about half that of the coefficient of weight for the response of SM to temperature (0.5590, equations 5.6 to 5.9). Variations in SM with weight involve only the effectiveness of temperature regulation below the zone of thermo-neutrality; variations in EM with weight include part or all of the zone of thermo-neutrality and the additional factors of feeding and cage activity.

Existence metabolism at 30 and 0 °C

Kendeigh (1970) showed for a limited number of species that passerine species had a higher level of EM on weight than non-passerines at 30 °C, but at 0 °C the rates were essentially the same. With many more data now available (Appendix 5.2), a more detailed comparison is possible and the new set of equations have greater accuracy. Equations are given separately for the two photoperiods, although the differences are small. Different species are included in each regression.

30 °C, 10± hour photoperiods (*winter, migration*)

$$\text{Passeriformes } (N = 71): M = 1.544 \, W^{0.6601} \pm 1.130 \qquad (5.23)$$

$$\text{Emberizidae } (N = 15): M = 2.030 \, W^{0.5314} \pm 1.205 \qquad (5.24)$$

$$\text{Non-passerines } (N = 40): M = 1.455 \, W^{0.6256} \pm 1.187 \qquad (5.25)$$

30 °C, 15+ hour photoperiods (nesting, moulting)

Passeriformes ($N = 70$): $M = 1.462\ W^{0.6880} \pm 1.108$ (5.26)

Fringillidae ($N = 9$): $M = 1.565\ W^{0.6930} \pm 1.048$ (5.27)

Non-passerines ($N = 70$): $M = 1.068\ W^{0.6637} \pm 1.320$ (5.28)

0 °C, 10± hour photoperiods (winter, migration)

Passeriformes ($N = 71$): $M = 4.437\ W^{0.5224} \pm 1.067$ (5.29)

Emberizidae ($N = 15$): $M = 4.158\ W^{0.5291} \pm 1.077$ (5.30)

Non-passerines ($N = 40$): $M = 4.235\ W^{0.5316} \pm 1.178$ (5.31)

Galliformes ($N = 11$): $M = 2.344\ W^{0.5970} \pm 1.165$ (5.32)

Passeriformes & non-passerines ($N = 111$): $M = 4.374\ W^{0.5266} \pm 1.114$ (5.33)

0 °C, 15+ hour photoperiods (nesting, moulting)

Passeriformes ($N = 70$): $M = 4.969\ W^{0.5105} \pm 1.070$ (5.34)

Non-passerines ($N = 70$): $M = 4.142\ W^{0.5444} \pm 1.127$ (5.35)

Psittaciformes ($N = 5$): $M = 5.765\ W^{0.5020} \pm 1.081$ (5.36)

At 30 °C and at both photoperiods, there are no significant differences between passerine and non-passerine species in the exponents of weight, or the slopes of the regression lines, but there are differences in their elevations, or Y-intercepts (Fig. 5.3). At 0 °C and 10± hour photoperiod, there is still no significant difference between passerine and non-passerine species in either the slope or elevation of the regression lines so that they may be combined, but at 15+ hour photoperiod there is a significant difference in both slopes and elevations. Passerines generally have higher metabolic rates than non-passerines.

The differences between these equations for Passeriformes at 30 °C do not differ in the weight exponent from those given by Kendeigh (1970) but the Y-intercepts are higher; at 0 °C the equation for the 15+ hour photoperiod is also higher but otherwise the equations for both passerines and non-passerines are not statistically different.

Of the six families of Passeriformes (Muscicapidae, Emberizidae, Fringillidae, Estrildidae, Ploceidae, Corvidae) where the data are adequate for detailed analysis, only the Emberizidae has a significantly different elevation in the regression, averaging a lower metabolic rate at 10± hour photoperiod and both 0 and 30 °C. At 15+ hour photoperiod and 30 °C, only the regression of Fringillidae is statistically different, giving higher rates. None of the weight exponents or slopes of the regression lines is significantly different.

Of five orders of non-passerines (Anseriformes, Galliformes, Gruiformes, Charadriiformes, Psittaciformes), only the Galliformes have a sig-

S. C. Kendeigh, V. R. Dol'nik & V. M. Gavrilov

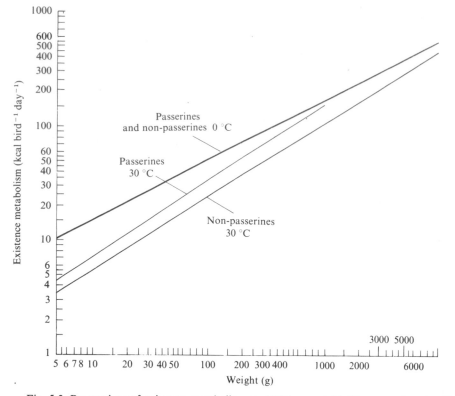

Fig. 5.3. Regressions of existence metabolic rates (EM) on weight. The regressions for 10± and 15+ hour photoperiods have been combined.

nificantly different elevation or lower metabolic rates at 0 °C, 10± hour photoperiod, and only the Psittaciformes have significantly higher rates at 0 °C, 15+ hour photoperiod. The slopes of the regressions for all individual orders are essentially similar to those for all orders combined.

Here again, a regression equation for EM on temperature may be calculated from the above equations for EM on weight at 30 °C and 0 °C for any species where it has not been obtained by measurement. Use may also be made of Fig. 5.3 for approximate values.

The combined average of the weight exponents for passerines and non-passerines at 30 °C and both photoperiods (0.6594) is comparable with the combined average for the weight exponents of BM in the zone of thermo-neutrality (0.6938, equations 5.1 to 5.4) and both come close to the theoretical value of 0.67. At these high temperatures, the slope of EM on weight is affected primarily by the factors involved in the surface law of Rubner (1883) and Richet (1885). The combined average of the weight exponents for EM at 0 °C is 0.5272, which is comparable with 0.5590 for

144

weight exponents of SM at 0 °C and close to the theoretical value of $W^{0.50}$. At low temperatures, heat production for body temperature regulation takes priority as the modifying factor.

Adaptation to local climate

Measurements of the EM of local populations of the house sparrow from a wide range of localities in North America were generally higher than expected at northern, cooler latitudes and lower at southern, warmer latitudes (Kendeigh & Blem, 1974). Using the data for this species, corrections for physiological adaptation to local climates for passerine species of approximately 25–30 g in weight may be made with the following equations. *PC* is the percentage that should be added or subtracted from values obtained as described in the above section, and *T* is the mean temperature for summer (June through August) and winter (December through February).

$$\text{Summer: } PC = 10.870 - 0.7403\,T + 0.0046\,T^2 \pm 3.6073 \qquad (5.37)$$

$$\text{Winter: } PC = 2.6982 - 0.6848\,T - 0.0094\,T^2 \pm 0.4097 \qquad (5.38)$$

Energy control of size limits

Kendeigh's (1972) argument that minimum size in Passeriformes may be limited somewhat above that weight at which the regressions of SM and EM at 0 °C intersect is not supported by the present data. The coefficient of SM at 0 °C on weight of 0.5277 and 0.5313 in equations 5.13 and 5.14 are almost identical to the coefficients of EM at 0 °C on weight of 0.5224 and 0.5105 in equations 5.29 and 5.34, and hence the two regression lines do not intersect. A coefficient of 0.42, based on a much fewer number of data, was used for SM at 0 °C on weight in the earlier study.

There is little question, however, that a decrease in size brings an increase in energy stress (Table 5.1). Birds in temperate zones and tropical birds during the breeding–moulting period can tolerate a drop of ambient temperature to only about 0 °C, and at this temperature EM is maximum. The smallest Passeriformes are 5–6 g in weight, the smallest non-passerines about 3 g. It may well be that an ability to maintain a metabolic rate of between 2.2 and 2.5 kcal g^{-1} day^{-1} over an extended period of time may limit the evolution of small size.

Kendeigh (1972) also postulated that the upper limit in size may be determined by the ability of birds to reduce their metabolism so as to tolerate high ambient temperatures. BM is the lowest level to which it can be reduced. The regression of BM on weight (equation 5.1) intersects that of SM at 0 °C on weight (equation 5.13) in Passeriformes at 4.6 kg which

145

Table 5.1. *Metabolic rates in terms of specific weight ($kcal\ g^{-1}\ day^{-1}$). EM at minimum weights found in Passeriformes (5 g) and non-passerines (3 g) are italicized*

Weight of bird (g)	BM	SM at 0 °C	EM at 0 °C
	Passeriformes		
50	0.26	0.54	0.72
25	0.33	0.76	1.03
10	0.43	1.16	1.61
5	0.54	1.62	*2.26*
3	0.63	2.06	2.90
	Non-passerines		
50	0.20	0.49	0.70
25	0.24	0.66	0.96
10	0.30	0.98	1.45
5	0.37	1.31	1.99
3	0.42	1.64	*2.51*

is somewhat too high. The largest passerine species is the raven *Corvus corax tibetanus* Hodgson, which reaches only 2 kg. In non-passerine species, the intersection of the two regressions (equations 5.3, 5.15) comes at 16 kg which is too low. The maximum size of living non-passerines is approximately 100 kg, as found in the flightless ostrich *Struthio camelus* L. The capacity, however, of non-passerine species to attain both larger and smaller size than passerines is clearly indicated. If the regression of SM on weight at a temperature higher than 0 °C for passerines and at a temperature lower than 0 °C for non-passerines were taken, intersection with the regressions of BM on weight would come closer to the actual maximum weights attained in the two groups. The capacity of non-passerines to attain greater size is correlated both with their lower BM and with their generally greater ability to increase the rate of heat loss at high temperatures by evaporative cooling.

Heat increment of feeding (SDA)

Animals digesting and assimilating food have a higher rate of metabolism than do fasting animals. The difference is the heat increment of feeding, commonly but erroneously called the *specific dynamic action* (SDA). At high ambient temperatures, this extra heat is not all needed for body temperature regulation and most of it is lost from the body. With decreasing ambient temperatures, progressively more and more, and eventually all of it, is retained, replacing by that amount the metabolism required for temperature regulation in fasting animals. This is

Rubner's (1910) law of compensation as worked out for mammals. We are assuming that it holds also for wild birds although it has not been tested.

Cage locomotor activity

Locomotor activity in cages is concerned in part with feeding, and increases moderately at low temperatures as the rate of feeding increases. Activity concerned with drinking, on the other hand, decreases at low temperatures, and the two may more or less balance each other. However, as temperatures rise, there is in some species a conspicuous and appreciable increase in spontaneous or 'frivolous' activity that apparently serves no useful purpose. The energy cost per unit of activity varies inversely with temperature (Kontogiannis, 1968), so that here again the actual cost of cage locomotor activity may be essentially a constant at all except very high temperatures where heat stress comes into play (Kendeigh, 1974).

Component percentages

Until the energy cost of SDA and cage locomotor activity are measured at different temperatures in various species, we can only conjecture as to their relative importance as components of EM. In the house sparrow (Fig. 5.4), it is probable that the difference between SM and EM (2.7 kcal bird^{-1} day^{-1}) at the lowest temperature of $-15\,°C$ is due entirely to locomotor activity. If the same amount of energy is expended in locomotor activity at the upper critical temperature (37 °C), then the EM at that temperature (13.9 kcal) minus 2.7 kcal minus the BM during the daytime (9.3 kcal) should be SDA (1.9 kcal), all of which is certainly lost. The 20% relation of SDA to BM (1.9/9.3) is within the range of values compiled by Ricklefs (1974). If SDA constitutes the same percentage of EM (1.9/13.9 or 13.7%) at all temperatures, it will increase in actual amount at lower temperatures because more food is assimilated. The amount retained for temperature regulation may be calculated as 0.137 EM minus the amount lost (EM$-$2.7$-$SM). Of the amount left (EM$-$2.7$-$SDA), all but the BM (9.3 kcal) would be for temperature regulation by tissue metabolism. Thus at the following two temperatures, the component percentages during May would be:

	0 °C	*37 °C*
BM	36	67
Temperature regulation	40	0
SDA retained	12	0
SDA lost	2	14
Cage locomotor activity	11	19

S. C. Kendeigh, V. R. Dol'nik & V. M. Gavrilov

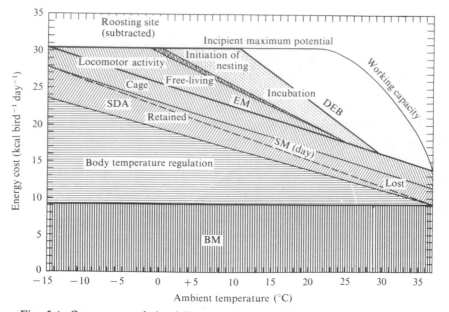

Fig. 5.4. Components of the daily energy budget (DEB) of the house sparrow while incubating during May in central Illinois. The regression line for SM during the daytime is indicated only to show the dividing line between SDA retained and SDA lost. Existence metabolism (EM) is calculated from the equation for 15+ hour photoperiods. Free-living locomotor activity is shown for its full potential and how energy is conserved by the birds using enclosed roosting sites. Since the energy available for free-living locomotor activity decreases below −1.0 °C, continued existence under natural conditions without acclimation would not be possible, although existence in cages could continue to −15 °C. If the regression for SM is extended, it is evident that the birds could live, perhaps a few hours, even down to −22.5 °C. Successful incubation would not be possible below about 11 °C, although nesting activities could be initiated.

In an analysis of the components of EM for the chaffinch *Fringilla coelebs* (L.), at 0 °C, Dol'nik (1974*b*) assigns a somewhat smaller percentage to temperature regulation (32%) and SDA (11%) and a larger percentage to cage activity (19%). At 37 °C, the birds become inactive and feeding so reduced that SDA constitutes only 7% and all the rest (93%) is BM. There is no evidence that cage activity diminishes in the house sparrow in the zone of thermo-neutrality (Eyster, 1954), but it becomes greatly reduced in the American tree sparrow *Spizella arborea* (Wilson) (West, 1960) and redpolls *Acanthis* sp. (Brooks, 1968), both of which are northern in distribution.

148

Productive energy (PE)

Productive energy (PE) is the amount that a bird mobilizes over and above what it requires for existence as measured in caged birds (EM). At its lower limit of temperature tolerance the maximum energy that a bird can metabolize is required for existence and there is no surplus available for other activities. At higher ambient temperatures, EM decreases and progressively more PE becomes potentially available, at least until the bird comes under heat stress. The determination of the amount of PE available is important since it is a controlling factor in the initiation, peaking and ending of such activities as reproduction, moulting and migration.

Maximum potential metabolism

To calculate the amount of PE a bird has available at any temperature, it is necessary to determine its maximum capacity for metabolism as exhibited at its lower limit of temperature tolerance. The *incipient* lower limit of temperature tolerance is determined by placing newly caught, locally-acclimatized birds, as soon as they become accustomed to caged conditions, at various low temperatures and observing at what temperature only 50% can survive (LD_{50}). For the house sparrow population in east-central Illinois, there is a progressive change in this tolerance between $-25\,°C$ in January and $0\,°C$ in August and September (Barnett, 1970). There is also a progressive change in low temperature tolerance with latitude, being greater northward at all seasons (Blem, 1973a). The *ultimate* lower limit of temperature tolerance is determined by acclimating caged birds step by step to progressively lower temperatures until the level for LD_{50} is reached. For winter house sparrows this may be as low as $-36\,°C$ (Kendeigh, 1973b). Even during the summer, the birds can be acclimated to temperatures nearly as low. The ultimate lower lethal temperature is of interest in showing the physiological capacity of the species but is of less importance than the incipient lethal temperature in relating daily energy budgets to natural conditions.

The maximum capacity for metabolism of a species is calculated with its equation for EM using its lower limits of temperature tolerance. For the house sparrow, the incipient maximum is 31.5 kcal in the winter and 25.7 kcal bird^{-1} day^{-1} in the summer using equations for $10\pm$ and $15+$ hour photoperiods, respectively. An ultimate potential maximum of 34.5 kcal bird^{-1} day^{-1} is indicated. A bird can mobilize energy on a sustained basis up to its incipient maximum under stress of needs other than temperature tolerance, such as locomotor activity and reproduction (Kendeigh 1973b; Dol'nik, 1974b). The incipient maximum rate of metabolism does not, however, represent the capacity of body tissues for energy output as it

149

can be exceeded for short periods, even up to several hours, as in prolonged flight, provided there is ample time later to replace the energy dissipated.

Available productive energy and working capacity

Although a permanently resident species, such as the house sparrow, can attain a higher level of metabolism with acclimatization to cold, as in winter, the actual amount of PE available does not vary in a corresponding manner. The house sparrow is subjected to a mean monthly temperature of -2.8 °C in January which would require an EM of 25.5 kcal bird^{-1} day^{-1}. Its available PE would be 6 kcal bird^{-1} day^{-1} (31.5 − 25.5 kcal) although with acclimatization it could muster 9 kcal bird^{-1} day^{-1} (34.5 − 25.5 kcal). In July, at a mean monthly temperature of 24.1 °C, its EM would be 18.0 kcal bird^{-1} day^{-1} and its PE, 7.7 kcal bird^{-1} day^{-1} (25.7 − 18.0 kcal).

Since PE is the difference between the amount of energy a bird is mobilizing and what it needs for existence, the potential amount that a bird has available increases proportionally as EM decreases with a rise in temperature. At high air temperatures, however, the bird eventually comes under heat stress and its working capacity, or ability to use all of its available PE, decreases. In the house sparrow (Fig. 5.4), maximum working capacity on a sustained basis comes at approximately 22 °C (Kendeigh, 1969a). This temperature is also the lower critical temperature of the zone of thermo-neutrality, and until empirical data can be obtained on other species, we may assume that maximum working capacity comes generally at the lower critical temperature. This correlation is useful as we know how the lower critical temperature varies in different species related to weight and season (p. 136).

In the zone of thermo-neutrality, the metabolism of the inactive, fasting bird is reduced to the minimum, BM. Feeding and activity of any kind bring an increase in heat production that must be eliminated if body temperature is held constant. Above the lower critical temperature this can only be done by increasing the rate of heat loss. The insulative properties of the plumage become reduced so that heat transference is maximized. However, as ambient temperatures approach or surpass body temperature, heat dissipation becomes dependent primarily on evaporative cooling (Shilov, 1968). The stress on the bird is evident in its panting and gular fluttering (Lasiewski & Seymour, 1974). A point is reached, the upper critical temperature, above which evaporative cooling is insufficient to prevent a rise in body temperature. Hypothermy may be tolerated for short periods, but sustained activity is virtually eliminated. In Fig. 5.4, we have indicated the decline in working capacity as a mirror image of the curvilinear increase in the rate of evaporative cooling (Kendeigh,

150

1969*a*; 1973*b*). The decrease in work capacity at high temperature may well be a factor in the decline of reproductive activities in mid-summer and the inability of some species to carry on normal activities in hot desert conditions.

Weight change

The availability of energy to the bird fluctuates with increase and decrease in body weight. While feeding, the bird adds to the energy reserves on its body, chiefly in the form of fat deposits, and when not feeding it utilizes these reserves. There is thus a daily cycle of weight with the maximum attained in late afternoon and the minimum just before onset of feeding in the morning (Kendeigh, Kontogiannis, Masac & Roth, 1969). In cold weather, greater fat reserves are deposited during the day to compensate for their faster loss at night, so birds commonly gain overall weight during the winter. Migrant species increase in weight, often to a considerable extent, preceding both spring and autumn migration, and this provides for their long flights. There is commonly also an increase in weight during the annual autumn moult. Changes in weight, therefore, provide an index of the energy balance of the bird and the rate at which it is being stored or utilized. Fluctuations in weight, however, also involve changes in the water content of body tissues and, to a lesser extent, carbohydrates and proteins. Several studies have made possible the assignment of calorific values to body weight changes.

The calorific value of overnight weight loss (q), corrected for the food content of the alimentary tract, is affected by ambient temperature (Dol'nik & Gavrilov, 1971*b*; Dargol'ts, 1973). In the house sparrow (Fig. 5.5) it varies as ($q = \text{kcal g}^{-1}$):

$$(N = 20)\, q = 5.74 - 0.124\, T \pm 0.34 \tag{5.39}$$

The lower calorific value of weight loss at high temperatures is due to a larger proportion of water being evaporated for body temperature regulation. At low temperatures, calorific values may sometimes approach or exceed that of dry fat (9.5 kcal g^{-1}) because of storage of the metabolic water produced when fat is oxidized.

During the period of moulting values are lower, varying from about 4.4 kcal g^{-1} at 0 °C to 1.1 kcal g^{-1} at 30 °C, probably because of high supply of blood to and water loss from the papilla of the regenerating feathers.

There are taxonomic differences. In determinations on some 60 passerine species, Gavrilov found that at the lower critical temperature of the zone of thermo-neutrality the calorific value of overnight weight loss varied from 2.7 to 3.9 kcal g^{-1}. At 0 °C, in those species with lower critical temperatures of 4 to 17 °C, the value of q varied from 3.8 to 5.5

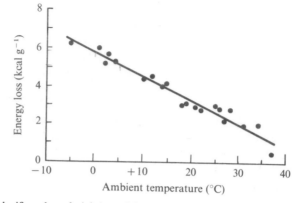

Fig. 5.5. Calorific value of nightly weight loss in non-moulting house sparrows (Dol'nik & Gavrilov, 1975).

kcal g^{-1}; in species with lower critical temperatures of 23 to 30 °C, q was much higher, 8.5 to 9.3 kcal g^{-1}.

Exercise, however, does not affect q in any consistent manner. Birds given forced exercise in a chamber rotated at two different speeds (12.5 and 24 revolutions min^{-1}) gave values not significantly different from birds at rest although there was a tendency for exercised birds to have lower values (Dargol'ts, 1974).

In migratory flight, the calorific value of weight loss was found to average 6.05±0.17 kcal g^{-1} (Dol'nik & Gavrilov, 1971b), and calculations from data in the literature on the energy cost of flight and weight loss during flight give values of 5.3 to 6.3 kcal g^{-1} (Dol'nik, 1969). According to analyses of body composition before and after non-stop migratory flights, calorific values ranged from 6 to 9 kcal g^{-1} (Dol'nik, 1971a). Perhaps these latter high values are caused by temporary retention of metabolic water during flight.

The calorific cost of adding a gram of weight is greater than the energy derived when that weight is lost because of the added work cost of feeding, assimilation and storing not only fat but also water and other materials. In the daily fluctuations in the weight of a house sparrow, approximately 6 kcal g^{-1} are required for a gain in weight during the daytime. With about 3 kcal g^{-1} obtained in the loss overnight, the gain/loss ratio is two (Kendeigh et al., 1969).

Changes in mean weight from day to day are caused by a disturbance in the balance between weight gain during the daytime and weight loss at night. In experiments with chaffinches, the gain/loss ratio during periods of weight gain (9.38 kcal g^{-1}) and weight loss (3.67 kcal g^{-1}) was 2.6 (Dol'nik, 1968; Dol'nik & Gavrilov, 1971b). In resident house sparrows, Dol'nik & Gavrilov (1975) found a mean value for weight loss of 4.0 kcal

152

g^{-1}, but in the migratory race *P. domesticus bactrianus* Zar. & Kud., it was 5.0 kcal g^{-1}. Assuming 6.0 kcal g^{-1} to be the cost of adding weight, the ratios would be 1.5 and 1.2, respectively. With the blue-winged teal *Anas discors* L., at 30 °C, the ratio between weight gain (6.7 kcal g^{-1}) and weight loss (3.2 kcal g^{-1}) was 2.1, but at lower temperatures the ratio decreased considerably (R. C. Owens, personal communication). It is obvious, therefore, that the use of fluctuations in weight as quantitative indices of energy exchanges requires caution. The calorific value per gram change differs not only between weight gain and weight loss, but also with ambient temperatures, the activity or physiological state of the bird and taxonomically.

Energy-conserving conditions and activities

Most activities of free-living birds under natural conditions are energy-demanding. There are some conditions and activities however, that are energy-conserving and these will be considered first. They generally decrease the cost of existence and increase the amount of PE that a bird has available. Behavioral thermo-regulation takes a variety of forms many of which are listed by Calder & King (1974).

Insolation

Birds lose heat by radiation to all objects in their surroundings as long as they maintain surface temperatures higher than those of their surroundings. This radiation would be considerable, because of the high body temperature of birds, were it not for the plumage that intervenes as a layer of insulation between the skin and radiating surface. Likewise, birds will receive heat from surrounding objects that have a higher temperature. Solar radiation could be excessive, even lethal, to birds were it not for the plumage layer. Thus plumage may function in the summer to keep excessive heat out of the body, as it functions in the winter to reduce the rate at which it is dissipated (Marder, 1973).

Radiation directly from the sun commonly varies between 0.8 and 1.4 gcal cm^{-2} min^{-1}. In addition, radiation from the sky varies between 0.2 and 0.3 gcal cm^{-2} min^{-1}, infra-red radiation from the atmosphere from 0.4 to 0.6 gcal cm^{-2} min^{-1}, and radiation from the vegetation and ground, 0.4 to 0.8 gcal cm^{-2} min^{-1}, depending on their surface temperatures. Most animals, except for white ones, have a dorsal absorptance to high sun of 70 to 90%, the rest being reflected. A white swan has 35% absorption, the pileated woodpecker *Dryocopus pileatus* (L.) 84%, and owls 44–75%. Generally, the absorptances for low sun are about 8% lower than for high sun (D. M. Gates, personal communication).

153

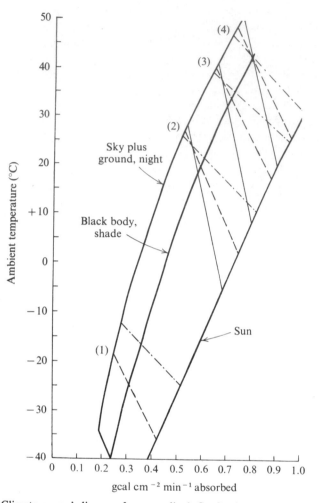

Fig. 5.6. 'Climate space' diagram for a cardinal *Cardinalis cardinalis* showing relations between air temperature, radiation absorbed and wind speed (——— 10 cm s^{-1}; – – – 100 cm s^{-1}; –·– 1000 cm s^{-1}) at lower limit of temperature tolerance (1), lower critical (2) and upper critical (3) temperatures of the zone of thermo-neutrality, and upper limit of temperature tolerance (4). At (4), the cardinal could not tolerate air temperatures exceeding 20 °C in full sunlight and still air (10 cm s^{-1}) but could tolerate up to 31 °C with wind speed of 1000 cm s^{-1}. At night, the bird could tolerate up to 50 °C in still air but only 46.5 °C with wind speed of 1000 cm s^{-1}. At the other extreme, birds are subjected to daily minimum temperatures early in the morning before they can benefit from solar insulation. If protected by trees or shrubs, they could survive down to −40 °C in still air but only −34 °C if exposed to the open sky. With higher wind velocities, tolerance to extreme cold would be even less. Within the zone of thermo-neutrality (2) and (3), birds could maintain the BM at much lower air temperatures if exposed to the sun and in still air (Porter & Gates, 1969).

154

Heat absorbed by the bird from insolation cannot be used as energy for the carrying on of activities. It is of value, however, in reducing heat loss from the body through the plumage and in reducing thermo-regulation costs. The SM of the cardinal, at -5 °C in still air and in shade on a winter day, for instance, would be 23.0 kcal bird^{-1} day^{-1}, according to its equation in Appendix 5.1. With the bird fully exposed to the sun, however, it is only 12.2 kcal, the same as at its lower critical temperature of 18 °C in the shade (Fig. 5.6). White-crowned sparrows *Zonotrichia leucophrys* (Forster), exposed to full sunlight for only three hours in early morning at an air temperature less than 10 °C, had the total calorific food intake for the day reduced by 12.6% (Morton, 1967).

At air temperatures above the lower critical temperature, absorption from the sun may produce an unneeded surplus of heat that must be dissipated to prevent a rise in body temperature. However, sun-bathing may occur at relatively high air temperatures and may have other functions besides heat absorption.

In an experiment with zebra finches *Poephila guttata castanotis* (Gould), exposed to an artificial source of radiation, black birds absorbed 3.1 gcal min^{-1} more than white birds and economized on SM to the extent of 28%. With larger birds, the energy conservation of exposure is less, however, since there is decreased surface area in proportion to the total body mass (Heppner, 1970). There is not much difference in effectiveness between black, brown or gray in dark-colored birds (Lustick, 1969, 1971). Some birds deliberately expose underlying dark coloration at temperatures below the zone of thermo-neutrality (Ohmart & Lasiewski, 1971; Marder, 1973).

An activity of birds which is of potential energy-conserving importance, therefore, is moving into and out of the sunlight for periods of time. With species occurring in open country where there is almost continuous exposure and clear, sunny days are the rule, conservation of energy by this means must be included in the analysis of daily energy budgets. This item becomes less important in biotopes such as forests where species are only occasionally exposed or where cloudy days are more prevalent. It is not practical at the present time to quantify the amount of energy conserved per unit time exposure to solar radiation in birds of different size. Aside from characteristics of the bird with respect to surface/mass ratio, coloration and orientation, the amount of heat incurred by a unit surface varies with time of day, season, latitude, elevation and the amount of water vapor and dust in the air. It should be noted, however, that probably all measurements of SM and EM have been taken from birds protected from incoming radiation. Birds under natural conditions, exposed even intermittently to solar radiation, will have lower metabolic rates than those predicted by the given equations. Nevertheless, it is worthwhile to calculate daily energy budgets without including the benefits of solar radiation, as birds must be able to survive and carry on their normal activities regardless of sky conditions.

Roosting site

Birds roosting in the open on clear nights are exposed to rapid radiation of heat to the sky. This radiation is reduced when the bird can obtain overhead protection from deep grass, snow, shrubs, vines, or trees (Moore, 1945; Shilov, 1968). The best roosting sites are in cavities, particularly in trees or buildings, where the walls have low rates of heat conductance. 'Northern' grouse and ptarmigan roost in snow burrows during periods of cold weather which sometimes last for several days (Volkov, 1968). Even redpolls and other northern passerine species may roost in snow cavities (Sulkava, 1969). Protected roosting sites become progressively more advantageous as ambient temperatures fall. For instance, house sparrows roosting singly in wooden boxes conserve energy at ambient night temperatures below 25 °C at an hourly rate (M_h = kcal bird^{-1} hour^{-1}) (Kendeigh, 1973b) of:

$$M_h = (3.29 - 0.133 \ T) \times W \times 10^{-3} \tag{5.40}$$

Huddling

Birds commonly roost in groups. This is advantageous at low ambient temperatures since the exposed surface available for radiation is reduced (Gerstell, 1942; Brenner, 1965; Case, 1973; Weiner, 1973). The amount of energy conserved varies with the closeness of contact between the birds and the number of individuals involved. The coal tit *Parus ater* L. commonly roosts in a loose linear group. With only two birds in the group, SM at 0 °C is reduced 18%, with three birds 26% (Gavrilov, V. M., 1972). The long-tailed tit *Aegithalos caudatus* (L.) forms a more compact linear group. With two birds, SM is reduced 27%, with three birds 39%. Further increase in number of birds did not improve the benefits appreciably. A closed circle is more effective in conserving energy than a linear order (Gavrilov, V. M., 1972).

When huddling occurs under the protection of cover, as discussed in the preceding section, the effect is additive and considerable conservation of energy occurs. Sociable weavers *Philetairus socius* (Latham), closely related and similar in weight to the house sparrow, roosting and huddling in groups within their well-insulated nest chambers, were calculated to conserve 43% of the energy that would have been expended roosting in the open when night temperatures dropped to near 0 °C (White, Bartholomew & Howell, 1975). Huddling, on the other hand, is disadvantageous at high ambient temperatures as the problem here is to dissipate heat rather than to conserve it.

Winter plumage

The EM immediately after the completion of the annual autumn moult is lower than it is preceding moulting because of the denser plumage insulation that is provided. In the house sparrow, the decrease is about 16% at 3 °C, 10% at 22 °C, and 9% at 32 °C (Blackmore, 1969); in the chaffinch, it is about 10% at lower temperatures (Dol'nik & Gavrilov, 1975).

Although the equations given in this chapter for summer and winter may be used uncorrected for these seasons taken as a whole, monthly values for September through December of both EM and SM should be lowered. In the house sparrow, we use the following corrections (Kendeigh, 1973*b*); September, -1.1; October, -2.3; November, -1.5; and December, -0.7 kcal bird^{-1} day^{-1}.

Fluffing of plumage

Metabolic rate is affected by the thickness of the plumage as well as by its density, as they both control the rate of heat loss from the body. When shifted from 20 °C to -5.5 °C, the yellowhammer *Emberiza citrinella* L. fluffed out its feathers and increased the volume of its plumage by 40%; the closely related ortolan bunting *E. hortulana* L., which is more sensitive to cold, increased the volume of its plumage 85% (Wallgren, 1954). Feathers are fluffed gradually and progressively, correlated with the drop in temperature to which they are exposed (Fig. 5.7). Fluffing of the plumage occurs most commonly as birds are roosting during the night but also occurs during the day with exhausted birds, even at high temperatures (MacFarland & Baher, 1968). Activity interferes with full erection of the feathers.

The chaffinch with its feathers fluffed and exposed to 0 °C lost heat at the rate of 0.523 kcal bird^{-1} hour^{-1}; with feathers prevented from fluffing, 0.712 kcal; and stripped of feathers entirely, 0.990 kcal (Gavrilov, Dol'nik, Keskpaïk & Yu, 1970). In two other species of small passerines at 0 °C, SM was increased 8.6 and 15.9% (Gavrilov, V. M., 1972). Similar increases in rate of SM have been demonstrated in the pigeon when prevented from fluffing its plumage (Burckard, Dontcheff & Kayser, 1933).

Fluffing of the plumage is more effective in conserving heat when the feathers are long than when they are short (compare Galliformes and Anseriformes, for instance). Body size is important here as the feathers

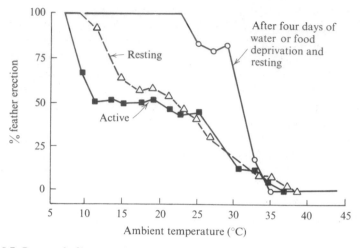

Fig. 5.7. Percent feather erection on the ring dove *Streptopelia decaocto* (Friv.) at various ambient temperatures (after McFarland & Baher, 1968).

are longer in large birds and can trap a greater volume of air. The lower critical temperature is only 1–2 °C lower in the goldcrest *Regulus regulus* L. (*ca* 6 g), with the plumage normally fluffed than when not fluffed; in the raven *Corvus corax* L. (*ca* 1000 g) it is 14–16 °C lower.

Concealing the head under the wing commonly accompanies fluffing the plumage, especially at night. Heat loss through the head is high because of the short feathers that cover it (Shilov, 1968).

Hypothermy and hyperthermy

Any reduction of the level of body temperature maintained at low ambient temperatures reduces by that much the amount of heat production required by the bird. Hypothermy is an adaptation to conserve body energy reserves during long cold periods and/or when food is scarce or feeding is not possible (Ohmart & Lasiewski, 1971). Hypothermy has been reported in the house sparrow, tree sparrow and several other passerine species (Steen, 1958).

The long, cold winter nights are often critical, especially with small birds. The willow tit *Parus montanus* Conrad, Siberian tit *P. cinctus* Boddaert (Haftorn, 1972) and black-capped chickadee *P. atricapillus* L. (Chaplin, 1974) may have a lowering of their body temperature as much as 8–9 °C below the 40 °C commonly found at night in passerine birds. Hummingbirds, being so small, are not capable of storing much reserve fat on their bodies so that even during the summer they are often hypothermic at night (Calder & Booser, 1973). A reduction of 10 °C in their body temperature may decrease SM by 25% (Calder, 1974).

158

Aerial species are frequently unable to feed adequately during inclement weather. During one four-day period of bad storms with air temperatures falling to 12–14 °C, sand martins *Riparia riparia* (L.) had a drop of 8 °C in body temperature even though they huddled together in compact groups. It was estimated that EM was reduced 62% (Keskpaïk, 1972). Hypothermy occurs under similar conditions in swifts (Koskimies, 1961) and in those goatsuckers that hibernate over winter (Dawson & Hudson, 1970).

Hyperthermy, on the other hand, is advantageous under stress of high ambient temperatures. Raising the body temperature increases the radiation of heat and puts less stress on the evaporative cooling mechanism. The extent to which body temperature can rise without becoming lethal is more limited than is the extent to which body temperature can be lowered.

Energy-demanding conditions and activities

Wind and rain, in addition to low temperatures, are often draining on a bird's energy, and locomotion, reproduction and moulting are all energy demanding.

Wind and rain

Body heat, conducted to the surface of the plumage, warms the adjacent air which then rises to be replaced by cooler air from the surroundings. This convective loss of heat becomes accelerated with air movements or wind (Fig. 5.6). Birkebak (1966) calculated the increase in rate of heat loss at about 10% m^{-1} s^{-1} increase in wind speed. In the snowy owl *Nyctea scandiaca* (L.) SM increases as a function of the square root of the air speed. A drop in temperature from −20 to −30 °C increased the cooling property of the wind by 20–30% (Gessaman, 1972). To minimize this wind-chill factor, birds commonly face into the wind to avoid the ruffling of their plumage, which would increase the rate of heat loss, and seek the lee side of woods, valleys or other obstructions that provide protection.

Rain may have relatively little influence unless it wets the plumage, as may occur when it is wind driven onto exposed birds. Water is a good heat conductor so that heat loss from the bird becomes accelerated.

Free-living locomotor activity

Birds under natural conditions move around more or less freely in their search for food and cover, social contacts with other birds, escape from enemies, etc. Flights are usually of short duration lasting only a few seconds. In caged birds, some of these motivated activities are replaced spontaneously by other kinds of activities, much of which is 'frivolous' in that it serves no apparent useful purpose (p. 147). Hops accompanied by wing movements replace the short flights of free-living birds. Increased locomotor activity often seen in caged birds before onset of darkness replaces the search for cover under natural conditions. Birds are disturbed by other individuals in adjacent cages and by the experimenter. During the migratory period there is nightly unrest (*Zugunruhe*) in caged birds. For instance, migratory unrest in caged chaffinches persisted for two months and was such that 1.2 million hops above the previous level of activity were recorded. The full distance of these hops was about 600 km, compared with 1500 km of their natural migratory route (Dol'nik, 1967*a*). In consequence of these substitute activities, the energy required for free existence exclusive of reproduction, moulting, sustained flights, etc. may not be much greater than that of caged birds. Some experimental data confirm this point.

Small groups of three to five house sparrows confined in a fairly large cage, $5 \times 2.5 \times 2.5$ m, that permitted short flights and social interactions at 13–19 °C, metabolized food 24% faster than solitary birds in the small laboratory cages used for measuring EM. This percentage difference decreased at lower temperatures probably because of energy conserved with huddling at night (Weiner, 1973).

Owen (1969) correlated the rate of heart beat, determined telemetrically, with metabolism in the blue-winged teal and found that non-breeding and non-moulting birds on a pond maintained a level of metabolism above EM, varying with prevailing temperature as:

$$M = 16.15 - 0.635\,T \qquad (5.41)$$

The birds had their wings clipped so that they were unable to fly, but they could swim freely. They were also exposed to solar radiation part of the day, which is not true of the caged birds with which they were compared. The increased cost of the free-living motor activity varied from 17% of EM at -10 °C to 0.5% at 25 °C. The heart rates of quietly foraging birds were not statistically different from resting birds, the heart rates of preening birds were about 30% higher, and of rapidly swimming birds, about 60% higher. The activities of small passerine species are obviously of a different character than those of ducks, but quantitatively they may be comparable (Kendeigh, 1973*b*). The measurements of free-living activity of the blue-winged teal were made at photoperiods averaging 12 hours a day. Total activity would doubtless be greater during the longer days of summer and less during the shorter days of winter. The lesser amount of activity during the winter may be offset, however, by the greater cost per unit of activity at low temperature (Kontogiannis, 1968) and the greater amount of activity per unit time required to satisfy their larger appetites.

160

Sixteen measurements of 'free existence' by means of telemetry of heart rate on four American kestrels *Falco sparverius* L., confined in a large quonset hut at 22.5 °C, gave a mean slightly lower than the value for EM calculated from equations 5.28 and 5.35, but the difference was not statistically significant (J. A. Gessaman & S. W. Sawby, personal communication).

It is of interest that in winter at temperatures of −20 to −30 °C, the blackcock *Lyrurus tetrix* (L.), a gallinaceous species, feeds for 30–60 minutes in the morning and spends the rest of the day under snow cover at about −4 °C. It has been possible to measure gross energy intake of free-living birds from stomach contents of individuals killed after a morning's feeding, and excretory energy from collections of excrement made around the snow burrow (Potapov & Andreev, 1973). The energy cost of free existence of blackcocks varied up to only 17% of the EM at −4 °C. Using a procedure that also involved analysis of stomach contents, collections of bird droppings and metabolized energy coefficients, the free-living willow ptarmigan *Lagopus lagopus* (L.), during the winter at −20 °C, was estimated to have a rate of metabolism about 28% higher than caged birds. In the rock and white-tailed ptarmigans *Alectoris groeca* (Meisner) and *Lagopus leucurus* (Richardson), however, the extra cost of free living was negligible (West, 1968; Moss, 1973).

Flight, swimming, running

For flying machines having the same shape and density but differing in size, the power required for horizontal progression, according to von Helmholtz's theory, is proportional to $W^{1.167}$ (Wilkie, 1959). In birds, however, a number of factors may modify this relation, such as wing span and flapping speed, as well as drag, velocity, air resistance, air density, etc., and there is also the energy expenditure for BM (Dol'nik, 1969; Pennycuick, 1969; Berger & Hart, 1974; Tucker, 1974). Assuming 'standardized' wing spans and flight velocities where the cost of transport is minimum, Tucker (1973) gives an equation for power input which varies with weight as $W^{1.00}$. Empirical data on the energy expenditure of birds in flight, measured by a variety of procedures, not all equally reliable (Dol'nik & Gavrilov, 1971a; Berstein, Thomas & Schmidt-Nielsen, 1973; Berger & Hart, 1974), provide values that are generally higher than those expected from Tucker's equation for small birds and lower than those expected for large birds (Fig. 5.8).

Tucker's (1973) equation does not allow for differences in aerodynamic quality in birds of different size. This varies with the lift/drag ratio. Lift is equal to body weight and drag to body surface. Aerodynamic quality ($W^{1.00}/W^{0.667}$) will therefore increase with weight as $W^{0.333}$ (Dol'nik, 1969).

161

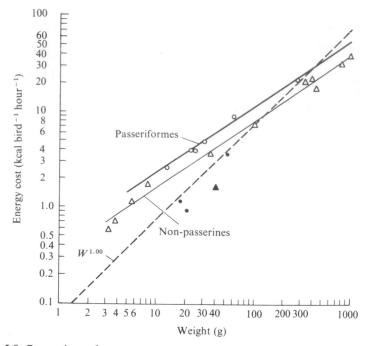

Fig. 5.8. Regressions of energy cost of flight on weight: passerine birds, equation 5.42; non-passerines, equation 5.43; Tucker's (1973) regression, $M_h = 0.0729\ W^{1.00}$. The solid dots are data for swallows, the solid triangle for a swift.

Using this correction, power input during flight should vary with weight as $W^{0.667}$ ($W^{1.00}/W^{0.333}$).

Excluding aerial feeders, the empirical values fit the following regressions (M_h = kcal hour^{-1}):

$$\text{Passerines } (N = 6): M_h = 0.4645\ W^{0.6911} \pm 1.076 \qquad (5.42)$$

$$\text{Non-passerines } (N = 11): M_h = 0.3157\ W^{0.6980} \pm 1.159 \qquad (5.43)$$

The exponents of weight come close to the predicted values and agree closely with those for BM on weight (Wilkie, 1959; Berger & Hart, 1974) but energy expenditure averages about 12.6 times higher in passerines and 12.1 times higher in non-passerines.

Aerial feeders, three swallows and a swift, were excluded from the above analysis because they have values well below those of other species (Fig. 5.8). These species have special adaptations for flight (Lyuleeva, 1970). Four hummingbirds are included at the lower end of the regression line for non-passerines. Values for these species agree fairly closely with expected ones based on their weights. Lasiewski (1963) and Hainsworth & Wolf (1972), however, give 215 gcal g^{-1} hour^{-1} as the cost of hovering

162

flight in hummingbirds, regardless of weight. Values for large birds fall
below the regression line of $W^{1.00}$ not only because of their better aerodynamic qualities but probably also because of their more frequent use
of passive flight, such as soaring and gliding.

The energy costs analyzed above are for sustained horizontal flight of
more than one or two minutes duration. Energy expenditure during take-off,
acceleration and gaining altitude may be up to twice as great. This is
reflected in the rapid rise in body temperature that occurs at the beginning
of flight. The rate of heat flow through the plumage also rises at the
beginning of flight but then declines to a lower level, well above that of
resting, that is maintained throughout the flight (Hart & Roy, 1967; Aulie,
1971). The flight muscles generate excess heat that must be dissipated. This
excess heat eliminates the need for additional temperature regulation
(Tucker, 1968; Keskpaïk & Khorma, 1972; Berger & Hart, 1974) so that
the energy expenditure of birds in sustained flight is independent of
ambient temperature. The energy expended in the short flights of foraging
and nesting birds, lasting only a few seconds at a time, although requiring
extra energy to perform (Teal, 1969), may be compensated for only in part,
if at all, by a decrease in the amount of heat produced for temperature
regulation after the bird comes to rest. Experiments by Kontogiannis
(1968) and Pohl & West (1973) showed that forced locomotion of small
birds in rotating cages required energy over and above EM at low as well
as at high temperatures.

Equations 5.42 and 5.43 give the energy expenditure of birds during
sustained flight. If we disregard the SDA of any food being digested, since
it will be small or absent, the energy cost of flying itself may be obtained
by subtracting BM. If we consider that both vary with weight as $W^{0.67}$
the equations become (M_f = kcal hour^{-1}):

$$\text{Passerines: } M_f = 0.4592\ W^{0.67} \qquad (5.44)$$
$$\text{Non-passerines: } M_f = 0.3333\ W^{0.67} \qquad (5.45)$$

These equations may be converted into energy cost per unit distance
flown (M_L = kcal bird^{-1} km^{-1}) by using Tucker's (1973) equation for
variation of flight speed with weight (v(km hour^{-1}) = 13.2 $W^{0.20}$):

$$\text{Passerines: } M_L = 0.0348\ W^{0.47} \qquad (5.46)$$
$$\text{Non-passerines: } M_L = 0.0252\ W^{0.47} \qquad (5.47)$$

The energy cost of flight, whether obtained by equations 5.44 and 5.45
or 5.46 and 5.47, needs to be added to EM to obtain the total energy
expenditure for the period. The energy saved in temperature regulation
during flight will be more than offset by the extra feeding activity required
to replace the energy used in the flight.

163

S. C. Kendeigh, V. R. Dol'nik & V. M. Gavrilov

At various swimming speeds under experimental conditions, mallards *Anas platyrhynchos* consume less energy at 0.50 m s^{-1}, which is their usual swimming speed, than at either higher or lower rates. At this speed, energy expenditure is about 2.2 times the resting level (Prange & Schmidt-Nielsen, 1970). Allowing for their threefold difference in weight, this level of energy expenditure agrees well with that which Owen (1970) found for locomotor activity in the free living, blue-winged teal.

The energetics of sustained running are similar to those of flying and there is a similar increase in body temperature. If either flying or running is excessive, especially at high ambient temperatures, the body temperature may reach the maximum that can be tolerated, then the activity must stop (Taylor, Dmi'el, Fedak & Schmidt-Nielsen, 1971).

Migration

In most species the amount of reserve energy within the body available for daily routine activities is insufficient for prolonged non-stop migratory flights. Many investigations have shown that before migration begins, and during stops along the migration route, there are periods of days or weeks of surplus fat deposition on the body to provide the energy required (Ward, 1963; King & Farner, 1965; Blyumental', Gavrilov & Dol'nik, 1967; Dol'nik & Blyumental', 1967; Fry, Ash & Ferguson-Less, 1970). In some species, one week of fat accumulation is required for one night of migratory flight (Dol'nik, 1971a). Fat deposition is made possible partly by increasing efficiency in the assimilation of food (Zimmerman, 1965b), partly by reduction of energy use in normal activities, but principally by increased food intake, or hyperphagia (King & Farner, 1959, 1965; Odum, 1960; Dol'nik & Blyumental', 1964; Dol'nik, Keskpaïk & Gavrilov, 1969). In both caged and free-living birds during vernal and autumnal pre-migratory periods, hyperphagia may increase the daily energy intake by 25 to 40%. Changes in body weight of recaptured, banded birds give information on the relative rate of fat deposition.

The maximum amount of fat deposited varies between species and populations within a species, is proportional to the length of the non-stop migratory flight, and varies with potential power output in flight. Among species with similar, lean body weight, the upper limit of fat deposition is inversely proportional to the rate at which energy is expended during flight, and this in turn is correlated with wing dimensions, especially length (Ward, 1964; Blyumental' & Dol'nik, 1970).

The energy cost of the migratory flight itself may be calculated with equations 5.46 and 5.47, when the distance between the breeding and wintering areas and the weight of the bird is known. The total energy expenditure of migration must, however, also include the cost of de-

164

Table 5.2. *Relative costs of migration and overwintering in birds of different weight*

Taxon	Weight (g)	Migration (kcal)	Overwintering (kcal)	Difference (kcal)
Passerines	10	342	265	+77
	100	1009	463	+546
Non-passerines	10	248	358	−110
	100	731	759	−28

positing the energy (pre-migratory fat) on the bird that is used to power the flight. If we assume a ratio of 1.5:1 for the cost of adding pre-migratory fat and the energy gained from its use, then the two equations above may be transformed to give the total energy expenditure of migration as follows (EEM = kcal bird^{-1} km^{-1}):

$$\text{Passerines: } EEM = 0.0522 \ W^{0.47} \tag{5.48}$$

$$\text{Non-passerines: } EEM = 0.0378 \ W^{0.47} \tag{5.49}$$

Obviously values thus calculated can only be approximations. Actual values of EEM will vary with the energy ratio for fat gained and used, velocity of flight, altitude, favorable or adverse winds, direct or round-about routes, etc. However, they permit some preliminary evaluations of the cost of migration in the annual energy budget.

Although migration requires a considerable amount of energy, this expenditure is partly or wholly compensated for by avoiding the energy stress of overwintering (90 days) in the colder latitudes (Dol'nik, 1971b). If we assume a round-trip migration of 2220 km over 10° of latitude with a 10 °C difference in mean daily temperatures between the two latitudes, the relative costs would be as given in Table 5.2. The cost of overwintering was calculated by finding the temperature coefficients for EM (equations 5.17, 5.18), multiplying by 10 °C, 90 days and 1.07 (cost of free existence, p. 180). It costs non-passerine species less to migrate than it does passerine species and, as the former are more sensitive to cold, they actually save energy by migrating. Small birds profit by migration more than do large birds.

Egg-laying and nest building

The growth of ova within the follicles of the ovary begins early in the year under the stimulus liberated by lengthening photoperiods. This development is slow, however, until copulation with the male when the ova undergo a period of rapid growth. Egg-formation involves the deposition

165

S. C. Kendeigh, V. R. Dol'nik & V. M. Gavrilov

not only of the nuclear material of the oöcyte but also proteins, carbo-hydrates and lipids that furnish building material for the embryo and energy for its growth and maintenance. In addition, calcium and phos-phorus are deposited to form the hard shell. These elements come from the food consumed by the female during this period and from its body reserves. Most of the energy deposition into the egg occurs in the yolk during the final period of rapid growth of the ovum, beginning seven to eight days before it is laid in large precocial species and three to four days in small altricial species. Ovulation (passage from ovary to oviduct) occurs the day before the egg is laid, and the albumen and shell are deposited around the ovum as it passes down the oviduct (King, 1973; Ricklefs, 1974).

There have been only a few direct determinations of the calorific value of whole, fresh eggs (Table 5.3). On a specific basis, eggs of altricial birds average 1.10 ± 0.039 kcal g^{-1} and of precocial and semi-precocial birds, 1.67 ± 0.105 kcal g^{-1}. The difference is significant and the higher value for eggs of the latter group is caused by their relatively higher content of yolk.

The calorific value of the whole egg ($q =$ kcal egg^{-1}) varies with its weight ($w =$ grams) in altricial passerine species as:

$$q = 1.124 \ w^{0.9438} \pm 1.088 \tag{5.50}$$

and for precocial and semi-precocial non-passerines as:

$$q = 1.910 \ w^{0.9574} \pm 1.189 \tag{5.51}$$

The two regressions have the same slope, close to $w^{1.00}$, but the difference in their Y-intercepts indicates higher calorific values for the precocial species.

The average clutch of the house sparrow (4.7 eggs in Illinois) is equi-valent to 14.4 kcal. If rapid egg formation begins three days before the first egg is laid and continues until the clutch is complete (7.7 days), there would be an energy deposition of 1.872 kcal bird^{-1} day^{-1}. If the house sparrow has the same efficiency of converting productive energy into energy deposited in the egg as occurs in the domestic fowl (77%, Brody, 1945), then the energy requirements for egg-laying would be 2.4 kcal bird^{-1} day^{-1}. The remaining 23% is apparently lost as SDA. A general equation for cost of egg-laying ($q =$ kcal egg^{-1}, $n \times$ number of eggs, $d =$ number of days before laying of first egg) may be formulated as:

$$M = (q \times n)/0.77 \ (n+d) \tag{5.52}$$

This cost to the bird would be reduced below the zone of thermo-neutrality to the extent that SDA compensates for temperature regulation. The energy cost of egg formation itself is probably the same at all ambient temperatures.

166

Table 5.3. *Calorific value (q) of fresh eggs, including shell*

Species	Egg weight (g)	Kcal egg[-1]	Authority
Altricial species			
Zebra finch *Poephila guttata* (Vieillot)	0.97	1.193	El-Wailly (1966)
Long-billed marsh wren *Cistothorus palustris* (Wilson)	1.14	1.179	Kale (1965)
Great tit *Parus major* L.	1.57	1.675	J. A. L. Mertens (personal communication)
Tree sparrow *Passer montanus* (L.)	2.2	2.244	Pinowski (1967 *b*)
House sparrow *Passer domesticus* (L.)	2.7	3.067	Tangl (1903)
Precocial and semi-precocial species			
Bobwhite *Colinus virginianus* (L.)	8.7[a]	16.2[a]	Case & Robel (1974)
Baird's sandpiper, *Erolia bairdii* (Coues)	10.0	16.0	Norton (1973)
Dunlin *Erolia alpina* (L.)	12.3	19.0	Norton (1973)
Laughing gull *Larus atricilla* (L.)	38	57[b]	Schreiber & Lawrence (1976)
Black-bellied tree duck *Dendrocygna autumnalis* (L.)	41	100	Cain (1976)
Domestic fowl *Gallus gallus* (L.)	56	89	Tangl (1903)
Domestic fowl *Gallus gallus* (L.)	58	92	Brody (1945)
Brown pelican *Pelecanus occidentalis* (L.)	92	127[b]	Lawrence & Schreiber (1974)
Herring gull *Larus argentatus* Pontoppidan	95	144	Drent (1967)

[a] Weighted average for birds at 15 and 25 °C.
[b] Some of the eggs measured contained embryos.

The cost of egg-laying may, however, vary with the season and locality within the same species. Fresh eggs of the house wren *Troglodytes aedon* Vieillot, for instance, are larger in size and probably have a higher calorific content when formed at medium ambient temperatures than at either higher or lower temperatures and for early clutches than for later ones in the year. Likewise, the number of eggs laid per clutch generally decreases with the onset of warmer weather (Kendeigh, 1941). As a general rule, species lay more eggs per clutch in northern, cooler latitudes than towards the tropics.

During the egg-laying period, the female is usually also involved in building or completing the nest structure and applies an increasing amount of heat to the eggs daily until the clutch is complete and full incubation begins. In a study with the zebra finch (El-Wailly, 1966), an efficiency of 77% was assumed for egg formation at ambient temperatures of 29–34 °C where the energy cost of the extra activities was minimal. This efficiency decreased at lower temperatures to only 49% at 14.5 °C, probably because of the relatively greater use of energy for these associated activities.

167

S. C. Kendeigh, V. R. Dol'nik & V. M. Gavrilov

Incubation

The work of incubation is maintaining the eggs at a relatively constant temperature well above that of the environment so that embryonic growth proceeds rapidly to hatching. In the zebra finch, El-Wailly (1966) has shown clearly that the energy cost of incubation increases with a drop in temperature. The lower the ambient and nest temperatures the faster the rate of heat loss from the eggs and the greater the amount of heat that the incubating bird must apply to maintain the egg temperature at its proper level. The energy cost of incubation is proportional to the rate of heat loss from the eggs in the nest which may be calculated from the following equation (Kendeigh, 1963):

$$M = n \times w \times c \times b(T_e - T_{na}) \times i(1 - s \times pc) \times 10^{-3} \qquad (5.53)$$

M = kcal bird^{-1} day^{-1}

n = number of eggs per clutch

w = mean weight of eggs (g) during incubation

c = specific heat capacity of eggs: 0.8 gcal g^{-1} °C^{-1}

b = rate of cooling of the eggs (°C hour^{-1}°C^{-1})

T_e = mean egg temperature (°C)

T_{na} = mean nest-air temperature (°C)

i = interval in hours (24)

s = percentage of egg surface covered by incubating bird

pc = percentage of i that the bird is on the eggs

10^{-3} = conversion factor, gcal into kcal

Substituting known values for the factors in equation 5.53, it may be simplified for the house sparrow (T_a = ambient temperature) as:

$$M = 8.09 - 0.241\, T_a \qquad (5.54)$$

Equation 5.53 measures the energy cost of maintaining the eggs at a constant mean temperature and is used most appropriately where both adults share incubation duties or one adult incubates more or less continuously. In many species, however, where only one parent incubates intermittently, or where the adults do not sit continuously on the eggs, the eggs cool during periods of inattentiveness, when the adult is away feeding, and must be warmed on each return for a period of attentiveness. This repeated raising of temperature of the clutch requires an expenditure of energy in addition to that calculated above. This additional expenditure may be calculated from the following equation:

$$M = n \times w \times c(T_e^{\text{max.}} - T_e^{\text{min.}}) \times n_{\text{att}} \times 10^{-3} \qquad (5.55)$$

$T_e^{\text{max.}}$ is the average maximum temperature attained by the eggs during the periods of attentiveness, $T_e^{\text{min.}}$ is the average minimum temperature attained by the eggs during the period of inattentiveness, and n_{att} is the

168

average number of attentive periods during the day (Kendeigh, 1952) plus the last visit to the nest at night. In the house wren, where values for these factors have been measured, this additional cost amounts to 35% of the value obtained by equation 5.53 at $T_a = 20\,°C$. The necessary information to calculate this cost for the house sparrow is lacking, but it is probably lower as the larger eggs cool less rapidly during periods of inattentiveness, and there are fewer inattentive periods per day.

Egg weights (w) are known for a number of species (Heinroth, 1922; Schönwetter, 1960–1972; Lack, 1968; King, 1973), and Rahn, Paganelli & Ar (1975) have developed a series of equations for different orders and families that estimate egg weights, when not known, from body weights. Egg weight varies with body weight approximately as $W^{0.67}$.

The cooling rate ($b = °C$ hour^{-1} $°C^{-1}$) is a constant independent of the stage of incubation (Kendeigh, 1963; Drent, 1967) but varying between species (Kendeigh, 1973a) in relation to egg weight (w) as:

$$b = 6.204\,w^{-0.3965} \pm 1.080 \qquad (5.56)$$

This equation is for isolated eggs in still air and includes heat loss due to radiation, convection and evaporation of moisture through the egg shell. It would be more accurate to use the composite cooling rate of the entire clutch, which would be lower (D. W. Norton, J. A. L. Martins personal communication). The cooling rate of eggs of the mallard *Anas platyrhynchos* L., according to equation 5.56, would be $1.31\,°C$ hour^{-1} $°C^{-1}$, but in the nest during periods when the incubating bird was absent it measured only $0.64\,°C$ hour^{-1} $°C^{-1}$ (Caldwell & Cornwell, 1975).

The internal egg temperature fluctuates as the incubating bird is attentive and inattentive but averages for different species between 29 and $38\,°C$, with highest values occurring in the Galliformes (Huggins, 1941; Drent, 1967). In most species it is around $35\,°C$. In the house wren (Kendeigh, 1963), where only the female incubates with short periods on and off the eggs during the daytime, the mean egg temperature varies with ambient temperature as:

$$T_e = 33.9 + 0.045\,T_a \qquad (5.57)$$

This equation should be adequate also for other small passerines nesting in wooden cavities. The temperatures measured were of the egg located at or near the middle of the clutch. It would be desirable in large clutches to have the mean temperature of all the eggs, as those around the periphery are often lower. In the semi-precocial species of herring gull, the internal egg temperature at the beginning of incubation averages about $32\,°C$, but as the embryo develops it rises to a level of 37.6–$39\,°C$ during the latter half of the incubation period. Since the two adults alternate on the eggs for 98% of the time, the internal egg temperature is negligibly influenced by variation in ambient temperatures (Drent, 1967).

The mean nest-air temperature surrounding the eggs is difficult to determine accurately since gradients in temperature occur from the top to the bottom of the eggs and from side to side. Nest-air temperature is also different when the incubating bird is attentive and inattentive and varies with ambient temperature. In the house wren (Kendeigh, 1963), the nest-air temperature is taken as the average of the temperature at the bottom of the nest under the eggs and near the top surface and varies with ambient temperature as:

$$T_{na} = 24.0 + 0.34 T_a \qquad (5.58)$$

Both Drent (1967), Rolnik (1970) and Caldwell & Cornwell (1975) have compiled nest-air temperatures for several non-passerine species.

The percentage of the egg surface covered by the incubating birds was estimated at 25% in Kendeigh (1963). In the herring gull, Drent (1967) measured the area of the brood patches, calculated what percentage of the surface area of the eggs this was, and obtained a factor of 0.18. Using a similar procedure with the great tit *Parus major*, J. A. L. Mertens (personal communication) also obtained a factor of 0.18. The agreement between these two species may be coincidental but the value is adopted here for the house sparrow. This percentage should be measured in other species. The area of the brood patch must be obtained from the bird. The surface area of an egg ($S = $ cm^2) may be obtained from its volume ($V = $ cc) by the use of Shott & Preston's (1975) equation:

$$S = 4.95 \, V^{0.67} \qquad (5.59)$$

The volume of the egg may be determined by water displacement, from its linear dimensions (Tatum, 1975), or from its weight ($w = $ g) (Drent, 1967) using the Baten-Henderson equation:

$$V = 0.933 \, w, \qquad (5.60)$$

or it may be estimated from the weight of the adult (Ricklefs, 1974). Actually, the rate of heat loss from the eggs when the incubating bird is on is doubtless reduced over the entire upper surface of the eggs not in direct contact with the brood patch (J. A. L. Mertens, personal communication). On the other hand, in large clutches, the percentage of total egg surface in contact with the brood patch is probably less than in small clutches.

Incubating behavior, i.e. whether one or both parents incubate, average lengths of attentive and inattentive periods, etc., is species-specific (Kendeigh, 1952). The amount of time that the adults spend incubating the eggs is closely correlated with maintaining a relative constant mean egg temperature. Caged village weaverbirds *Ploceus cucullatus* (Müller) increased their attentive time on the eggs proportionately with increase in cooling rate of the eggs as affected by the amount of insulation provided by the nest and air temperature (White & Kinney, 1974). Above 37 °C, egg and air temperatures became the same and attentiveness ceased. At 18 °C, on the other hand, the adult was providing the maximum heat of which it was capable. Below this air temperature, the incubating bird could not prevent a critical drop in egg temperature.

The cost of incubation should be added to EM and the cost of free existence for obtaining the daily energy budget in those species where the incubating bird is frequently inattentive throughout the day foraging for food, preening, etc. In those species where the incubating bird leaves the eggs for only a brief period during the day or is supplied with food by its mate, there is no significant cost of free existence and the cost of incubation should be added directly to EM.

The total cost of incubation, as analyzed above, is not necessarily a liability for the incubating adult. Some of the heat required for successful incubation comes from the embryo itself. In the semi-precocial herring gull, embryonic heat production is small during the first half of the 27-day incubation period but may provide up to 75% of the amount required during the latter days (Drent, 1967). In three species of precocial shore-birds *Calidris* spp., embryos nearly ready to hatch contributed 35–40% of the total incubation cost for the day. For the entire 19.5–22-day incubation periods in the different species, the embryonic heat contribution was 4.4–5.6% of the total required (Norton, 1973). Somewhat similar data are available on a few other species (Kashkin, 1961; Spellerberg, 1969; Rolnik, 1970).

Physiological regulation of body temperature becomes established in precocial species shortly before the time of hatching. This is not true in altricial species, yet the contribution of heat by the embryo may still be considerable. Calculations from equations of oxygen absorption by eggs of the house wren at different temperatures (Kendeigh, 1940), for a mean egg temperature of 34.8 °C and converting these values to kcal clutch^{-1} day^{-1}, indicate that on the second day of incubation at an ambient temperature of 20 °C embryonic heat production may amount to only 1.1% of the total required but that it increases rapidly, until on the twelfth day of the 14-day incubation period it may amount to 35%.

Nests may be located in such a position as to receive heat from the sun for a portion of the day. Desert larks *Ammomanes deserti* (Lichtênstein), for instance, often place their nests on north-facing low slopes or on the north side of bushes where they receive solar radiation during the cool mornings, are shaded during the hot mid-day hours, and receive cool northwestern breezes in the late afternoon (Orr, 1970). When house wrens nest in boxes exposed to full sunlight for much of the day there is an increasing amount of energy saved below air temperatures of 26 °C (Kendeigh, 1963), which in terms of percent (*PC*) is:

$$PC = 34.63 + 0.752T - 0.064T^2 \pm 2.830 \qquad (5.61)$$

At temperatures above 26 °C, however, such exposure may bring excessive temperatures in the nest and may cause mortality of the embryos.

Nests in cavities are protected from losing heat by radiation to the open sky. Such exposure of open nests, especially at night, adds to the cost of incubation as calculated by the above equations. Most nests, however, are placed in sheltered situations where they are protected by overhanging vegetation (Calder, 1974), or the nest itself may be constructed with a cover overhead and with an entrance from the side. The location and structure of the nest are also adjusted in relation to their exposure to wind or air movements that produce cooling (Shilov, 1968).

When the bird crouches deep in the nest cavity, as it incubates the eggs, it benefits from the nest insulation so that heat loss from its body is conserved. With only one-quarter of their body exposed above the nest walls, hummingbirds at 0 to 4.6 °C were estimated to save 60% of the heat loss of fully exposed birds and birds with one-half of their body exposed saved 40% (Calder, 1973a). J. A. L. Mertens (personal communications) has developed a theoretical model of heat exchange for the great tit nesting in boxes that shows that the incubating bird deep in its nest thereby conserves much of the extra energy required for maintaining the eggs at the proper temperature. Smith, Roberts & Miller (1974) have similarly developed a model for the hummingbird.

Because of these many variables, we are obviously still far from quantifying accurately the net energy cost of incubation to the bird. The equations so far developed are useful if one assumes that the unquantified parameters that conserve energy compensate for parameters that increase energy expenditures, but these need to be investigated in detail.

Equation 5.53 could be simplified if the cooling rate of the entire clutch in the nest was used. This composite cooling rate, b_n, would involve not only heat provided by the eggs but also the insulation provided by the nest structure, and would be a function of air temperature (T_a), which is easily measured. The factor $b_n(T_e - T_a)$ could then be substituted for $n \times b(T_e - T_{na})$. The above equation disregards nest insulation because use is made of the nest-air temperature (T_{na}), to which the eggs respond directly and which is the result of interaction between ambient temperature and nest insulation.

The importance of nest insulation to the cooling rate of the eggs is demonstrated in an experiment by F. N. White (unpublished). An artificial steel 'egg', weighing 24.16 g, was cooled in still air from 40 to 25 °C in 37 minutes or at an average rate of 0.4054 °C min^{-1}. When placed in the nest of a village weaverbird with the dome removed to simulate an open nest, the 'egg' cooled at the average rate of 0.3260 °C min^{-1}; in the normal domed nest of the species but without the feather nest-lining, it cooled 0.3000 °C min^{-1}; and with the feather lining, 0.2586 °C min^{-1}.

Two factors are of concern in evaluating the insulation of nests of different species: the temperature gradient between the egg and its environment and the extent to which egg and nest-air temperatures are affected by changes in the ambient temperature. In the herring gull, a ground nester and a species where the adults alternately incubate almost continuously, there is a difference of 7 °C between the nest-air temperature and the temperature of the nest-bottom in contact with the ground (Drent, 1967). In the house wren, on the other hand, where the nests are in wooden boxes above the ground and the female covers the eggs intermittently, the difference is only 0.5 °C (Kendeigh, 1963). However, in the house wren, the nest-air temperature varies 0.32 °C per degree change in ambient temperature while in the herring gull there is practically no change (Kendeigh, 1973*a*).

Brooding young

Newly hatched, altricial young respond to changes in ambient temperature as do ectotherms (poikilotherms). Regulation of body temperature is gradually attained during the next week or 10 days, varying with the species (Dunn, 1975*c*). During this period the adult bird must brood the young in a similar manner to the way it incubated the eggs, and the amount of heat that it applies must compensate for the heat that is lost from the nestlings. Mertens (1972) has developed a physical model for the rate of heat loss to the environment for nestling broods (W_b = weight of brood, grams) of the great tit in their wooden boxes, which may be transformed into energy requirement of the brooding adult to replace this heat loss, by the following equation (T_b = mean body temperature of brood, T_a = ambient temperature):

$$M = 0.0719 \, W_b^{0.613} \, (T_b - T_a) \qquad (5.62)$$

As the nestlings develop the ability to regulate their own body temperature, they will contribute to this heating requirement and will thereby relieve the brooding parent by a corresponding amount until eventually, brooding is no longer required. Heat loss per specific weight decreases as the size of the brood increases, related to the decrease in the exposed surface area of the whole brood in proportion to its total mass. This is particularly important for economizing on the energy expenditure of the adult in species with large broods, such as tits. In species with small broods, such as the house sparrow, this economy is less important but is compensated for by good insulation of the nest structure (O'Connor, 1975).

With nests exposed to the sun, the problem is often to prevent overheating of the young. With open nests, the parent bird may stand with partially outstretched wings to shade the young. Orientation of the covered nest of cactus wrens *Campylorhynchos brunneicapillus* (Lafresnaye), so

S. C. Kendeigh, V. R. Dol'nik & V. M. Gavrilov

that the entrance faces into the cooling southwest winds during the hot part of the breeding season, is correlated with higher percentages of the young successfully leaving the nest (Austin, 1974).

Feeding young

Under natural conditions, chaffinches commonly raise four to five nestlings, each parent expending about half of the productive energy it has available, 2 kcal bird^{-1} day^{-1}. With this expenditure of energy they provide the young with 32 kcal gross energy, or at a ratio of 1:8. Experiments with caged birds have shown that either the male or female is capable of feeding and caring for 9, but not 10, young, 9–11 days old. Under these conditions the parent would need to expend its near-maximum available energy (Dol'nik, 1972, 1974a).

Except for a study with ring doves, which represents a special situation, since they feed their young with 'pigeon milk' (Brisbin, 1969), there is very little other information available on the energy cost for feeding and care of nestlings and fledglings until they are able to take care of themselves. Since quantitative data for the house sparrow were lacking, the average energy cost per adult for brooding and feeding the nestlings was considered to be the same as the energy cost of incubation. In comparison with incubating, the female may spend more energy during the daytime feeding the young but less energy at night brooding them.

With fledglings out of the nest, the attention of the parents to the young continues at about the same rate for several days but then decreases gradually until it ceases altogether, about 10 days to two weeks after fledging occurs. When a later brood is raised, the parents may initiate activity for it as their responsibility for the first brood diminishes.

Singing, courtship, territorial defense

Spermatogenesis in the male is not very energy demanding and is spread over a relatively long period of days (King, 1973; Ricklefs, 1974). There have been no direct measurements, to our knowledge, of the energy cost to the male for establishing and defending territories, taking the aggressive role in courtship, and in his other activities around the nest. From a detailed analysis of time spent in these different activities and estimates of the energy cost of each, T. W. Custer (unpublished) has shown that the total energy expenditure of the male lapland longspur *Calcarius lapponicus* (L.) is about equal to the females during the breeding season. This is what we would expect. In several species, the male shares in incubation and in even more species he helps to feed the young, in and out of the nest.

174

Moulting

As old feathers are lost, new feathers begin growing immediately. The energy requirements for moulting continue until the new feathers are fully grown. There may also be a larger number of feathers present after moulting than there was before moulting. Difficulty may be experienced in examining live birds in the field in determining the exact time moulting starts and finishes. In the caged house sparrow, the moulting period lasts 91 days at 3 °C, 109 days at 22 °C, and 126 days at 32 °C. It normally begins gradually in late July but at low temperatures it may not start until the very end of August. However, it reaches a peak in early October, regardless of temperature, and is completed at about the same time at the end of November (Blackmore, 1969). Dol'nik & Gavrilov (1975) have also observed in the chaffinch that regardless of the time moulting begins in different birds, it is completed on about the same date.

In considering the energy cost of moulting, it is necessary to distinguish between the cost of plumage replacement and the actual cost to the bird because of compensating factors. Blackmore (1969) was able to obtain the cost of plumage replacement by a computer program that eliminated the compensating factors. The net energy cost of moulting (NEM = kcal bird^{-1}) has also been calculated using the difference in chemical composition of the feathers and the kind of food consumed. Feathers have a higher concentration of sulphur-containing amino acids (cystine, cysteine) than is found in either plant or animal foods (Newton, 1968; Dol'nik & Gavrilov, 1975). The synthesis of feathers (w_{pl} = weight of new feathers, g) therefore needs an increase in food intake proportional to the difference in concentration of these amino acids in the plumage (A_{pl}) and in the food proteins (A_{pr}). The equation used is:

$$NEM = (w_{pl} \times A_{pl}) \times \frac{GEI \times MEC}{A_{pr} \times PR} \qquad (5.63)$$

where GEI is the specific energy content of the dry food eaten, MEC is the metabolizable energy coefficient (p. 191), and PR is the total protein content of the food. Assuming a ratio of A_{pl} to A_{pr} in plant seeds of 5:1, GEI = 4 kcal g^{-1}, MEC = 0.8 and PR = 0.15, then NEM = 107 kcal g^{-1} of feathers in granivorous birds. The ratio of A_{pl} to A_{pr} in animal food is about 2.5:1, and if GEI is about 5.7 kcal g^{-1}, MEC is 0.8, and PR is 0.20, then NEM for carnivorous (insectivorous) birds is 57 kcal g^{-1} of feathers.

In four studies (Table 5.4) where measurements were made directly, NEM varies almost proportionally with the weight of the plumage (Dol'nik, 1965; Blackmore, 1969; Dol'nik & Gavrilov, 1975). The weight of plumage, not dried and taken at various times of the year, also varies

Table 5.4. *Total energy cost of moulting in passerine species (Dol'nik & Gavrillov, 1975; Blackmore, 1969)*

| | | | | | Kcal bird[a] | | |
| | | | | | | Calculated | |
Species	Age	'Lean' weight (g)	Weight of plumage moulted (g)	Direct measurement	Equation 5.64	107×weight of feathers (g)
Chaffinch, *Fringilla coelebs*	juvenile	19.2	0.8[a]	86.7	81.7[b]	86
	adult	19.2	1.4	140	143	150
House sparrow (Illinois)	adult	25.2	1.7	185	185	182
House sparrow (Leningrad)	adult, juvenile	25.8	1.9	218	189	203

[a] Moulting is incomplete in juveniles.
[b] *NEM* (equation 5.64) ×0.8/1.4.

almost proportionally (5–6%) to bird weight ($W^{0.9591}$) (Turček, 1966; Kendeigh, 1970). Knowing the energy cost of feather replacement in the house sparrow (185 kcal bird^{-1}), the equation which can be established for the complete moulting period in passerine species is approximately:

$$NEM = 8.377\ W^{0.9591} \tag{5.64}$$

Values obtained from this equation are reasonably close to measured values (Table 5.4).

Previous to and in the early stages of moulting, metabolized energy is high because the loss of feathers decreases the insulation provided by the plumage (Blackmore, 1969; Lustick, 1970; Dol'nik & Gavrilov, 1975). As the new feathers develop, insulation becomes greatly improved so that even at the peak of moulting the full cost of plumage replacement is not realized, and during the latter part of moulting the metabolized energy drops well below the pre-moult level (Blackmore, 1969; Dol'nik & Gavrilov, 1975).

The full energy cost of moulting (*NEM*) is reduced in other ways. Feathers are largely proteinaceous, and the assimilation of proteins has a high SDA. Measurements of the SDA of feather growth are not available but the heat liberated could significantly replace that which would need to be produced for temperature regulation.

It is a general observation that birds become exceptionally quiet under natural conditions during the period of moulting. There is also a decrease in locomotor activity of caged birds while moulting. In three species of Passeriformes, this decrease varied from 31 to 80% during the daytime

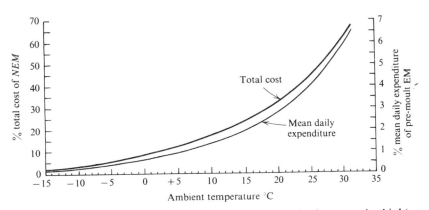

Fig. 5.9. Variation with temperature in net total cost of moulting in a passerine bird (upper curve) in percentage of values (net energy cost of moulting, *NEM*) obtained with equation 5.64 (Table 5.4), and net average cost per day throughout the moulting period (lower curve) as percentage of pre-moult existence metabolic rate (EM).

and from 48 to 91% at night with birds previously exhibiting *Zugunruhe* (Eyster, 1954). Any decline in activity would be energy conserving.

These compensating factors are such that the mean daily energy expenditure for the whole period of moulting was increased in the house sparrow only 0.2 kcal or 0.8% above the pre-moult EM at 3 °C, 0.6 kcal or 3.2% at 22 °C, and 1.0 kcal bird^{-1} or 6.5% at 32 °C (Fig. 5.9). Even at the peak of moulting the increase above the pre-moult level was only 22, 22 and 30% at these three temperatures. The total energy expenditure for the whole moulting process over the pre-moult level of EM was 19.9, 68.9 and 123.2 kcal bird^{-1} at 3, 22 and 32 °C, respectively (Blackmore, 1969). The percentages of these values of *NEM* reported in Table 5.4 (185 kcal bird^{-1}) are 11% at 3 °C, 37% at 22 °C and 67% at 32 °C (Fig. 5.9). The energy cost of plumage replacement is temperature independent, but percentage-wise the actual cost of moulting appears greater at high ambient temperatures because the compensating effect of SDA on temperature regulation is less and EM itself is lower.

To obtain the actual energy cost of moulting to the bird, therefore, values obtained with equation 5.64 need to be reduced to a percentage depending on the ambient temperature. The exact percentages to be used, based on studies with the house sparrow, can be approximated from Fig. 5.9.

Total metabolism or daily energy budget (DEB)

The total energy metabolized by the bird during a day is its daily energy budget (DEB) and is calculated as the sum of the EM plus the cost of energy-demanding conditions and activities minus the savings of energy-

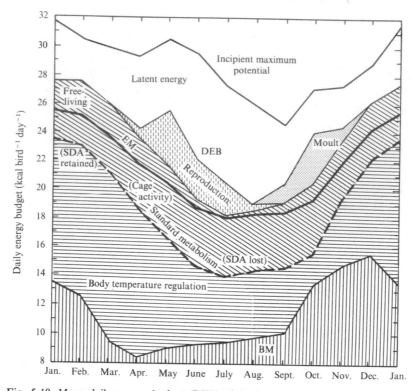

Fig. 5.10. Mean daily energy budget (DEB) of the house sparrow each month throughout the year in central Illinois. No attempt is made to separate the energy expended for cage activity from SDA lost nor to separate SDA retained and body temperature regulation. The abrupt variations, April to May and October to November, in both DEB and incipient maximum potential are artifacts caused by shifting from equations for 10± to 15+ hour photoperiods in the spring and the reverse in the autumn.

conserving processes. The mean DEB for each month has been calculated for the house sparrow in Illinois using the normal mean monthly temperatures and the equations developed in this chapter (Fig. 5.10). The mean monthly DEBs average nearly one-third higher than those calculated by Weiner (1973) for southern Poland where winter temperatures are comparable but summer temperatures are cooler.

In Illinois, the DEB is higher in the winter when only the cost of free existence is added to EM than in the warm summer months when reproductive and moulting activities are also included. During the 'winter' months, the DEB is only 8% higher than the EM during December and January but averages 10% higher for the entire period (November through April). During the 'summer' months (May through October) there is greater monthly variation but the average DEB is 16% higher than the EM. The seasonal changes in the DEB parallels a similar change in the incipient

178

maximum potential. The low point in the latter curve came in September in the particular year it was determined but may frequently come in August as does the low point in the curve for the DEB.

Latent energy

The difference between the DEB and the incipient maximum potential is *latent energy* or an energy reserve available in times of need. Since the curves in Fig. 5.10 were calculated from the *mean* monthly temperatures, on approximately half of the days each month temperatures are lower and during half of the years on record mean monthly temperatures are lower and energy demands higher than indicated. Latent energy allows the bird to survive these fluctuations in temperature as well as other emergencies. It is of interest that latent energy is least, i.e. the DEB is closer to the incipient maximum potential that the bird can mobilize, during autumn and winter and that there is the greatest potential reserve of energy during spring and summer when reproduction is underway.

Adaptation to local climate

The DEB for any locality will vary with the prevailing climate, particular temperature and photoperiod and the timing and extent of reproductive, moulting and other activities. Table 5.5 gives the mean DEB each month for seven localities in North America, extending from the tropics to the subarctic regions and calculated in the same manner as for Fig. 5.10. The house sparrow is found at all localities except Balboa Heights, but there is no reason, as far as energy resources are concerned (Kendeigh & Blem, 1974), why the species should not disperse to this and other tropical localities in the future.

Daily energy requirements during late spring and summer are clearly less at all localities than during the winter months, but the difference between summer and winter energy requirements decreases from northern to southern localities. Likewise, the general level of the total energy requirements decreases from northern to southern latitudes.

Since both seasonal and geographical trends vary inversely with temperature, a general relation between energy requirements and temperature was determined by averaging all monthly data for intervals of 10 °C. There may be a question about averaging data for different months at different localities, but there is justification for so doing. For instance, egg-laying in Illinois begins in April when temperature reaches 8 °C and attains a peak when temperatures rise to 14 °C. These temperatures and presumably the beginning of egg-laying do not occur at Winnipeg until early May but occur at more southern latitudes by at least March. Such a relation does not hold

Table 5.5. *Normal mean monthly temperature and 'measured' DEB* (M = kca.
bird⁻¹ day⁻¹ for the house sparrow at various localities. All monthly data for
intervals of 10 °C, regardless of the time of year or location, are bracketed

	Locality and latitude (N)													
	Churchill, Manitoba 58° 40′		Winnipeg, Manitoba 49° 54′		Champaign[a] Illinois 40° 8′		Amarillo, Texas 35° 14′		Tucson, Arizona 32° 14′		Veracruz,[b] Mexico 19° 10′		Balboa Heights,[b] Panama Canal Zone 8° 58′	
Month	°C	M	°C	M	°C	M	°C	M	°C	M	°C	M	°C	M
January	−27.5	36.0	−18.3	36.0	−2.8	28.3	2.6	26.4	9.9	21.2	21.9	18.4	26.6	16.0
February	−26.4	37.8	−15.7	35.8	−1.5	28.3	5.2	25.5	11.6	20.6	22.5	18.1	26.9	15.9
March	−19.8	37.0	−8.1	33.3	4.1	26.3	8.7	24.6	14.4	21.4	23.0	19.1	27.6	16.5
April	−10.7	34.8	3.3	28.3	10.6	24.2	14.2	25.7	18.8	21.1	25.3	18.7	27.9	17.0
May	−2.3	32.9	10.6	29.5	16.5	25.5	19.2	24.3	23.0	18.2	26.9	15.0	27.2	14.8
June	5.8	29.2	16.5	27.8	21.8	21.6	25.0	19.9	28.1	15.1	27.2	14.8	26.8	15.0
July	12.0	29.4	19.7	25.8	24.1	19.8	27.0	18.9	30.1	14.1	26.9	15.1	26.9	15.1
August	11.6	27.3	18.6	25.3	23.0	18.6	26.4	18.5	28.4	14.1	26.9	14.6	26.8	14.6
September	5.7	27.2	12.5	24.8	19.3	20.2	22.1	19.5	26.9	13.2	26.6	13.8	26.6	13.8
October	−1.1	31.2	6.5	27.6	12.8	24.2	15.8	23.6	21.1	14.9	26.1	13.8	26.2	13.8
November	−11.7	30.3	−4.4	29.3	4.9	24.7	7.8	23.3	14.5	18.8	23.6	17.2	26.2	15.8
December	−21.9	32.7	−13.7	32.9	−1.0	26.9	4.1	24.9	11.0	19.9	22.5	17.4	26.6	15.3
Mean/ Total (Mcal)[c]	−7.2	11.7	2.3	10.8	11.0	8.8	14.8	8.4	19.8	6.5	25.0	6.0	26.9	5.6

[a] The metabolic measurements were actually made on a population sample from Monticello, about 30
km from Champaign.
[b] No metabolic measurements were made for birds at Veracruz and Balboa Heights and the data given
are calculated from equations derived at the other localities.
[c] Mcal = Mcal (1000 kcal) bird⁻¹ year⁻¹.

in the autumn, however, as nesting mostly terminates with the onset of
moulting at about the same time in late summer at all localities. There is
also some tendency for daily energy values to be higher from January
through August than from Septemper to December when calculated for
particular temperatures. Perhaps these differences are due to poorer
feather insulation preceding the autumn moult, but they are small and thus
disregarded.

A significant regression between the mean monthly DEB and temperature follows the equation:

$$M = 28.6 - 0.426\,T \pm 1.07 \qquad (5.65)$$

The data for the interval −20 to −30 °C, all obtained during winter months
at Churchill, are not included in this calculation, as the birds in this

180

locality are confined during the winter in and around grain elevators and are not fully exposed to the prevailing ambient temperatures. An inverse regression of the DEB on temperature is, of course, to be expected since each component varies significantly with temperature. However, this equation is of special interest as it expresses the composite relation of all components of the DEB to temperature when taken as a unit.

When values calculated with equation 5.65 are compared with the average 'measured' values of the DEB at each locality, most of the calculated values are lower than the 'measured' ones at northern localities and higher at southern localities. Although the metabolic adaptations of populations to local climate are included in the values of the DEB in Table 5.5, in combining data for the same temperatures at different localities, much of these adjustments were lost and need to be restored. When the average differences between expected and 'measured' values for the DEB at each locality are related to the mean normal yearly temperatures (T_{yr}), the following significant regression is obtained:

$$M = 3.86 - 0.236 T_{yr} \pm 0.971 \tag{5.66}$$

Using equations 5.65 and 5.66, predicted values of the DEB averaged within 6% of the 'measured' values, with the greatest discrepancy being 11.4%.

Relation of the daily energy budget to existence metabolism

As an alternate procedure, the DEB could be predicted from the EM, if the relation between the two was established. At the four localities (Winnipeg, Champaign, Amarillo, Tucson) where EM was measured and the DEB calculated for mid-winter (December, January, February), the DEB averaged $107.1 \pm 0.4\%$ of the EM with no consistent difference between localities or temperatures. During mid-summer (June, July, August), however, the DEB, expressed as a percentage (PC) of the EM, increased progressively northward with decrease in mean summer temperature (T_s) as:

$$PC = 134.2 - 0.923 T_s \pm 0.9 \tag{5.67}$$

Blem (1973a) measured the EM of house sparrow populations from four other localities (Duluth and St. James, Minnesota; Flagstaff, Arizona; and Vero Beach, Florida). These localities were not included in the above analysis, since their ambient temperatures were similar to localities used. The DEB was calculated at these localities for two winter seasons and three summer seasons, both by using equations 5.65 and 5.66 and the percentage method. The chi-square test showed no significant difference between the two sets of values.

S. C. Kendeigh, V. R. Dol'nik & V. M. Gavrilov

Relation to species weight

An exact determination of how the DEB varies in different species in relation to weight must wait until measurements of the DEB can be made on a variety of species. However, a tentative relationship can be established based on the data available for the house sparrow, and by assuming that the relation between the DEB in species of different weight varies in the same way as does the theoretical relation between EM and weight.

Since the DEB during the *winter season* averages 107.1% of the EM, regardless of ambient temperature, when allowances are made for physiological adaptation to local climates, then equations 5.23 and 5.29 for *Passeriformes* may be changed to show the relation of the DEB to weight as follows:

$$(30\ °C) \quad M = 1.588\ W^{0.67} \tag{5.68}$$

$$(0\ °C) \quad M = 5.208\ W^{0.50} \tag{5.69}$$

During the *breeding–moulting season*, the DEB (expressed as a percentage of the EM, according to equation 5.67, is 106.5 at 30 °C and 134.2 at 0 °C. Using these percentages, equations 5.26 and 5.34 may be converted to show the relation of the DEB to weight as:

$$(30\ °C) \quad M = 1.676\ W^{0.67} \tag{5.70}$$

$$(0\ °C) \quad M = 6.961\ W^{0.50} \tag{5.71}$$

If we assume that during the *winter season*, the DEB of *non-passerines* also averages 107.1% of the EM in birds physiologically adapted to local climates, then equations 5.25 and 5.31 may be changed to show the relation of the DEB to weight as follows:

$$(30\ °C) \quad M = 1.075\ W^{0.67} \tag{5.72}$$

$$(0\ °C) \quad M = 5.908\ W^{0.50} \tag{5.73}$$

During the *breeding–moulting season*, the closest approach to measurement of the DEB in a non-passerine species, using procedures similar to those in this chapter, is that by West (1968) on the 574 g willow ptarmigan *Lagopus lagopus* (L.). His caged birds were exposed to fluctuating temperatures and photoperiod out-of-doors in an approximately natural environment and underwent normal moulting (102.0 kcal bird^{-1} day^{-1}) and egg-laying (126.0 kcal bird^{-1} day^{-1}), but not incubation or care of young. The mean temperature, from April through September, was 8.2 °C. At this temperature, the EM would be 98.8 kcal bird^{-1} day^{-1} (Appendix 5.2) and the cost of free living (10.8%), about 10.7 kcal. If we assume that the cost of egg-laying was representative of the whole reproductive period of West's birds and that reproduction and moulting divided the period equally, then the DEB averaged 126% of the EM [(102+126)/2+10.7]/

98.8] or what it would also be in Passeriformes at this temperature. Assuming that DEB varies with temperature, as in Passeriformes, we may convert equations 5.28 and 5.35 to give the DEB for nonpasserine species:

$$(30\ °C)\quad M = 1.079\ W^{0.67} \tag{5.74}$$

$$(0\ °C)\quad M = 8.059\ W^{0.50} \tag{5.75}$$

Values from equations 5.68 to 5.75 will be somewhat too low for northern birds and too high for southern birds. However, it is not possible to provide additional equations to correct for adaptations to local climate. Barely adequate data on this point are available only for the EM of the house sparrow (equations 5.37 and 5.38), and we do not know how this adaptation may vary for species of different weight.

Growth of young

The energy cost of embryonic growth before hatching is charged to the egg-laying item of the adult female's DEB. After hatching, often both the male and female feed and care for the young and this also is charged to their DEB. However, the energy equivalent of the food given to the young by the adults or eaten by the young in precocial species requires a separate estimate. The energy cost of growth may be calculated from the total amount of food metabolized by the young birds until they reach the stage of independence from their parents. This is considered to be the time of first flying in precocial species.

Energy requirement

Energy utilization during the period of growth has been measured in laboratory-reared birds (Table 5.6). The house sparrow nestlings were maintained at approximately 36 °C (Blem, 1975), and the precocial young of the black-bellied tree duck *Dendrocygna autumnalis* (L.) at 32 °C (Cain, 1976). These high temperatures approximate nest or brooding temperatures and, in the house sparrow, render the metabolic rates of the 'cold-blooded' early stages comparable with the later stages after body temperature regulation is acquired. Ducklings have their temperature regulation fairly well established at hatching. The EM was determined for each age group by calculating the regression of total metabolized energy on gain in weight and solving for zero growth. The DEB minus the EM gave the amount of PE used to produce growth.

Regression lines for the EM and DEB of young in the house sparrow have different slopes but those for the black-bellied tree duck are similar (Fig. 5.11). The slopes and elevations of the regression lines for the EM,

S. C. Kendeigh, V. R. Dol'nik & V. M. Gavrilov

Table 5.6. *Statistics for young birds; the data for the house sparrow derived from Blem (1975), the data for the black-bellied tree duck from Cain (1976)*

Age	Weight[a] (g)	EM	PE	DEB	MEC (%)	GEI
		(kcal bird^{-1} day^{-1})				(kcal bird^{-1} day^{-1})
House sparrow						
(Days)						
1	3.0	1.6	1.6	3.2	55.7	5.7
2	5.3	2.5	2.6	5.1	59.5	8.6
3	6.6	3.8	3.0	6.8	58.5	11.6
4	10.0	4.8	2.7	7.5	62.3	12.0
5	12.9	5.8	3.8	9.6	68.1	14.1
6	14.0	10.4	3.4	13.8	70.2	19.7
7	16.6	11.7	2.9	14.6	70.3	20.8
8	18.0	12.4	2.0	14.4	62.6	23.0
9	19.3	13.2	3.5	16.7	67.7	24.7
10	21.3	14.1	2.5	16.6	67.1	24.7
11	21.6	12.9	2.7	15.6	67.8	23.0
12	22.7	13.5	2.8	16.3	67.0	24.3
13	23.8	14.2	3.6	17.8	68.9	25.8
14	24.5	14.8	2.8	17.6	67.9	21.2
Adult[b]	25.2	14.2	0	—	67.0	21.2
Total:						
Nestling		135.7	39.9	175.6	—	263.9
Fledgling[c]		196.0	—	196.0	—	292.5
Hatching to independence		331.6	39.9	371.6	—	556.4
Black-bellied tree duck						
(Weeks)						
1	26[d]	(12.8)[e]	(3.6)	(16.4)	(69.3)	(23.7)
2	54	26.9	7.6	34.5	69.3	49.8
3	109	47.7	14.3	62.0	66.7	93.1
4	149	55.8	19.3	75.1	64.7	115.3
5	199	84.5	33.7	118.2	78.2	150.4
6	275	96.8	63.7	160.4	82.9	195.6
7	380	101.1	40.9	142.0	85.8	172.3
8	478	(144.4)	(34.6)	(179.0)	(87.3)	(205.0)
9	576	187.7	28.2	215.9	89.1	242.2
Adult	685	108.7	0	108.7	86.5	125.5
Total:						
Hatching to flying[f]		5304	1722	7026	80.5	8732

[a] Laboratory-reared birds but closely similar to natural growth weights.
[b] From equation in Appendix 5.2.
[c] 14.0 kcal bird^{-1} day^{-1} × 14 days and excluding extra cost of free existence, if any.
[d] Cain (1970), weight of three-day old ducklings; fly at eight to nine weeks.
[e] Data in parentheses are interpolated.
[f] Total for young in columns × seven (days in week).

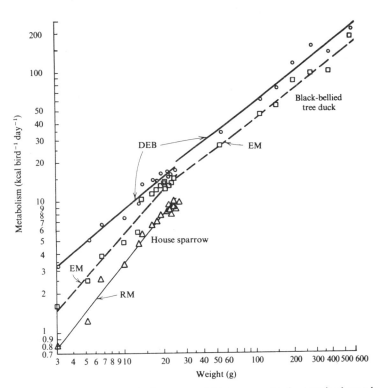

Fig. 5.11. Regression of metabolism on weight in young birds: 'resting' metabolism RM, equation 5.79; existence metabolic rate (EM), equations 5.76, 5.77; and the daily energy budget (DEB) of the house sparrow, $M = 1.305\ W^{0.8274} \pm 1.098$ and the tree duck *Dendrocygna autumnalis*, $M = 1.638\ W^{0.7784} \pm 1.160$.

but not for the DEB, are different in the two species, so that the data for the DEB may be pooled. The relevant equations are as follows:

$$\text{EM, house sparrow: } M = 0.4264\ W^{1.1263} \pm 1.151 \qquad (5.76)$$

$$\text{EM, black-bellied tree duck: } M = 1.230\ W^{0.7749} \pm 1.125 \qquad (5.77)$$

$$\text{DEB, the two species combined: } M = 1.353\ W^{0.8140} \pm 1.114 \quad (5.78)$$

'Resting' metabolism (RM) of inactive house sparrows undergoing food assimilation and growth was determined from measurement of gas exchanges (Blem, 1975) and has the regression:

$$M = 0.2040\ W^{1.2249} \pm 1.119 \qquad (5.79)$$

This regression has a lower elevation but essentially the same slope as that of the EM (Fig. 5.11). Measurements of RM in the house sparrow made by Myrcha, Pinowski & Tomek (1973) give a regression ($M = 0.0749\ W^{1.6361} \pm 1.105$) with a lower Y-intercept but a steeper slope which intersects that of Blem's (1975). The difference between the levels of RM and EM is due to the activity of the birds and is further complicated by the fact that 'resting' birds are growing while the birds at 'existence' metabolism are not.

In an attempt to measure the daily energy requirement for growth, Myrcha *et al.* (1973) studying the house sparrow, Diehl & Myrcha (1973) the red-backed shrike *Lanius collurio* L., and Tomek (1975) the rook *Corvus frugilegus* L., obtained 'assimilation' values by adding together the RM and the calorific equivalent of daily gains in weight. The regression lines obtained generally lie intermediate between those for the EM and the DEB in Fig. 5.11 (see also Fig. 5.12), because of the omission of the energy cost of locomotor movements of the young bird in the nest. Of the total energy metabolized by the house sparrow during the period of growth, Blem (1975) assigns 24% to this activity, 57% to RM and 19% to the increase of biomass. Westerterp (1973) also ascribes a significant amount of energy expenditure in the starling to nest activities. The values obtained by Ricklefs (1974) for 'total energy requirement' of growth in five altricial and precocial species do not represent the DEB because they do not include either the energy cost of locomotor activity or the heat increment of feeding (SDA).

The young house sparrows leave the nest when 14–15 days of age and are cared for by the adults for approximately another 14 days. Of the total DEB from hatching to independence (371.6 kcal bird^{-1}), 89% is used for existence and 11% for growth. In the black-bellied tree duck the corresponding figures are 75 and 25%.

The energy required by altricial nestlings after endothermy is established (6–10 days) varies with the size of the brood. Brooding by the adult is greatly reduced or absent, except often at night, so the young are more responsive to the ambient temperature. In larger broods, the relative rate of heat loss is decreased, thus energy production for existence varies as a function of the total mass, approximately as $W^{0.67}$ (Mertens, 1969). This is particularly true at low ambient temperature; at high temperatures, large broods are at a disadvantage when excess heat must be dissipated to prevent hyperthermy. Offsetting the greater energy conservation with large broods is a decrease in the amount of food each bird receives compared with that which it obtains in small broods. In the house sparrow, the adults provide more food for broods with two and three young than broods with one or two, but above three the amount of food provided remains constant, so the share of each individual declines (Seel, 1969).

The DEB of nestling starlings was estimated by determining the calorific value of samples of the meals given to a brood of four young in the nest by the adults and the number of meals per day. Likewise, the calorific value of samples of faecal sacs was determined and the rate at which they

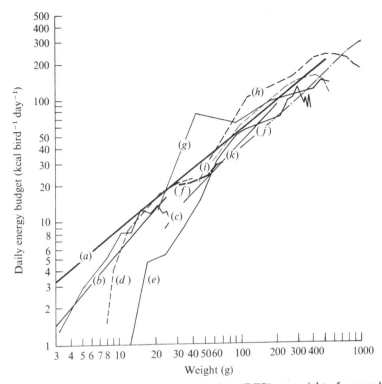

Fig. 5.12. Regression of the daily energy budget (DEB) on weight of young birds: (*a*) house sparrow–tree duck (equation 5.78); (*b*) house sparrow (Myrcha, Pinowski & Tomek, 1973); (*c*) red-backed shrike *Lanius collurio* (Diehl & Myrcha, 1973); (*d*) dunlin *Erolia alpina* (Norton, 1973); (*e*) rook *Corvus frugilegus* (Tomek, 1975); (*f*) starling *Sturnus vulgaris* (Westerterp, 1973); (*g*) coot *Fulica america*, and (*h*) black duck *Anas rubripes* (Penny & Bailey, 1970); (*i*) lesser scaup duck *Nyroca affinis* (Sugden & Harris, 1972); (*j*) double-crested cormorant *Phalacrocorax auritus* (Dunn, 1975*a*); and (*k*) blue grouse *Dendropagus obscurus* (Stiven, 1961).

were produced noted (Westerterp, 1973). Values thus obtained are below the sparrow–duck regression line (Fig. 5.12) probably because growth in these birds was retarded (Westerterp, personal communication). Low values were also obtained for nestling double-crested cormorants *Phalacrocorax auritus* (Lesson) by Dunn (1975*a*) using a similar procedure.

Norton's (1973) data on the DEB for the precocial young of the dunlin *Erolia alpina* (L.) fit the sparrow–duck line very closely from the eighth day through the twenty-second day. The birds were artificially reared indoors at 25–35 °C, and for the first few days, their growth was abnormally slow compared with that of the outdoor or wild birds. The total energy metabolized was 57% of the total ingested.

Penny & Bailey (1970) measured the DEB during the first eight weeks of growth in laboratory-reared young black ducks *Anas rubripes* Brewster, and American coots *Fulica americana* Gmelin, and Sugden & Harris (1972) measured both the DEB and what was approximately the EM during the first 12 weeks in young lesser scaup ducks *Nyroca affinis* (Eyton). The procedures and calculations were somewhat different and the temperatures at which the young birds were reared after leaving the brooder are uncertain, but the values obtained for much of the pre-flight period fall fairly close to the regression line in Fig. 5.12.

Data obtained on the amount of food consumed by semi-altricial young wood storks *Mycteria americana* L. during their first nine weeks of growth are not comparable with the above. After the first week, the young were maintained in outdoor cages at unspecified temperatures, and metabolized energy was calculated assuming a very high and constant metabolized energy coefficient of 90% (Kahl, 1962).

It is apparent that to obtain the daily amount of energy used by young birds, equation 5.78 is suitable for both altricial and precocial species. Equations for the DEB and EM of adult birds of different weight are not applicable to young birds. The slopes of the regression lines for passerine adult birds are too flat; that of the EM for non-passerine species has the same slope but is too low. Sugden & Harris (1972) also noted this for the lesser scaup duck. The EM of passerine adult birds is higher than that of young birds before they acquire regulation of body temperature but becomes lower after they acquire temperature regulation, some days before leaving the nest. Ricklefs (1974) discusses possible reasons for these differences.

After full weight in passerine birds and flying ability in non-passerines is attained, there tends to be a decline in the DEB of young birds. This is evident for several species in Fig. 5.12 and was also noted in the double-crested cormorant (Dunn, 1975a). Blem (1975) found that his hand-reared house sparrows showed a decrease from 17.6 kcal on the fourteenth day to 13.7 kcal bird^{-1} day^{-1} during the next three days when they would normally have left the nest. This decrease suggests that the metabolic rate of young birds drops to the level of adult birds within a few days after they acquire independence of their parents.

The measurements of the DEB made with the nestling altricial passerine birds were at approximately nest temperatures. When these birds leave the nest and become fledglings, along with precocial birds immediately after hatching, they are subject to the influence of ambient temperatures which generally are lower than those at which the measurements were made. During this period, brooding by the female parent may continue as needed, especially at night, but whenever they are exposed to low environmental temperatures they must normally increase their energy consumption as is required of any homoiotherm. To obtain the DEB of young birds out of the nest, minimum values may be calculated using equation 5.78 and additions made which are proportional to the extent that ambient temperatures are below 30 °C and the time that the birds are exposed to these temperatures. Diehl (1971) obtained a regression equation for EM on temperature in fledgling red-backed shrikes which have a higher temperature coefficient (*b*) than would be expected in adult birds. The juvenal plumage of young birds is much less dense than that of adults. However, until studies can be made on a variety of species, the regression equations obtained for adult birds will provide us with the approximate additions needed to correct values for fledgling birds.

Although precocial young have established a degree of temperature regulation at hatching, it is not perfect. In Baird's sandpiper, *Erolia bairdii* (Coues) and some other shorebirds, the body temperature of the young is thermolabile, and they commonly forage and are active without stress even when body temperatures drop from 37 °C, which they attain with brooding, to 30 °C. Lowering the body temperature reduces energy requirements for free existence and is therefore adaptive (Norton, 1973). One can expect that the same situation occurs in many other precocial species.

Respiration–production ratios

Growth efficiency is the ratio of the amount of energy expended, 'respiration', to the amount of new tissue produced and may be calculated in terms of lean dry weight, dry weight, or live weight. The calorific value of lean dry weight in the house sparrow decreases slightly with age, as it also does in the Japanese quail *Coturnix coturnix* (L.) (Brisbin & Tally, 1973), but averages 4.3 kcal g^{-1}. Dry weights, which include the lipids, average about 5.2 kcal g^{-1} in the house sparrow and several other species. The calorific value of fresh tissue (live weight), which also includes water, increases with growth in the house sparrow from 0.854 to 1.567 kcal g^{-1} (Blem, 1975), largely because the percentage of water in the tissues decreases with age, and the same is true in other species (Sugden & Harris, 1972; Brisbin & Tally, 1973; Myrcha *et al.*, 1973; Westerterp, 1973; Ricklefs, 1974; Dunn, 1975*b*; Tomek, unpublished).

Of the total energy required from hatching to fledging in the house sparrow (DEB), 77% is used for existence (EM) and 23% for growth (PE), and in the black-bellied tree duck it is 75 and 25%, respectively. The two percentages are similar in other species. The energy used for growth includes both that anabolized as new tissue and that required to bring about this increase in biomass.

The increase in calorific equivalents of house sparrows from the first day, 2.56 kcal, to the fourteenth day, 38.37 kcal, is 35.81 kcal bird^{-1}. This is 90% of the total PE (39.9 kcal), 20% of the DEB (175.6 kcal) and 14% of the GEI (263.9 kcal) (Table 5.6). For the black-bellied tree duck, the increase in calorific value of the duckling from the second to the ninth week is 1192 kcal and the percentages of growth efficiency are 69% of the PE, 17% of the DEB, and 14% of the GEI. Growth efficiency in the duck is lower in terms of the PE and DEB than in the house sparrow but is the same for the GEI. The efficiency of 20% of the DEB in the house sparrow is the same as the percentage of assimilated energy obtained in this species by Myrcha *et al.* (1973).

When calculated on a daily basis, growth efficiency has been shown to decline progressively with age in all species studied. In the house sparrow, this decline in terms of the PE is from 97% on the first day in the nest to 67% on the last day and in terms of the DEB from 50 to 12%. In the tree duck, the decline in relation to the PE is from 93 to 19% and in relation to the DEB from 20 to 2.5%.

In terms of specific weight (Fig. 5.13), both the RM and EM increase in the house sparrows as they establish body temperature regulation. The RM then stabilizes but the EM declines. Correlated with the ducklings having body temperature regulation established at hatching, the EM declines progressively from hatching. In the sparrow, the PE nearly equals the EM at first but then rapidly declines, reaching the same general level as in the duck during the last days in the nest. The DEB declines in both species as the birds increase in size.

Efficiency may also be calculated in terms of energy expenditure for reproduction by the adult bird. In Table 5.5, the total energy cost for reproduction per adult house sparrow was estimated at 300 kcal or for both adults approximately 600 kcal. In Illinois, an average of 8.94 eggs are laid in successive clutches in the same nest throughout the season. If we assume that these constitute the entire layings of one pair of birds and that of these, 3.14 hatch and fledge (Will, 1973), then the total production to successful fledging averages 120 kcal pair^{-1} (3.14 nestlings × 38.37 kcal nestling^{-1}) or an efficiency of 40%. However, if we add to the cost of reproduction, the cost of free existence for the two adult birds during the breeding season, late April through early August (4219 kcal), the efficiency is reduced to 2.5%.

In order to obtain the 'production'/'respiration' ratio of a pair of birds in the ecosystem, we need first to add to the total GEI of the adults through the year (26269 kcal), the GEI of the fledged young (3.14 nestlings × 263.9 kcal nestling^{-1} = 828.6 kcal) and the hatched young that did not fledge and are assumed to have died half-way through the nestling period (2.74 nestlings × 263.9/2 kcal nestling^{-1} = 361.5 kcal). This totals 27459 kcal, of which 9061 kcal is lost in excreta, leaving 18398 kcal which is metabolized or 'assimilated', part of which goes into 'production', and

190

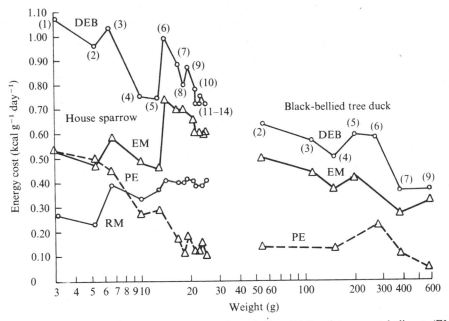

Fig. 5.13. Changes with growth in 'resting' metabolism (RM), existence metabolic rate (EM), productive energy (PE), and the daily energy budget (DEB) in relation to specific weight. Numbers in parentheses for the house sparrow represent age in days; for the tree duck, age in weeks.

part of which is lost as 'respiration'. Production consists of the successful fledglings (120 kcal), unhatched eggs (3.06 eggs×3.067 kcal egg^{-1} = 9.4 kcal), nestlings that died (2.74 nestlings×38.37/2 kcal nestling^{-1} = 52.6 kcal), and new feathers after moulting (1.7 g bird^{-1}×2×4.5 kcal g^{-1} = 15.3 kcal), totalling 197 kcal pair^{-1}. 'Respiration', therefore, is 18201 kcal (18398 − 197 kcal) and the net production efficiency is only 1.07%.

Food consumed

In order to translate the DEB into the amount of food consumed, the efficiency with which the food is digested, absorbed and utilized must be known. The metabolizable energy coefficient (MEC) is calculated by:

$$MEC = \frac{GEI - \text{energy in excrement}}{GEI} \times 100 \qquad (5.80)$$

The *MEC* has been measured in adult birds of only a few species and on only a few diets. On the standardized University of Illinois chick mash, no. 521, 12 species of granivorous passerine species varied between 62

191

and 80% in their *MEC*. On other chick mash compositions, the *MEC* in four species varied between 81 and 87%. One species when fed with the first diet had a *MEC* of 67%, but on other diets it varied between 81 and 87% (Kendeigh, 1974). Long-billed marsh wrens on a mixed insect diet had an average *MEC* of 75.9% (Kale, 1965).

S. N. Postnikov and V. R. Dol'nik (personal communication) found that during January at temperatures of −10 to −15 °C, 10 species of passerines feeding on the seeds of the sunflower *Helianthus annuus* had a *MEC* varying between 82 and 91%, but for the house sparrow it was 76%. It is 84% in the evening grosbeak feeding on sunflower seeds (West & Hart, 1966). Coefficients are known to vary not only with species of bird but also with kinds of food consumed and with temperature (Tables 5.7, 5.8).

In non-passerine species, willow ptarmigan feeding on a commercial diet had a *MEC* of 68.4% in winter, 73.4% during the summer moult, and for laying (non-moulting) females a *MEC* of 73.6% was obtained (West, 1968). During the winter, wild birds of the willow ptarmigan, feeding mostly on willow, had a *MEC* of 44%; rock ptarmigan, feeding mostly on birch, a *MEC* of 39%; and white-tailed ptarmigan, feeding on a mixture of birch, alder and some willow, a *MEC* of 41% (Moss, 1973). Individually caged bobwhites, feeding on a balanced mash, had a *MEC* of 76%, but birds in coveys showed an increase in the *MEC* from 77% at 5 °C to 81% at 35 °C (Case, 1973).

In young house sparrows, feeding on a mixed diet, the *MEC* was low during the first two or three days, reached a peak about the seventh day, and then declined (Table 5.6). Excluding the first three days when the newly hatched bird may still be using yolk absorbed from the egg, the *MEC* averaged 67%. Myrcha *et al.* (1973) also found in both the house sparrow and tree sparrow that the *MEC* was low at first, increased about mid-way in the growth period, and then declined. The *MEC* is highest during the period of most rapid growth. For the entire nestling period, Myrcha *et al.* found the average *MEC* to be much higher than did Blem (1975), being 82 and 76% in the house and the tree sparrow, respectively. Blem (1973 *b*) found it to average 75.3% in six other passerine species. Diehl (1971) gives it as 70% for nestling red-backed shrikes. The overall average for the starling (Westerterp, 1973) was 64%, but it decreased from 80% on the fourth day to 63% on the eighth day to 60% on the sixteenth day.

In contrast with the passerines, the *MEC* increased with age in the black-bellied tree duck (Table 5.6) as it also did in the black duck and coot (Penney & Bailey, 1970), and in the double-crested cormorant (Dunn, 1975*a*). In the dunlin, the *MEC* remained more constant and averaged only 57% (Norton, 1973).

The amount of food consumed varies strongly with the calorific concentration of the food, as birds eat primarily to obtained energy, although

Table 5.7. *Metabolizable energy coefficients (%) (Willson & Harmeson, 1973)*

	26+ °C	0 °C
Cardinal *Cardinalis cardinalis* (L.)		
Sunflower *Helianthus annuus* L.	80	67
Ragweed *Ambrosia trifida* L.	77	69
Hemp *Cannabis sativa* L.	73	73
Foxtail *Setaria faberii* Herm.	68	78
Smartweed *Polygonum pensylvanicum* L.	65	77
Song sparrow *Zonotrichia melodia* (Wilson)		
Foxtail *Setaria faberii*	89[a]	88
Hemp *Cannabis sativa*	80	86
Pigweed *Amaranthus retroflexus* L.	80	57
Smartweed *Polygonum pensylvanicum*	—	55

[a] Difference between temperatures not significant.

Table 5.8. *Metabolizable energy coefficients (%) at 0.5–15.5 °C October–February (Gibb, 1957)*

	Scots pine seed	Sunflower seed	Ground nuts	Meal-worms	Mixed insects
Great tit *Parus major* L.	78.3	81.0	81.3	—	—
Blue tit *Parus caeruleus* L.	74.8	—	76.9	84.1	—
Coal tit *Parus ater* L.	80.6	—	81.3	86.1	66.8

there is evidence that under special circumstances birds may also eat for nutrients, not just energy. A fairly large number of measurements on the calorific equivalents of different kinds of wild food, especially seeds, have been made. Useful compilations are those of Kendeigh & West (1965), West & Meng (1966), Cummins (1967), Grodzinski & Sawicka-Kapusta (1970) and West (1973). Kale (1965) gives the average calorific value of several species of insects as 5.5 kcal g^{-1}, and Custer (personal communication) as 5.3 kcal g^{-1}.

The total energy requirement of adult house sparrows throughout the year in central Illinois (Champaign) is a little less than 8.8 Mcal per bird (Table 5.5). Assuming a *MEC* of 67% and a mixed grain diet averaging 4.3 kcal g^{-1}, an individual adult bird would consume about 3.0 kg year^{-1} or about 7.3 g bird^{-1} day^{-1} in the summer and 9.0 g bird^{-1} day^{-1} in the winter (Kendeigh, 1973*b*). At Churchill, Manitoba, to survive and reproduce out-of-doors, a bird would require one-third more grain; in the tropics less than two-thirds as much.

According to Summer-Smith (1963), the food given to the young consists

S. C. Kendeigh, V. R. Dol'nik & V. M. Gavrilov

of only 32% grain or plant food, the other 68% being insects. If we assume 5.5 kcal g⁻¹ as the calorific equivalent of insects and 4.3 kcal g⁻¹ for grain, then the total consumption by a nestling would be 35.1 g of insects and 16.5 g of plant food. Total consumption of a young bird from hatching through the fledgling stage to independence would be 57.2 g of insects and 18.3 g of plant food if the same percentages persist through the fledgling stage.

If 3.14 nestlings are fledged per pair each year, and another 2.74 nestlings survive half-way to fledging, the total consumption to the point of leaving the nest would be 117.2 g of insects and 55.1 g of plant food, and to independence 186.6 g of insects and 60.8 g of plant food. These amounts are relatively small compared to 6000 g, mostly grain, consumed by the adults through the year.

Evolution

Birds, as a taxonomic group, probably first evolved in a tropical climate (Darlington, 1957). The primitive bird, already covered with feathers, doubtless had some physical control over body temperature by increasing the rate of heat loss while active during mid-day heat and reducing the rate of heat loss while at rest during the cooler night. Seasonal changes in climate were small and temperature stresses could largely be avoided by basking in the sun during cooler weather and seeking shade during warm periods or by other appropriate behavioral responses. The zone of thermo-neutrality may reflect the range of ambient temperatures to which these primitive birds were subjected.

As bird populations grew and pressure was exerted to disperse out of the tropics into climates having lower temperatures, especially at night and during the winter season, selection favored those individuals that could increase their heat production at temperatures below the zone of thermo-neutrality. This increase would have to be proportional to the increase in rate of heat loss which in turn was controlled by the effectiveness of the birds' insulation and the drop in ambient temperature. A premium was placed on maintaining a relatively uniform high body temperature for greater activity, reproduction and competitive effectiveness. As birds increased their ability to regulate a constant body temperature by meta-bolic as well as physical means, they then expanded their breeding range during the warmer months into the temperate climatic zone. Competition for space and food is especially great at this time, and the longer photo-periods of the temperate zone are advantageous. Many modern species that nest in temperate regions but migrate into the tropics for the winter are in this stage of evolution, such as the dickcissel (Zimmerman, 1965a).

Before a species could become a permanent resident outside of the warm

194

tropics, it had to develop seasonal acclimatization. This required replacement of worn plumage in the autumn before the onset of cold weather, increase in daily fat deposits on the body to provide energy to survive the longer nights, and ability to increase tissue heat production for tolerating lower ambient temperatures. Behavioral responses, such as huddling, seeking sheltered roosting sites, sun-bathing, etc., reduced physiological stress and became better established. Some species came to tolerate hypothermy or even dormancy to save energy during cold periods. A lowering of the zone of thermo-neutrality at both its lower and upper critical temperatures and a rise in the BM resulted from these adjustments. During the spring and summer, however, just the opposite set of responses occurred. Metabolic rates decreased, as there was no longer a need to tolerate extreme low ambient temperatures, and the reduced BM saved energy. Likewise, a higher upper critical temperature was advantageous in tolerating exposures to high ambient temperature. This seasonal acclimatization is well shown in the responses of the house sparrow. Other species, such as the field sparrow *Spizella pusilla* (Wilson) (Olson, 1965), show intermediate steps in this evolution of permanent residency in that they appear early in the spring, leave late in the autumn, migrate only part-way to the tropics, and some hardy individuals survive the northern winter.

With continued differentiation of species and saturation of available niches in the environment, selective pressures favored those species that acquired tolerance of the still colder climates of the sub-arctic and arctic. The American tree sparrow *Spizella arborea* (Wilson), is a good example (West, 1960). This species overwinters in the temperate zone and in the spring, instead of losing its acclimatization to cold as do the permanent residents, it maintains this ability and migrates to the sub-arctic to nest. The final step in this evolution is permanent residency in the far north, where there needs to be seasonal acclimatization but at a higher level than that found in temperate zone species. The redpolls have nearly reached this goal, although they still migrate short distances, at least during many years (Brooks, 1968). All this, of course, supposes that food of the right sort is available at all times and other conditions are at least tolerable.

The total energy that a bird can mobilize sets a limit to its activities. A bird's activities are normally regulated so that this limit is seldom reached as otherwise there is risk of mortality. A potential reserve capacity of 'latent energy' allows for daily, monthly, or yearly fluctuations in temperature. For the house sparrow at 40° N, this latent energy averages 16% of the incipient maximum potential throughout the year and 19% of their average DEB. It appears, therefore, that the DEB is limited by physiological capacity which includes a reserve for environmental fluctuations and emergencies. We do not know whether the bird continually

195

extends its activities to the limit set by these restrictions, but it would seem to its competitive advantage to do so, especially for reproduction. Since other metabolic levels and capacities have been shown to vary with weight, at any particular temperature, this gives justification to believe that the DEB does also.

There are many compensations in a bird's physiology, anatomy and behavior that function to keep its energy requirements safely within the set limits. The SDA relieves the body of extra tissue metabolism in temperature regulation so that a feeding animal can tolerate low ambient temperature better than a starving one. A greater loss of weight at night due to low temperature or unrest results in overcompensated feeding the following day so that there is an extra gain in weight. Overnight tolerance of cold weather is aided by selection of protected roosting sites, the development of huddling behavior and ability to fluff out the plumage effectively. Tolerance of cold is also aided by sun-bathing and evolution of dark pigmented plumage to absorb solar energy.

Sustained flying is more energy demanding than intermittent activity in ground feeding, preening, singing, etc. This is partly compensated for by the use of the increased heat production by the flight muscles for temperature regulation. More or less continuous fliers, such as swallows, swifts and others, have evolved adaptations in wing and body structures that reduce the amount of energy required. Migrants commonly take advantage of favorable tail winds in long flights.

Moulting and renewal of the plumage are highly energy demanding. However, the new feathers, as they grow, give increased insulation to the body, there is generally a compensating decrease in other energy-demanding activities, and the process extends over a period of weeks or months so that the increase in daily energy requirements is kept at a tolerable level.

Reproduction, moulting and migration do not overlap as a general rule, although there are adaptive exceptions (Foster, 1975). This temporal spacing permits most efficient use throughout the year of limited energy resources.

Birds nesting in cavities have better nest insulation than those in open situations. This permits one parent to do all the incubation with frequent inattentive periods for feeding and/or larger clutches of eggs. Open nesters have similar advantages when the nests are well made and the adult crouches well down in the nest when sitting on the eggs. Ground-nesting birds, often with little nesting structure, conserve energy by having only one or two relatively short inattentive periods for filling their crops with food. Species with flimsy nests in exposed situations generally have small clutches. There is less stress on the incubating bird, which may permit larger clutches, if she is fed by her mate or if both adults share in the

196

incubation duties. In cold climates only one brood may be raised per year because a greater share of available energy must be used for existence; in warm climates where existence is less strenuous, more broods are raised. There thus appear to be dependent interrelations between nest site, nest structure, clutch size, number of broods, role of sexes, incubating behavior and distribution based on the efficient use of limited energy resources that need much more study. Conservation of energy by one means makes possible its utilization in another way. The different ways in which the energy mobilizing ability is apportioned doubtless has selective value in evolution for the particular species in the particular situation.

All of these adjustments are related to the size (weight) of the bird. There is doubtless an optimum size for a bird to occupy any particular niche so as to make most effective use of energy resources available in that niche. Selection of an appropriate size must therefore have interacted and been coordinated in these evolutionary processes.

Discussion and summary

The ability of an animal to secure ample energy supplies from its environment and to mobilize energy physiologically for useful purposes determines to a large extent much of its life-history, such as where it occurs, its abundance, the season and extent of such activities as migration, reproduction, and moulting, and its interactions with other individuals of its own and other species. A knowledge of bioenergetics is necessary for an understanding of behavior as well as for evaluation of the impact of a species in the ecosystem where it lives. In this chapter we have been concerned with the analysis and quantitative measurement of the various components of the DEB of birds and with providing means for estimating these energy requirements in species generally.

The DEB is a complex balance between energy intake, energy-conserving and energy-demanding processes. When the energy equivalents of these various processes are known, they may be summed or integrated to get the total daily energy requirements of the bird, and hence this procedure may be designated the *integrative method*.

Integrative method

The measurement of the various components of the DEB has been of concern in animal husbandry for many years, especially the energy cost of performing physical work, the production of meat, milk, eggs, etc. and the economics of providing domestic animals with the best feed for obtaining the highest rates of production. The concepts and terminology used in this chapter are patterned after the studies of the animal nutri-

S. C. Kendeigh, V. R. Dol'nik & V. M. Gavrilov

Table 5.9. *Comparative terminology of the components of the daily energy budget* (*DEB*)

Animal nutrition (Harris, 1960)	Avian energetics (this chapter)	Secondary productivity (Petrusewicz, (1967)
Food intake gross energy (IGE)	Gross energy intake (GEI)	Consumption (C)
Fecal energy (FE)	Excretory energy (EE)	
Digestible energy (DE)		Rejecta (FU)
Urinary energy (UE)		
Gaseous products of digestion (GPD)	(Insignificant, not measured)	
Metabolizable energy (ME)	Metabolized energy	
Heat increment (HI)	Heat increment of feeding,	
Heat of nutrient metabolism (HNM)	specific dynamic action (SDA)	
Heat of fermentation (HF)	(Insignificant, not measured)	
Net energy (NE)		
Maintenance energy (NE$_m$)	Existence metabolic rate (EM) (with constant weight)	
Basal metabolism (BM)	Basal metabolic rate (BM)	Respiration (R)
Heat to keep warm (HBW)	Body temperature regulation standard metabolic rate (SM)	
Energy to keep cool (EBC)		
Voluntary activity (VA)	Cage activity	
Production energy (NE$_p$) (excludes HI)		
Work		
Energy storage	Productive energy (PE)	
Semen	(includes SDA)	Production (P)
Eggs		(excludes SDA)
Growth		
Feathers		

tionists but are modified and adapted for analyzing the energy balance of wild birds under natural conditions. The components of the DEB may be combined and simplified in still a different way for estimating the role of birds, compared with that of other organisms, in the secondary productivity of the ecosystem (Table 5.9).

For obtaining entirely precise quantification of the DEB, each component needs to be measured for each species under consideration. This is a time-consuming process, however, and cannot be carried out with equal ease on all species. Where direct measurements cannot be made, an estimate of the energy value of each component of the DEB may be obtained with the equations provided in this chapter, and then these values may be summed to give the DEB for the species. Finally, if only an approximation of the DEB is desired, equations 5.68 to 5.71 may be used for passerine species, equations 5.72 to 5.75 for non-passerine species, and equation 5.78 for young birds. To solve these equations the weight

198

of the bird must be known and, for the DEB of the adults, the prevailing air temperature.

Heavy isotopes

There are other means of estimating the DEB besides the integrative method. Each different method has certain advantages as well as disadvantages (Gessaman, 1973). The use of *doubly labelled water*, $^2H_2^{18}O$, is a direct measurement of total energy use over a period of time (Lifson & McClintock, 1966). It requires the capture of the animal to inject the isotopes of hydrogen and oxygen and again, later, to obtain a blood sample to determine their elimination. Where knowledge of only the total DEB is desired, the method has the advantage in that data may be obtained relatively quickly. The method does not provide information, however, about the energy cost of the various components in the DEB when different activities are involved. Likewise, the method is costly and only a few laboratories are able to process the samples.

Radioisotopes

The use of radioactive isotopes of such elements as iodine, zinc, phosphorus, caesium and others, depends on their rate of excretion being correlated with the rate of metabolism (Odum, 1961). A known amount of the isotope is injected into an animal or taken in with food and water, and the rate at which it is eliminated is determined by recapturing the animal after a period of time. As with doubly labelled water, only total metabolism is measured. The use of any large amount of some isotopes requires licensing the operator, and there is a restriction on applying it to wide ranging animals. Its potential as a tool for measuring the DEB on wild animals needs further development (Gessaman, 1973).

Biotelemetry

Biotelemetry is also a promising procedure for measuring DEBs (Gessaman, 1973). This method requires a radio-transmitting device fastened on the animal and a receiver to record the radio signal at a distant location. At the present time, the rate of heart beat is commonly used as an index of metabolic rate. By variation in the character of the signals emitted, it is often possible to distinguish the type of activity the animal is undergoing. Interpretation of the rate of heart beat in terms of metabolism must, however, be done with care. Artificial eggs may be fitted with a transmitting device to record egg temperature and attentivity of adult birds to incubation (Varney & Ellis, 1974).

199

S. C. Kendeigh, V. R. Dol'nik & V. M. Gavrilov

Time–activity–energy budgets

Time–activity budgets are commonly converted to *time–energy* budgets. This procedure involves measurement in the field of the amount of time an animal spends in various activities. The energy cost of each activity is calculated and summed to obtain the DEB. Since reliable equations permit the calculation of the BM and the EM when the species weight is known, the energy costs of various activities are calculated as multiplying factors or coefficients of one or the other of these two rates. This procedure has a distinct advantage in that time–activity budgets can be readily compiled from field observations almost anywhere in the world with a minimum of equipment.

The critical factor in this procedure is calculating the energy cost of each activity. Although the approximate costs of egg-laying, incubation, moulting and sustained flight can be estimated from equations given in this chapter, very few measurements have been made on the energy cost of other free-living activities, such as short flights, perching, preening, singing, territorial defense, walking, hopping, swimming, roosting. Our discussion of the energy cost of all these activities taken together, i.e. 'free existence', is probably the weakest part of this chapter, and we will not attempt even to guess at the energy cost of each of its components. For instance, the energy cost of perching varies with air temperature, exposure to insolation, wind velocity, recent feeding, time of day and season. Quantification of the energy cost of specific activities under different situations must wait for future research using one or another or a combination of procedures. Meanwhile, we urge that students measuring time–activity budgets also record in as much detail as possible the micro-climate to which the birds are exposed when undergoing these activities. It is important, of course, that the micro-climate be measured whatever the procedure used in determining the DEB.

Birds in ecosystems and computer modelling

The energy requirements of individual birds for existence and for carrying on their various activities are high compared with those of other organisms, even mammals that are also 'warm-blooded'. They therefore consume a disproportionate amount of food and have an impact on lower trophic levels greater than that of other organisms of the same size. This impact would be even greater had not territorial and other behavior evolved which insures the better survival and reproduction of individuals but prevents the development of comparably high populations. Only about 1% of the total energy that birds take from their environment becomes anabolized into new tissues which may be used as food by predators in higher trophic

200

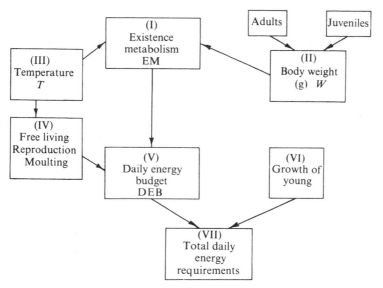

Fig. 5.14. Computer model for calculating 'individual energy demands' during the breeding–moulting season.

levels or, if not taken by them, may provide energy to the decomposers on their death. This percentage doubtless varies from one ecosystem to another but is very low. Birds do not appear to be very important organisms in ecosystems except where their food consumption helps to control population levels of organisms in lower trophic levels.

In order actually to assess the impact of birds in ecosystems, use needs to be made of computer simulation programs. Such computer modelling is the subject of Chapter 6, and included in the computer model is a sub-model for 'individual energetics' (Figs. 6.1, 6.2). On the basis of the new information provided in this chapter, we suggest that separate sub-models are required for the breeding–moulting and the non-breeding–migration seasons.

For the *breeding–moulting season* (Fig. 5.14):

(I) An equation ($M = a - b\,T$) for the EM of any species on temperature may be calculated from the following equations:

Passerines

$$30\,°C: \quad M = 1.462\, W^{0.6880} \qquad (5.26)$$

$$0\,°C: \quad M = 4.969\, W^{0.5105} \qquad (5.34)$$

Non-passerines

$$30\,°C: \quad M = 1.068\, W^{0.6637} \qquad (5.28)$$

$$0\,°C: \quad M = 4.142\, W^{0.5444} \qquad (5.35)$$

(II) For W, use the mean weight of the adults for the season (in grams). Preferably this should be the early morning weight before maximum deposition of fat.

(III) The equation obtained from (I) may be used for calculating the EM for any time period of one day or longer using the mean temperature to which the bird is exposed.

(IV) The combined costs of free living, reproduction and moulting may be approximated for passerines and probably also non-passerines using the equation ($PC = \%$ DEB of EM):

$$PC = 34 - 0.92\ T \tag{5.67}$$

(V) DEB = EM (I) plus the cost of additional activities (IV).

(VI) The growth of young each day requires energy proportional to their weight as:

$$M = 1.353\ W^{0.8140} \tag{5.78}$$

The total requirements from hatching to independence may be calculated from the growth curve of their weight. This should be multiplied by the number of young raised per pair of adults during the season and a fraction added for those young that die before attaining independence. If divided by the number of days during which the young are being cared for by the adults, the 'total daily energy requirements' (VII) could be obtained for this period only; otherwise divide by the number of days in the season.

After reaching independence, the energy requirements of juveniles approaches that of adults. (IV) may be calculated for juveniles using the same equation. Although not reproducing, this is offset more or less by their somewhat higher EM, and, of course, they also moult. Compute the total energy used during the juvenile stage and divide by the number of days in the season.

(VII) Total average daily energy requirements during the season is the DEB for adults plus the DEB for juveniles plus growth of young (VI).

For the *non-breeding–migration season* (Fig. 5.15):

(I) Calculate the equation for variation of the EM on temperature from the following equations:

Passerines

$$30\ ^\circ\text{C}: \quad M = 1.544\ W^{0.6601} \tag{5.23}$$
$$0\ ^\circ\text{C}: \quad M = 4.437\ W^{0.5224} \tag{5.29}$$

Non-passerines

$$30\ ^\circ\text{C}: \quad M = 1.455\ W^{0.6256} \tag{5.25}$$
$$0\ ^\circ\text{C}: \quad M = 4.235\ W^{0.5316} \tag{5.31}$$

(II) and (III): as above.

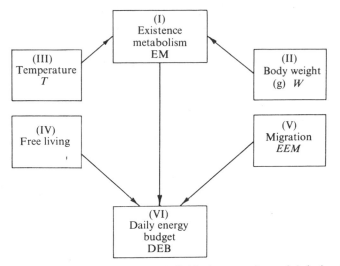

Fig. 5.15. Computer model for calculating 'individual energy demands' during the non-breeding–migration season.

(IV) Free-living costs over winter, allowing for physiological adaptation to local climates, appear to be independent of temperatures in the house sparrow and average 7% of the EM. Until such values can be determined for other species, 7% may be assumed for both passerines and non-passerines.

(V) The energy expenditure for migration (EEM = kcal bird^{-1} km^{-1}) is calculated per unit distance (km^{-1}) as:

$$\text{Passerines: } EEM = 0.0522\ W^{0.47} \qquad (5.48)$$

$$\text{Non-passerines: } EEM = 0.0378\ W^{0.47} \qquad (5.49)$$

Multiply EEM by the total distance covered in autumn and spring migrations and divide by total number of days in the season.

(VI) The average total daily energy requirements for the season is obtained by adding (I) and (IV) (and (V) if the species is migratory). The DEB for the migration period only may be calculated by dividing (V) and (VI) by the number of days involved in migration.

A number of people have contributed to this chapter in one way or another. We are particularly grateful to Charles R. Blem, Brian W. Cain, Erica H. Dunn, David W. Norton and K. Westerterp who provided unpublished data and read the section on the energetics of growth in young birds; to David M. Gates for comments on the section dealing with insolation, to Vance A. Tucker who helped with an early draft of the section on flight, to Fred N. White and J. A. L. Mertens for unpublished data and help with the energetics of nest insulation and reproduction, and to J. A. L. Mertens, James A. Gessaman, January Weiner, John A. Wiens and

S. C. Kendeigh, V. R. Dol'nik & V. M. Gavrilov

George C. West who gave us critical comments and suggestions on an intermediate draft of the whole manuscript. The senior author wishes especially to acknowledge the office, secretarial and computer facilities made available at the University of Illinois by Edwin M. Banks and Lowell L. Getz and the essential help in computer programming provided by David J. Moriarty. Without the help of these persons and the intensive studies of this topic by many graduate students, the breadth and depth of coverage attained in this chapter would not have been possible.

6. Assessing the potential impact of granivorous birds in ecosystems

J. A. WIENS & M. I. DYER

Granivorous birds have attracted the attention of laymen, the concern of farmers, and the study of scientists not only because they possess interesting biological attributes and are in some respects amenable to study and experimentation, but also because some species feed upon grain and seed crops which are of economic importance and the birds thus come into potential or direct competition with man for a limited and valuable commodity. These granivorous birds are often labelled 'pests', and if the agricultural impact is sufficient their populations may become the target of massive management or 'control' efforts. The impact which a granivorous species may have within a natural or an agricultural ecosystem is an expression of its biological attributes. Determination of this impact potential thus builds upon the features of granivorous birds considered in previous chapters in this volume – their evolutionary history, population dynamics, biomass and production rate and bioenergetics. There is considerable information available on these aspects of the biology of granivorous birds, and the task of these previous chapters has been to synthesize this existing knowledge. However, in this chapter the foundation of existing information on the impact of granivorous birds in ecosystems is scanty, and we instead emphasize developing an approach to the determination of 'impact' and discerning some priorities for further studies. We retain the emphasis on *Passer* populations, as in the previous chapters, but expand our coverage to consider several other granivorous species as well as some community-level aspects of avian granivory.

Determining the importance of granivorous birds

While the 'importance' of granivorous birds is most frequently thought of in terms of economic effects, there are several partially-overlapping dimensions in which one may consider a given population to be 'important'. Since the emphasis given these dimensions may influence our estimates of 'impact', and thus the nature of possible management responses, it is worthwhile to discuss them briefly, even though we will emphasize only one in this chapter.

J. A. Wiens & M. I. Dyer

Population dynamics

By virtue of various aspects of its population dynamics, a given population may assume a position of importance through its influence or dominance of community structure or organization or its role in the web of species interactions, such as competition, which determine community patterns. Thus a species which dominates the avifauna of a given biotope, either through numbers of individuals or proportion of the total avian biomass, may be of considerable importance in determining the overall structuring of the community and may influence such features as the territorial spacing or tactics of biotope utilization by other sub-dominant species. In many xeric or early successional environments, the dominant species are frequently granivorous during at least part of the year, and many granivorous species which have little direct economic impact may nonetheless be 'important' in community structuring.

At the opposite extreme, a population may assume importance if its abundance and/or distribution are so limited as to warrant the status of rare or endangered. This, of course, represents an administrative rather than a biological determination of importance, since the rare species are frequently of quite limited importance to the dominance patterns or structural composition of avian communities. In fact, the patterns of species-abundance distributions in communities (Preston, 1948; Williams, 1964; Whittaker, 1965) dictate that some species will inevitably be rare, and such populations should perhaps merit consideration as 'important' only if they become endangered, especially as a result of human activities.

Aesthetics and recreation

Birds are colorful, conspicuous creatures and are the objects of the attention of legions of bird-watchers throughout the world. Bird populations which lack 'importance' in other dimensions may nonetheless be important in a recreational sense, but evaluation of the 'impact' of this role in a quantitative sense is extremely difficult. In North America, economic expenditures associated with recreational 'bird-watching' may be considerable (Payne & DeGraaf, 1975), but since they are associated with sectors of the economy rather distantly related to utilization of the biotopes actually occupied by the birds, they are frequently ignored. Aesthetic and emotional judgements of the importance of bird populations potentially conflict with impact in other dimensions, and when this conflict is intense it renders the adoption of proper management options difficult. One recent case, involving granivorous bird populations, illustrates the problems.

206

Red-winged blackbirds *Agelaius phoeniceus* (L.), common grackles *Quiscalus quiscula* (L.), brown-headed cowbirds *Molothrus ater* (Boddaert) and the European starling *Sturnus vulgaris* L. are widely distributed and abundant throughout North America, especially the eastern portion. On their summer breeding grounds they either provide a source of recreation (especially the redwing) or are thought to be important insectivorous predators in the breeding-grounds ecosystems. However, these beneficial roles are counter-weighted by apparent deleterious roles during the winter when the birds gather in numerous, enormous roosts in southeastern USA. There, they are involved in depredations and possible disease roles. The question, then, is how do we allocate a qualitative or quantitative measure of impact to these birds or judge their value? In one case they are desirable and in the other they are definitely undesirable. Can there be a way to measure and weigh aesthetics, which are abstract qualitative projections of human thought, against depredations or disease, which are measured in economic or easily-defined terms dealing with quality of life? Currently, we know of none.

Role in ecosystem functioning

Natural ecosystems are integrated through the webs of interrelationships of populations and the holistic effects these generate. These interrelationships are usually considered in terms of energy flow or nutrient cycling, and it appears that bird populations rarely assume 'important' roles in either. In hardwood forests of northeastern USA, for example, the estimated annual energy flow through the avian community represents only 0.17% of the annual net primary production (Holmes & Sturges, 1973, 1975), while in Oklahoma tallgrass prairies the breeding season energy demands of birds were estimated to comprise 0.17% of the standing crop of living vegetation in mid-summer (Risser, 1972). The role of bird populations in nutrient cycling is virtually unstudied, although Sturges, Holmes & Likens (1974) documented that birds contained but a tiny percentage of the standing crop of nutrients in the northeastern USA hardwood forests and transported a small fraction from the system when contrasted with other transport mechanisms. Their apparently minor roles in the energy and nutrient dynamics of at least terrestrial ecosystems have suggested that birds are 'frills' in the ecosystem (Wiens, 1973) or 'evolution's ornaments on the tree of community function' (Whittaker, MS), subsisting off the ecosystem without influencing its energy or nutrient dynamics in any significant manner. However, this assumption may not invariably hold true. Dyer (unpublished data) estimates that 8 to 15% of the total fixed nitrogen in agricultural crops can be transported by blackbirds in the southeastern ecosystems of the USA.

207

It is possible, of course, that a bird population may play an important role in ecosystem functioning quite apart from its processing of energy or nutrients. Turček (1969) and others (e.g. Varley, 1967; McCullough, 1970; Chew, 1974) have suggested that some species may be important in governing or controlling flux patterns or rates in ecosystems without directly participating in these fluxes. This may be achieved most directly through control of the population dynamics of other species which are themselves 'important'. Such control may be most apparent among insectivorous birds, which may spread consumption over several age classes of their prey and thus potentially influence not only population size but age structure as well. While insectivorous birds are known to respond both functionally and numerically to increases in abundance of insect (especially larval) populations (Morris, Cheshire, Miller & Mott, 1958; Gage, Miller & Mook, 1970; Koplin, 1972), it is doubtful that they normally play a significant role in controlling the abundance of their prey, although they may reduce the growth rate of the prey populations (Thomas, Crouch, Bumstead & Bryant, 1975; Wiens, 1975). Granivorous birds exert their influence on a much more restricted age class (i.e. seeds) of their 'prey' populations, but this consumption is concentrated on relatively small seeds, which frequently are produced in large quantities by the plants. Birds may, in fact, consume a sizeable proportion of the seed production of a given year. In coniferous forests, juncos *Junco hyemalis* (L.) may consume 17 to 20% of the seedfall (Hagar, 1960; Gashwiler, 1970; Noble & Shepperd, 1973), and while this consumption has aroused concern in reforestation programs, the effects it may have upon the natural dynamics of the plant populations are unknown (Wiens, 1975). Pulliam & Enders (1971) suggested that wintering populations of five finch species consumed at least 40% of the available seeds in abandoned farm fields in North Carolina, but their analytic procedure is suspect, and potential effects upon the plant populations were not demonstrated. Some granivorous birds are known to play a role in the dispersal of seeds and may govern the spatial pattern of plant populations in that manner. Overall, however, it seems unlikely from the evidence available that birds, especially granivores, customarily play an important 'control' role in ecosystem dynamics. The evidence is meagre, however, and careful studies of such control effects should have a high priority.

It is worth noting that all of these avenues of 'control function' are through consumption or predation. It is thus important to understand the behavioral processes underlying predation in order to evaluate the potential for exerting a controlling influence in ecosystem processes. Unfortunately, such behavioural processes as diet selection (Ellis, Wiens, Rodell & Anway, 1976), 'switching' between preferred prey (Murdoch,

1969; Oaten & Murdoch, 1975), or formation of searching images (Gibb, 1962; Croze, 1970) are generally ignored in ecosystem studies, which contributes to the difficulty of demonstrating 'control functions' in ecosystems.

Food consumption

Perhaps the most obvious dimension of importance of bird populations is in the direct effects of their consumption of prey. Thus a population which consumes large quantities of, say, a pest insect species, may be 'important' in a positive or beneficial sense, while others which feed upon foods which are of economic value to man may be considered to be important through their negative or 'pest' impact. Those granivorous birds which achieve importance through their food consumption are generally included in this latter category, and it is this aspect of 'importance' or impact which we will emphasize in this chapter. It is important to note, however, that the determination of 'importance' in this framework rests heavily upon economic and political considerations – a population whose food habits bring it to exploit a commodity valued by man attains importance and may become the object of control programs, while a population consuming equal quantities of economically unimportant seeds is scarcely noticed. Ideally, of course, the evaluation of impact should be based upon overall cost–benefit considerations, but usually if a bird population incurs an economic cost to man, possible benefits are ignored or deleted from the cost–benefit analysis. Such problems are developed more fully in the following chapter, and our intention here is to analyze the component processes which determine the impact potential of granivorous bird populations.

The impact which a granivorous bird population can exert upon a resource of economic or agricultural importance is influenced by several components of the population–environment system (Fig. 6.1). Most directly, the total quantity of food consumed by the birds is an expression of the energy demands of the population, and this in turn is a function of the metabolic demands and efficiencies of individuals, the total size and composition of the population and the dynamics of the population biomass. But how energy demands are expressed in food consumption also depends heavily upon the composition of the diet, and their preferences for some foods over others are a product of the evolutionary adaptations of the birds to utilize food resources and of the relative availabilities of various food types (Ellis *et al.*, 1976). Thus the requirement of a given quantity of energy by a granivorous bird population creates the *potential* to affect human crops adversely. Whether that impact is 'important' depends upon the

209

J. A. Wiens & M. I. Dyer

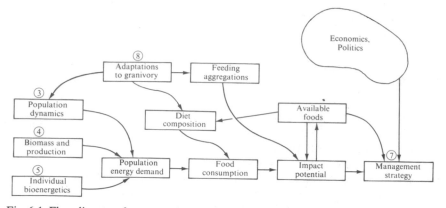

Fig. 6.1. Flow diagram of conceptual relationships among the major components contributing to the determination of impact potential of a granivorous bird population. Circled numerals refer to the chapters in which a component is considered in detail.

economic values attached to the food resources. If the impact is judged to be 'significant', management programs may be initiated, as described in the following chapter.

This is, of course, a simplistic view of what determines impact. While the remainder of this chapter will deal with impact within this framework, we should draw attention to three complicating considerations. First, there are important spatial features to potential impact. At a regional level, these may be most significant when a species or population has impacts of one sort on one part of its range and of another sort elsewhere. Starlings, for example, are considered beneficial to agriculture during the breeding season in northwestern USSR, where they eat insects which may be crop pests. Here, populations are encouraged by the placement of nest boxes adjacent to croplands. When the birds leave the USSR in the autumn they migrate west and then south; in Germany and France they can be serious pests in vineyards, where they destroy large quantities of ripening grapes. Farther south, in Tunisia, the birds occupy olive orchards and have substantial negative impacts on that crop (W. Keil, personal communication). In New Zealand, introduced starlings are considered to be beneficial by virtue of their consumption of noxious insects, but this favorable status is by no means secure (Bull, 1973). On a more local scale, the food consumption of many granivorous species is associated with flock feeding or colonial nesting and thus may be quite localized. Red-winged blackbirds aggregate during the non-breeding season in North America and feed locally in concentrated flocks. Their impact is thus highly localized, so that some crop fields are virtually unaffected (Dyer, 1967). Similarly, *Passer domesticus* (L.), *P. montanus* (L.) and *P. hispaniolensis* (Temminck) cause considerable damage to fields of ripening wheat, barley and other grains

210

in close proximity to the nesting colonies or shrubs and trees in which the sparrows find shelter. The birds associate with biotope edges, and margins of the fields attacked by sparrows generally suffer more than the middle parts (Kashkarov *et al.*, 1926; Akhmedov, 1949; Keleïnikov, 1953; Dawson, 1970; Havlín, 1974). The potential of granivorous birds to have significant agricultural impacts must therefore be determined at a local level, although the management responses to this impact may often necessarily be regional or international.

It is also apparent that food consumption alone may not be the sole measure of impact. Qualitative aspects of consumption may modify the quantitative effects: the impact of consumption of a given amount of seed from detritus on the ground is fundamentally different from that of the same amount of the same seed taken from the standing plants, where it is harvestable. Grain losses during the soft grain stages may be economically more severe than quantitatively similar losses suffered after the grain has hardened, although under some circumstances damage to the early phenological stages of ripening grain *may not* result in loss of grain to the whole plant or to the damaged field (Dawson, 1970; Dyer, 1975). More importantly, in some cases granivorous bird populations may have a greater impact on a crop through structural damage of the crop than through actual consumption of grain. Gavrilov's (1962, 1963) studies of *P. hispaniolensis* in Kazakhstan indicate that while the actual consumption of grain by the birds represents roughly 2% of the total production, losses due to damage may amount to as much as 40% of the crop. Sparrows cause as much damage by breaking the stalks of ripening wheat and barley as by picking out grains from the ears (Havlín, 1974). In the USA, blackbirds, mainly grackles, have been implicated in sprout pulling of wheat and rice. Clearly, if fields of newly sown grain are heavily attacked, subsequent production can be more greatly affected than if the invading birds fed only upon ripening grain. Thus, determinations of impact potential derived solely from estimates of food consumption, such as we will describe below, should not be considered a final assessment to determine if consumption is accompanied by damage.

Finally, it should be obvious that the various dimensions of 'importance' we have described are not independent. For any granivore population there are varying degrees of importance in each dimension, and that which we wish to emphasize as 'best' is contingent upon specific objectives, e.g. maximization of agricultural yield, preservation of natural ecosystems, protection of species populations, or whatever. Moreover, the importance or role of a population in each of these dimensions is determined by a variety of interacting population or environmental attributes (e.g. Fig. 6.1). A proper assessment of 'importance', or a true cost–benefit measure of impact, must consider these various components

211

and the complex interrelations among them. This, of course, is rarely (if ever) done; usually the impact or importance of a given granivorous bird population is evaluated in terms of limited objectives or single dimensions.

The estimation of potential impact: a simulation model approach

The potential impact of granivorous birds in ecosystems, in terms of food consumption, is not easily measured under natural conditions, partly because many of the variables which influence food consumption are not amenable to direct measurement and partly because so many variables are involved. Direct field estimates of impact are thus generally *post facto* assessments. While these may be quite useful if conducted properly, it is also desirable to be able to suggest conditions in which the *potential* for economic impact exists. The main failure in making field assessments of depredations to date has been that only estimates of removal are considered, instead of evaluating the differences in production of the crop being attacked in contrast with nearby areas where birds had not been feeding. Vastly different answers arise when these two approaches are compared (Dyer, 1975; Mattson & Addy, 1975). Nowhere has there been an effort to look at production figures over widespread areas when birds are involved in cereal crop depredations. To derive reasonable estimates we must employ various *indirect* means of approximating potential impact (Wiens & Dyer, 1975 *a*). The need for such an approach is evident when one considers energy demands, upon which the magnitudes of food consumption are largely founded (Fig. 6.1). Direct field measurement of the energetics of free-ranging birds are not feasible at present, especially when entire populations must be considered. Several investigators (Holmes & Sturges, 1973, 1975; Wiens & Innis, 1973, 1974; Hémery & Le Toquin, 1975 *a*, *b*; Weiner & Głowaciński, 1975) have employed models developed from basic information on individual bioenergetics and life history features to estimate the energy demands and/or food consumption rates of free-living birds. Such modelling approaches have the advantage of being able to integrate field data on readily measurable population attributes with mathematically derived estimates of less readily measured features, to produce initial projections of energy demand or food consumption, but they suffer the disadvantage of lacking the detailed resolution or insights which are necessary to make specific or highly precise statements about specific situations. It is important to understand that indirect approaches to estimating impact potential, such as those which we will develop here, are *not* alternatives to careful, intensive field studies. Some information on biological attributes of a given population is necessary to

212

undertake any indirect assessment of impact potential, but detailed field studies are time-consuming and expensive. Under such conditions, model-derived approximations of energy demand or food consumption rates may facilitate broad determinations of impact potential under a variety of conditions, provide preliminary indications of situations which warrant closer direct investigation, and suggest which of the various components contributing to impact merit close study. The modelling approach we describe here represents only a beginning in the attempt to generate realistic predictions of the impact of granivorous birds in ecosystems, but in developing this approach we should discern some of the requirements of a more complete approach.

The model we have employed in estimating the potential impact of granivorous bird populations is a computer simulation model (affectionately named BIRD) which was developed by Wiens & Innis (1973, 1974; Innis, Wiens, Chuculate & Miskimins, 1974) as part of the modelling efforts within the Grassland Biome Program of the IBP (USA). It was originally developed to provide estimates of energy flow through breeding bird populations in grassland ecosystems (Wiens & Dyer, 1975 *b*), but it has subsequently been applied to the determination of energy flow and food consumption patterns in bird populations of coniferous forests (Wiens & Nussbaum, 1975) and coastal marine environments (Wiens & Scott, 1975). The model can be conceptually divided into three sections, and while the model structure is described in detail in the publications of Wiens & Innis (1973, 1974), some aspects of these sections of the model will be discussed here with particular reference to granivorous bird populations and the information contained in Chapters 3 and 5 of this volume.

Population dynamics section

The model, as originally developed, deals with the dynamics and energetics of populations occupying a given location and the treatment thus deals with populations in terms of densities (individuals km^{-2}) rather than total size. Our emphasis is therefore based upon what happens in a specified area, and as individuals or populations depart from that area they are no longer considered. Information derived from direct field studies or from the literature is used to specify various features of the annual phenology of a population at a location: the timing and duration of arrival and departure for migratory populations; the onset of incubation for each brood as it is distributed among breeding females; the duration of incubation, nestling and fledgling stages; and the disappearance of juveniles from the area. These phenological features serve to specify the time patterns of fluxes of individuals between age classes in the population.

(*a*) Population dynamics submodel

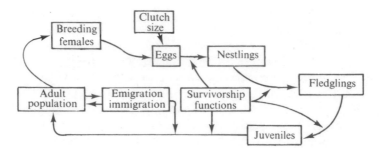

(*b*) Individual energetics submodel

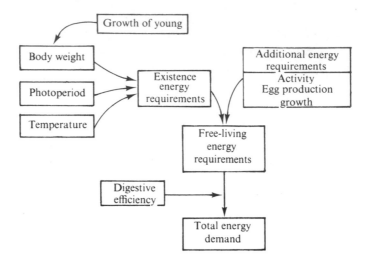

(*c*) Population energy demands and food consumption submodel

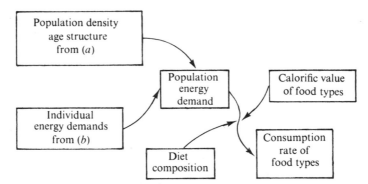

Fig. 6.2. Generalized flow diagrams of the three major portions of the BIRD simulation model of avian population bioenergetics. Part *a* depicts the components which determine the dynamics of population age classes through time. Part *b* indicates the interrelationships among features contributing to model calculations of individual bioenergetics. In Part *c*, the combination of population dynamics with individual energetics to estimate population energy demands is shown, with the further steps used to calculate food consumption rates. The arrangement of these components reflects the structure of the model and is not in total agreement with the more detailed treatments of population dynamics or avian energetics given in Chapters 3 and 5.

Processes operating during the non-breeding season, such as wintering mortality, are assumed to act uniformly over that period, even though this is an oversimplification.

Population densities of adults, derived from field estimates, must be specified for the onset of immigration at the initiation of the breeding season for that locality, during the main period of the breeding season, and at the beginning of the post-breeding period, following emigration. Obtaining density estimates for species which spend a considerable portion of the year in aggregations, such as many granivorous species, is especially difficult (Chapter 3, this volume). In fact, one assumption of the model is that the individuals present in an area, as measured by density, are evenly distributed over that area. Transients are ignored in the model calculations, since in most cases their influence on total population energy demands is probably slight.

Population density increases as reproduction proceeds. The major components of this recruitment process which are included in the model structure are depicted in Fig. 6.2 a. A certain portion of females in the population will be successful in initiating first broods, second broods and so on, and if the adult sex ratio in the breeding population is known we may then determine the number of females (= nesting attempts) involved in each brood. Given the mean clutch size for each brood, an egg age class of a certain size (density) is thus produced by these females. The timing of the entry of eggs into the population is determined by the various phenological values described above. Following a period of time (specified by the incubation period), the eggs in a clutch are transferred to the nestling age class; the proportion of eggs which become nestlings is determined by the mean hatching success for a given brood over the population as a whole. In a similar fashion, individuals remain in the nestling age class for a time determined by the nestling period, following which a portion (specified by fledging success) of this age class density transfers to the fledgling age class. Individuals remain in the fledgling age class until they attain adult body weight. The model thus generates a daily estimation of the densities of individuals in each of these age classes of the population. Survival of juveniles or adults during the non-breeding period may be included if input data are available.

Obviously, fairly detailed information is required to specify all of the input variables for a given population or locality. While such information exists for several granivorous bird populations for some variables (Chapter 3, this volume), data on others are limited and quite difficult to obtain. This is especially true of information regarding mortality rates following fledging. If one lacks values for a specific situation one may use values obtained from other populations of the species, or even from similar species, although this necessarily reduces the confidence that may be placed in the model estimates of population density dynamics. If the purpose of modelling, however, is to generate estimates of energy demand or food consumption over a year or a season, it is likely that precise values for all of these variables, especially those dealing with reproductive age classes, may be unnecessary, as sensitivity tests of the model have shown (Wiens & Innis, 1974).

Individual energetics section

The determination of individual energy demands within the model structure is founded upon calculations of existence metabolism (EM). As indicated in Chapter 5, EM measures the rate of energy utilization by caged birds undergoing no appreciable weight change, engaged in no 'special' activities such as moulting or migratory unrest, and given surplus food and water. It therefore integrates the individual bioenergetics associated with basal metabolism (BM), thermo-regulation, the heat increment of feeding and limited locomotor activity. The EM was originally selected as the metabolic basis for the model calculations because it more closely approximates the metabolic requirements of free-living birds than either BM or standard metabolism (SM), because it can be related to important environmental parameters such as temperature and photoperiod, and because it is measured over time periods of several days, a level of resolution similar to that of the model. The components which contribute to model determinations of individual energetics are shown in Fig. 6.2 *b*.

In Chapter 5, Kendeigh, Dol'nik & Gavrilov develop the bioenergetic foundations of EM. Some details of the model structure differ from relationships presented in Chapter 5, due to revisions incorporated into the chapter after the computer analyses were conducted. Following the treatment of Chapter 5, we calculate the EM in the model structure according to the following relationships:

For passerines:

$$30\ ^{\circ}\text{C, 10-hour photoperiod:}\ M = 1.544\,W^{0.6601} \tag{6.1}$$

$$30\ ^{\circ}\text{C, 15-hour photoperiod:}\ M = 1.462\,W^{0.6880} \tag{6.2}$$

$$0\ ^{\circ}\text{C, 10-hour photoperiod:}\ M = 4.437\,W^{0.5224} \tag{6.3}$$

$$0\ ^{\circ}\text{C, 15-hour photoperiod:}\ M = 4.969\,W^{0.5105} \tag{6.4}$$

For non-passerines:

$$30\ ^{\circ}\text{C, 10-hour photoperiod:}\ M = 1.455\,W^{0.6256} \tag{6.5}$$

$$30\ ^{\circ}\text{C, 15-hour photoperiod:}\ M = 1.068\,W^{0.6637} \tag{6.6}$$

$$0\ ^{\circ}\text{C, 10-hour photoperiod:}\ M = 4.235\,W^{0.5316} \tag{6.7}$$

$$0\ ^{\circ}\text{C, 15-hour photoperiod:}\ M = 4.142\,W^{0.5444} \tag{6.8}$$

where M = kcal bird^{-1} day^{-1}, and W = weight in grams. These equations are equivalent to equations (5.23), (5.26), (5.29), (5.34), (5.25), (5.28), (5.31) and (5.35). W is best taken as early morning weight before maximum daily deposition of fat. As such, it may be considered as relatively constant for a population, although varying among individuals. Differences in W between sexes are ignored unless they are substantial. Temperature is considered in a general manner, by specifying as input values mean daily air temperatures from a location (or, more often, a nearby weather station), averaged over weekly, 10-day, or occasionally monthly time intervals. This 'temperature', of course, is only a general approximation of the temperatures to which a bird is exposed in the micro-habitats it occupies, but such detailed temperature information is rarely available. The use of air temperatures to drive model estimations of the EM also ignores the ways in which variations in surface–air radiation exchanges may influence heat exchange and thus metabolic demands (Chapter 5, this volume). The resolution of the model is not sufficient to consider such details of individual bioenergetics, and since their effects are frequently short term, of the order of minutes or hours, they can perhaps safely be ignored in the calculations of the EM made on a daily basis. The other driving variable of EM, photoperiod, is considered in the model as a simple function of latitude. Given input specifications of weight, temperature and latitude, the model employs equations 6.1–6.8 to calculate the EM for an individual of a given weight on a daily basis, integrating the effects of temperature and photoperiod. The calculations of EM ignore day–night differences in metabolism (Aschoff & Pohl, 1970 *a*, *b*), which are built into the initial derivations of the EM. The seasonal and regional acclimation or adaptation of metabolism which are of demonstrated importance in some species (Chapter 5, this volume) are of necessity ignored, since we normally lack the information to make the appropriate adjustments.

Once the EM is calculated, it must be adjusted for the additional costs which may be levied by several 'special' functions (Fig. 6.2 *b*). Free-living birds, for example, are undoubtedly more active than the caged individuals upon which the estimations of EM are based, and this additional activity is not without metabolic costs. Unfortunately, we know relatively little about the actual activity 'levels' of free-living birds, or about their seasonal variations or true metabolic costs. Recorded time-budget observations of several species, however, have convinced us that the frequency of apparently 'high-cost' activities such as flight, active display, or aggression is greater during the breeding than in the non-breeding season. In the model calculations we elevate energy requirements by 0.25EM during the peak of the breeding season and by 0.10EM during the middle of the non-breeding season to adjust for activity costs. Actually, activity costs are varied between these two extremes in a sine-wave function. This

217

approach differs somewhat from that of Chapter 5, and resolution of these differences must await careful field studies of activity levels and their associated costs; such studies will be difficult, but they are essential.

During the breeding season, females must bear the additional metabolic costs of egg production and incubation. We assume that incubation costs are included in the general 'activity' costs, but the costs of egg production must be calculated separately. The model follows rather closely the approach detailed in Chapter 5; the metabolic costs (kcal) of producing a clutch of average size, EC, are given by:

$$EC = EW \times CS \times 1.1 \times 1.37 \tag{6.9}$$

where EW = egg weight (g), determined from relations of egg weight to adult body weight, which is specified elsewhere in the model; CS = clutch size; 1.1 = kcal per g of egg; 1.37 = egg production efficiency (73%). These costs are distributed over $CS+3$ days, assuming that one egg is laid each day and that the female begins to incur the costs of egg production three days prior to the laying of the first egg.

There are also costs associated with the growth of nestlings and fledglings. Growth itself is modelled following logistic functions (Ricklefs, 1968, 1972), and the EM is elevated by 20% in nestlings and 5% in fledglings to account for growth costs. These estimates are probably slightly conservative; the more detailed treatment of growth costs given in Chapter 5 was not available at the time of development of the BIRD model.

The requirements of the EM of individuals in each age class, after adjustment for these additional costs, reflect an approximation of the actual metabolic requirements of free-living birds engaged in normal activities. But since our objective is to estimate the energy or resource demands which are placed upon the environment by the birds, the estimated metabolic requirements must be converted into total energy demands, which incorporates not only the metabolic requirements, but also the unassimilated energy which is ingested. This adjustment is made using a measure of digestive efficiency, the metabolizable coefficient (MEC). As shown in Tables 5.7 and 5.8, the MEC varies with species, temperature and diet; since we generally have little specific information about these variations, we assume a standard value of $MEC = 70\%$ in the model analyses. Thus, the calculated free-living metabolic requirements are multiplied by 1.43 to obtain the energy equivalent of the food intake per individual for each age class in the population.

Two potentially important sources of additional energy demands are not considered in the model. Migration is accompanied by substantial energy costs, not only in the expense of the migratory flight itself, but in the increased activity, hyperphagia and fat deposition which customarily

218

precede migration (Dol'nik & Gavrilov, 1975). This omission can be justified to a degree, since we are concerned with estimating the energy demands of birds while they are associated with a specific location, and costs incurred after the birds have left that area are thus irrelevant. But some of the costs of pre-migratory fat deposition may occur at the site of interest, and to the degree that these are ignored by the model calculations we will underestimate actual energy demands. Energy demands may also be elevated during moulting (Chapter 5, this volume). However, since it is uncertain to what degree other energy-demanding functions, such as activity, are reduced during the moulting period, and since we have little information on the duration or timing of the moulting period, or on the manner in which the total costs of moulting are distributed over this period, we have omitted the costs of moulting from our calculations. This will also lead to a potential underestimate of energy demands, at least during the moulting periods.

Food consumption section

Once estimates of population density and age structure and of individual energy demands are calculated, they may be integrated to project a total population energy demand, on a daily basis, in kcal m^{-2} day^{-1} (Fig. 6.2 c). Information on the composition of the dietary intake of the population at various points in time is placed into an array in the computer program, along with values of the calorific content of each food type (kcal g^{-1}), and a linear interpolation function employed to generate estimates of diet composition on a daily basis. Given the calorific values of the food types, the energy demands of the population are converted into estimated rates of consumption of each food type (g m^{-2} day^{-1}).

One limitation on exercising this portion of the model is the frequent lack of availability of information on the dietary habits of a given population. Such data are customarily obtained from analyses of stomach contents, and these analyses are difficult, time-consuming and require the collection of at least moderately-sized specimen samples. This is often incompatible with ongoing studies of population dynamics, with the result that we rarely have an adequate knowledge of both the dynamics and the food habits of specific natural populations of birds. Since many birds, and perhaps especially granivorous species, are relatively opportunistic in their feeding, the diet composition of a given population at a point in time may reflect a variety of proximate circumstances. Diet information gathered over broad geographic areas in a variety of biotopes, over a number of seasons and years (such as that in the US Fish and Wildlife Service files) thus cannot be applied in a specific way to specific situations (e.g. Pulliam & Enders, 1971; Hespenheide, 1973) but can be used to indicate the

219

general dietary habits of a species. For this reason, the analyses of food consumption which we undertake in this chapter are restricted to a few general food categories.

There is a further complication in the use of diet information obtained from stomach analyses. Quantification of dietary composition is made from the occurrence of parts of food items of various types contained in the stomach (usually the gizzard) at the time of collection. But since food types of different sorts are digested at different rates (Coleman, 1974; Custer & Pitelka, 1975), those types which are relatively resistant to digestion will remain in the stomach longer, leading to an overestimation of their importance in the diet. Unfortunately, adjustments for such differential digestibilities of prey are not available for granivores, so we must use uncorrected results of stomach analyses to approximate dietary composition.

Constraints on model application

The advantages of employing this simulation model approach to the estimation of potential impact through food consumption stem primarily from its versatility and ease of application. Given basic information on the dynamics and breeding biology of a population, one may obtain preliminary estimates of measures such as energy demand or food consumption rates which are extremely difficult even to approximate under field conditions without intensive, time-consuming investigations. An indirect approach, such as the BIRD model, facilitates preliminary evaluations and comparisons of impact potential over a broad range of situations and may suggest those conditions in which the potential is greatest and in which additional direct studies are required. Further, in the process of constructing and applying a model, the components of impact are identified and compartmentalized, and this leads to a clearer definition of exactly what factors contribute to impact, and how they may interact. We believe this is an important result of our preliminary model investigations of granivorous bird impacts, and we will return to it at the conclusion of this chapter.

This is not to say that the model described here lacks constraints. There are several important limitations which must be kept in mind when considering the results of model simulations. First, the model presumes that the potential of a granivorous bird population to exert an economic impact on their food resources is entirely a function of their consumption of food, and that this in turn is entirely an expression of the satisfaction of *energy* demands of individuals. This is, of course, a simplification. Special nutritional demands, such as requirements for protein and calcium by females during egg formation (Jones & Ward, 1976) or for sulphur-containing amino acids (cystine and cysteine) during moulting (Newton, 1968), may increase food consumption levels beyond that required to satisfy metabolic energy demands or may lead to selection of food types on the basis of nutritional quality rather than energy content (Hughes, 1972; Moss, 1972; Emlen, 1973). In chaffinches *Fringilla coelebs* L., for example, food consumption rates with natural food types may be increased during moulting by 20 to 25% over the non-moulting levels. Birds rely heavily upon insects during this moulting period and if they are not available seeds must be used, and consumption rates are further increased to meet nutritional needs (V. R. Dol'nik, personal communication). The model, in ignoring such special nutritional requirements, underestimates impact potential, although not necessarily energy demand.

Second, the model and our approach in general deal with impact potential in only a *proximate* sense. The energy demands of individuals which determine the magnitudes of food consumption are determined by observed population densities, reproductive outputs and environmental conditions, and are *not* necessarily the energy requirements required to maintain the population at some equilibrium level or to maximize individual fitness. There are thus no feedback relations between energy demand or food consumption and reproductive performance in our approach. Such relations undoubtedly operate to some degree in natural populations, but they are more appropriately considered in the context of overall adaptive dimensions of granivory, which are discussed in Chapter 8 of this volume.

Finally, we must reiterate that the model estimates the energy demand or food consumption required to support a certain number of individuals with a certain reproductive output in some specified area and does not consider how these food demands may be distributed in space. Individuals may restrict their movements during moulting, for example, leading to localized concentrations of their food consumption, or individuals associated with breeding colonies of *Passer*, *Quelea*, or *Agelaius* may forage within a restricted radius about the breeding site. Such spatial variations in food consumption are not considered by the model, but must be included in the process of translating estimated impact potential to realistic projections of actual effects.

Population energetics and food consumption

Sources of data

Information which enabled us to conduct model simulations of population dynamics and energy demands was obtained from a survey of scientists throughout the world working on granivorous bird populations (primarily *Passer* spp.). This survey, conducted during 1974–1975 under the spon-

Table 6.1. *Features of study locations which provided information for the model analyses of Passer populations*

Number	Name	Country	Latitude	Longitude	Species	Years sampled	Contributors
1	Kangasala	Finland	61° 27' N	24° 03' E	*P. domesticus*	1967–1971	P. Rassi (personal communication)
2	Oxford	England	52° 00' N	01° 00' W	*P. domesticus*	—	D. G. Dawson (personal communication)
					P. domesticus, P. montanus	1961, 1963–1964	Seel (1968 *a, b*, 1969, 1970)
3	Schuilenburg	Netherlands	51° 57' N	05° 31' E	*P. montanus*	1971	J. H. van Balen (personal communication)
4	Loenen	Netherlands	51° 53' N	05° 46' E	*P. montanus*	1973	J. H. van Balen (personal communication)
5	Oosterhout	Netherlands	51° 53' N	05° 50' E	*P. montanus*	1973, 1974	J. H. van Balen (personal communication)
6	Marke	Belgium	50° 48' N	03° 13' E	*P. montanus*	—	de Bethune (1961)
7	Wolfsburg	GFR	52° 25' N	10° 47' E	*P. montanus*	1968, 1969	Scherner (1972, personal communication)
8	Yördenstorf	GDR	53° 50' N	12° 35' E	*P. domesticus*	1968–1972	A. Wendtland (personal communication)
9	Rottenau	GDR	52° 00' N	12° 50' E	*P. montanus*	1969–1971	Kaatz & Olberg (1975)
10	Steckby	GDR	51° 50' N	12° 00' E	*P. montanus*	1968–1970	Clausing (1975, personal communication)
11	Annaberg-Buchholz	GDR	50° 35' N	13° 00' E	*P. montanus*	1968–1971	S. Schlegel (personal communication)
12	Gdynia	Poland	54° 37' N	18° 30' E	*P. domesticus, P. montanus*	1969–1971	Pinowski & Wieloch (1973, personal communication)
13	Gdańsk	Poland	59° 28' N	18° 30' E	*P. domesticus, P. montanus*	1971–1972	M. Wieloch (personal communication)

No.	Location	Country	Latitude	Longitude	Species	Years	References
14	Wieniec	Poland	54° 20' N	18° 56' E	*P. domesticus*	1971–1973	B. Pinowska (personal communication)
15	Rzepin	Poland	52° 21' N	14° 51' E	*P. domesticus, P. montanus*	1968	Mackowicz, Pinowski & Wieloch (1970)
16	Turew	Poland	52° 25' N	16° 50' E	*P. domesticus, P. montanus*	1971–1973	Strawiński & Wieloch (1972); Wieloch & Strawiński (1976); Wieloch & Fryska (1975)
17	Dziekanów Leśny	Poland	52° 20' N	20° 50' E	*P. domesticus, P. montanus*	1960–1973	Pinowski (1968); Mackowicz et al. (1970); Pinowski & Wieloch (1973); Myrcha, Pinowski & Tomek (1973); M. Wieloch, B. Pinowska and J. Pinowski (personal communication)
18	Kraków	Poland	50° 04' N	19° 57' E	*P. domesticus, P. montanus*	1967–1973	Mackowicz et al. (1970); J. Pinowski (personal communication)
19	Nowy Targ	Poland	49° 28' N	20° 00' E	*P. domesticus, P. montanus*	1969–1970, 1972	Mackowicz et al. (1970); J. Pinowski (personal communication)
20	Zakopane	Poland	49° 18' N	19° 58' E	*P. domesticus*	1972	Novotný (1970)
21	Slezké Rudoltice	Czechoslovakia	50° 13' N	17° 43' E	*P. domesticus*	1963–1965	Ion & Valenciuc (1969); Ion (1971, 1973); Ion & Saracu (1971)
22	Iasi	Romania	47° 11' N	45° 11' E	*P. montanus*	1967–1968	
23	Kazakhstan	USSR	43° 05' N	77° 00' E	*P. hispaniolensis*	1959–1961	Gavrilov (1962, 1963)
24	Coldspring, Wisconsin	USA	42° 50' N	88° 40' W	*P. domesticus*	—	North (1969, 1973 *a*)
25	Portage des Sioux, Missouri	USA	38° 55' N	90° 22' W	*P. domesticus, P. montanus*	1968–1973	Anderson (1973, 1975)

223

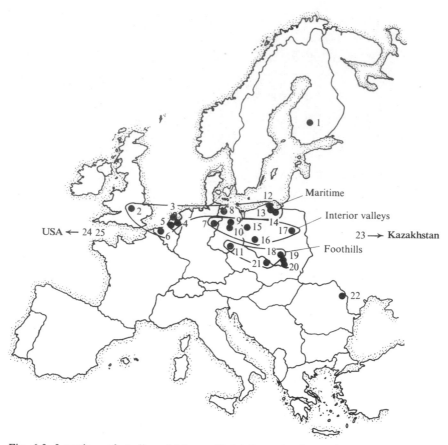

Fig. 6.3. Locations of studies which provided information for model analyses of *Passer* energetics. The numerals indicate the sites listed in Table 6.1.

sorship of the IBP Working Group on Granivorous Birds and the USA National Science Foundation, requested detailed information on reproductive biology and timing, growth of young, environmental features and population sizes of granivorous species. In all, 158 sets of data from 13 countries were obtained in sufficient detail to permit model analyses. The locations represented in this data base are indicated in Table 6.1 and Fig. 6.3. The information on population dynamics, a substantial share of it unpublished, was included in Chapter 3.

For many of the populations studied by these investigators, estimates of populations sizes or densities at various times of the year were not available. Since the BIRD model is based upon density values, we considered the dynamics of 'standardized' populations, rather than use rough approximations of density, at least in our treatment of *Passer* data. These

Table 6.2. *Model estimations of annual energy demands of standardized* Passer *populations*

Location	Number of years	P. domesticus Energy demand[a] (kcal m^{-2} year^{-1})	CV[b]	P. montanus Energy demand[a] (kcal m^{-2} year^{-1})	CV[b]
Kangasala, Finland	5	1.87 (0.15)	7.89	—	—
Oxford, England	1	1.65	—	—	—
	1	1.62	—	1.50	—
Schuilenburg, Netherlands	1	—	—	1.64	—
Loenen, Netherlands	1	—	—	1.79	—
Oosterhout, Netherlands	2	—	—	1.68 (0.18)	10.58
Marke, Belgium	1	—	—	1.39	—
Wolfsburg, GFR	2	—	—	1.77 (0.13)	7.40
Yördenstorf, GDR	5	2.23 (0.22)	9.94	—	—
Rottenau, GDR	3	—	—	2.09 (0.20)	9.58
Steckby, GDR	3	—	—	1.88 (0.10)	5.12
Annaberg-Buchholz, GDR	4	—	—	1.58 (0.12)	7.54
Gdynia, Poland	3	1.87 (0.14)	7.61	1.35 (0.11)	7.79
Gdańsk, Poland	2	1.89 (0.00)	0.015	1.37 (0.04)	2.88
Rzepin, Poland	1	2.01	—	1.40	—
Turew, Poland	3	1.86 (0.27)	14.25	1.41 (0.22)	15.27
Dziekanów Leśny, Poland	14	2.07 (0.18)	8.72	1.60 (0.08)	4.90
Kraków, Poland	7	1.87 (0.25)	13.58	1.51 (0.11)	7.32
Nowy Targ, Poland	2	2.23 (0.05)	2.06	1.56 (0.00)	0.23
Zakopane, Poland	1	2.05	—	—	—
Slezké Rudoltice, Czechoslovakia	4	2.07	—	—	—
Iaşi, Romania	2	—	—	1.71 (0.12)	6.85
Coldspring, Wisconsin, USA	1	1.41	—	—	—
Portage des Sioux, Missouri, USA	1	1.62	—	1.50	—

[a] Mean; standard deviation (S.D.) in parentheses.
[b] Coefficient of variation = s.D./$\overline{X}$ × 100.

analyses employed the available data on reproductive biology and other population attributes supplied by the various investigators, but specified a population density of 100 breeding adults km^{-2}. The estimates derived from calculations on such 'standardized' populations can be adjusted to the real density of a population, where it is known, by dividing the real density by 100 and multiplying the model estimates by the resulting value. Unless otherwise specified, all of the results we discuss here for *Passer* populations are derived from analyses of populations of 100 individuals km^{-2}.

225

Energy demands of Passer domesticus *and* P. montanus

Average estimated annual energy demand for standardized populations of
P. domesticus was generally lowest in the USA and Britain, and greatest
in northern GDR and southern Poland (Table 6.2). At the Polish sites, mean
annual energy demand of *P. domesticus* was lowest in the northern
coastal areas and at Turew and Kraków, intermediate at Rzepin, Dziekanów Leśny and Zakopane, and highest at Nowy Targ. There is no
general pattern to these variations, and they are probably related to
short-term local variations in climate or reproductive biology at the times
the studies were conducted.

Standardized populations of *P. montanus* generally exhibited lower
estimated annual energy demands than *P. domesticus*, which is not surprising given their smaller body size. Values ranged from a low of 1.35
(Gdynia) to a high of 2.09 kcal m^{-2} year^{-1} (Rottenau, GDR) (Table 6.2).
Mean values averaged low throughout northwestern and northern Poland
and in Belgium, somewhat higher at Dziekanów Leśny and sites in
southern Poland and at Annaberg-Buchholz, GDR, and highest in a cluster
of sites in the Netherlands and GDR. In the one USA site at which both
Passer species occurred, the *P. domesticus* population required somewhat
more energy than *P. montanus*.

There is considerable variation in annual energy demand estimates
between years at a given site, beyond the inter-site variation, as was also
described for certain population parameters noted in Chapter 3. A
preliminary approximation of the *relative* degree of variation in annual
energy demand for a given site may be obtained from comparisons of
coefficients of variation (CV) of the site values (Table 6.2). Since these
CV values are dimensionless, the numbers in themselves mean little except
in a comparative framework. It is apparent that annual variations are
greater at some locations than others. Estimated annual energy demand
of *P. domesticus* populations was most variable at Turew and Kraków,
Poland, and least variable at two other Polish sites, Nowy Targ and
Gdańsk. Estimates of *P. montanus* energy demand varied most at Turew,
Poland, and least at Nowy Targ, Poland. Of the six locations at which both
Passer species occurred, *P. montanus* populations had more variable
annual energy demands at three, and *P. domesticus* at three.

The studies by Polish scientists have provided information on *Passer*
population attributes at several locations over several successive years,
and thus prompt a more specific examination of annual variation in energy
demand estimates. The patterns of estimated annual energy demands of
standardized *Passer* populations at Dziekanów Leśny and Kraków are
shown in Fig. 6.4. The greater demands of *P. domesticus* populations are
clear, as is the greater degree of yearly variation in energy demand of this

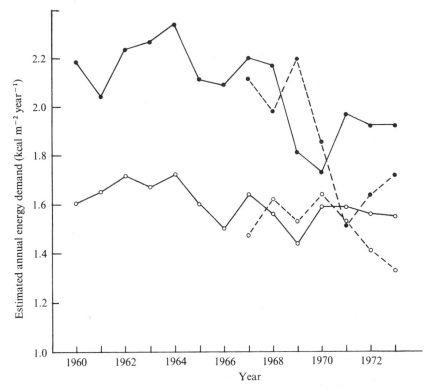

Fig. 6.4. Yearly variations in model-estimated annual population energy demand of *P. domesticus* (solid circles) and *P. montanus* (open circles) at Dziekanów Leśny (solid line) and Kraków (dashed line) in Poland.

species. It is also apparent that annual population energy demands are more variable at Kraków than at Dziekanów Leśny for both species. There are no close associations in the patterns of trends in energy demand for *P. domesticus* and *P. montanus* populations at the same location, however, nor are the yearly trends at Kraków very similar to those at Dziekanów Leśny. It is interesting to note that while no trends of energy demands are visible between the house and tree sparrows between areas, an analysis of nesting phenology pointed out considerable correlation (Chapter 3, pp. 61–5). At Kraków, *P. domesticus* had no third broods in 1970, 1971, 1972 and 1973, and in 1971 they had only first broods. This was coincident with high nestling mortality, and these factors undoubtedly contributed to the decrease in energy demands.

The energy demand of a *Passer* population also varies throughout a single year. The greatest energy demand occurs during the breeding season (Fig. 6.5), which is, of course, to be expected with a resident population in which young birds disperse out of the local area in the post-breeding

J. A. Wiens & M. I. Dyer

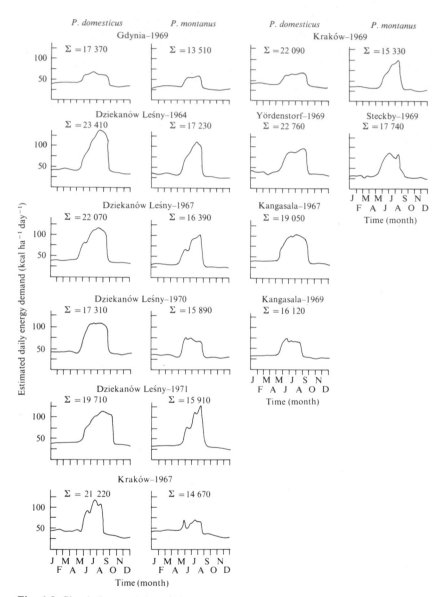

Fig. 6.5. Simulation model graphings of projected daily energy demands of *Passer* populations at selected locations and years. The total annual population energy demand is also given for each location.

228

period. It is also apparent that energy demands during the non-breeding season vary little between sites or years, but there is substantial variation in the pattern and magnitude of breeding season energy demands. This is an inevitable product of our procedure of standardizing adult population densities; in real populations there is often considerable year-to-year variation in adult density, and thus in non-breeding season energy demand.

The level of resolution of the model and the model inputs at this stage is not sufficient to justify analysis of weekly changes in energy demands, but we can consider the general patterns of energy demand during the breeding and non-breeding seasons for these *Passer* populations. At Gdynia, Poland, for example, the peak energy demands of *P. domesticus* and *P. montanus* populations in 1969 were similar, but the breeding season of *P. montanus* was shorter and production of young somewhat less. Even in situations in which peak daily energy demands during the breeding season were greater for *P. montanus* than *P. domesticus*, such as at Dziekanów Leśny in 1970, estimated total yearly energy demand was generally greater for *P. domesticus*. Under some conditions populations of the two species exhibited similar patterns of energy demand through the year (e.g. Dziekanów Leśny in 1964), but in other situations the two species differed markedly (e.g. Dziekanów Leśny in 1970). The considerable interlocality variation for a single year is apparent through comparisons of the 1969 analyses for Gdynia, Kraków, Yördenstorf, Steckby and Kangasala. These local and yearly differences in patterns and magnitudes of energy demand are produced by variations in the complex of features which drive the model analyses, i.e. temperature, reproductive timing, clutch size, reproductive success and so on.

Rather than attempt a detailed dissection of relations between variations in these input variables and variations in model output, we may examine the patterns of breeding and non-breeding season energy demands in a more general manner. The patterns of Fig. 6.4 suggested that the magnitude of annual energy demand might be associated with the total length of the breeding season for a local population. Breeding seasons of *P. domesticus* averaged from 27 to 29% of the year at Nowy Targ, Poland (where the low value is largely influenced by the abnormally cold weather of 1970), and Kangasala, Finland, to 55% at Slezké Rudoltice, Czechoslovakia (Table 6.3). Including the single USA population, *P. montanus* breeding season duration varied less between locations and was shorter than that of *P. domesticus* at the same locations. Annual variation in breeding season length (as measured by CV values) differed among the locations and between the species but in general seemed somewhat greater in *P. montanus* than *P. domesticus*; annual variation in breeding season length was greater for *P. montanus* at four of the six Polish locations at which both species were present. Correlation tests comparing breeding

Table 6.3. *Features of estimated* Passer *population energy demands during breeding season*

Locality	Number of years	P. domesticus Breeding season length (% year) X	s.d.	CV	P. domesticus % Annual energy demand in breeding season X	s.d.	CV	P. montanus Breeding season length (% year) X	s.d.	CV	P. montanus % Annual energy demand in breeding season X	s.d.	CV
Finland Kangasala	5	29.7	4.89	16.46	44.9	6.36	14.16	—	—	—	—	—	—
England Oxford	1	41.1	—	—	55.6	—	—	—	—	—	—	—	—
	1	46.7	—	—	56.9	—	—	35.6	—	—	51.5	—	—
Netherlands Schuilenburg	1	—	—	—	—	—	—	35.6	—	—	52.9	—	—
Loenen	1	—	—	—	—	—	—	38.4	—	—	59.2	—	—
Oosterhout	2	—	—	—	—	—	—	30.1	3.82	12.69	42.7	7.99	18.71
Belgium Marke	1	—	—	—	—	—	—	38.4	—	—	68.2	—	—
GFR Wolfsburg	2	—	—	—	—	—	—	30.1	3.89	12.92	44.2	10.25	23.20
GDR Yördenstorf	5	39.5	2.45	6.20	55.0	3.22	5.85	—	—	—	—	—	—
Rottenau	3	—	—	—	—	—	—	37.5	1.62	4.32	60.6	2.55	4.21
Steckby	3	—	—	—	—	—	—	34.7	1.56	4.50	60.7	2.44	4.02
Annaberg-Buchholz	4	—	—	—	—	—	—	32.9	7.43	22.58	48.3	13.53	28.00
Poland Gdynia	3	30.1	2.75	9.14	40.0	5.07	12.67	26.5	3.12	11.77	35.9	2.07	5.76
Gdańsk	2	38.4	11.67	30.39	52.8	12.02	22.77	35.6	3.89	10.93	48.9	3.89	7.95
Rzepin	1	43.8	—	—	61.1	—	—	32.9	—	—	50.9	—	—
Turew	3	44.8	5.68	12.68	57.3	6.35	11.08	27.5	7.46	27.13	45.1	17.40	38.58
Dziekanów Leśny	14	39.7	3.83	9.67	54.9	5.35	9.75	38.8	2.10	31.70	56.8	3.31	5.82
Kraków	7	33.7	3.80	11.28	43.4	7.32	16.87	30.9	3.77	12.20	44.7	5.04	11.28
Nowy Targ	2	27.4	11.60	43.34	39.2	19.52	49.78	26.1	9.69	37.13	35.8	19.23	53.72
Zakopane	1	38.4	—	—	52.7	—	—	—	—	—	—	—	—
Czechoslovakia Slezké Rudoltice	4	55.0	—	—	69.8	—	—	—	—	—	—	—	—
Romania Iasi	2	—	—	—	—	—	—	37.0	1.98	5.35	50.1	2.69	5.36
USA Wisconsin, Coldspring	1	35.6	—	—	36.1	—	—	—	—	—	—	—	—
Missouri, Portage des Sioux	1	43.8	—	—	53.8	—	—	38.4	—	—	47.5	—	—
Total		39.2 (7.27) n = 15	—	17.40 (12.85)	51.6 (9.14) n = 8	—	17.87 (13.84)	33.7 (4.29) n = 18	—	16.10 (10.96)	50.2 (8.57) n = 12	—	17.22 (15.98)

season duration for all locations and years with estimated total annual energy demand, however, suggest that these variations have relatively minor overall effect on annual energy demand (for *P. domesticus*, $r = 0.029$; $N = 48$; for *P. montanus*, $r = 0.304$; $N = 53$). In fact, correlation coefficients between annual energy demand and a wide array of single variables were generally moderate to low, although consistently somewhat higher for a given variable for *P. montanus* than for *P. domesticus*. The proportion of the annual energy demand incurred during the breeding season, however, was closely correlated with the length of the breeding season (for *P. domesticus*, $r = 0.904$, for *P. montanus*, $r = 0.870$). Breeding season energy demands accounted for 39 to 70% of total annual energy demand over all *P. domesticus* populations sampled and from 36 to 68% for all *P. montanus* populations (Table 6.3).

We may attempt to dissect these patterns further by considering the *Passer* populations of northern and central Europe in terms of general, environmentally-defined regions (Fig. 6.3). Over the entire region, *P. domesticus* annual population energy demands averaged almost 2 kcal m^{-2} $year^{-1}$, those of *P. montanus* 1.6 kcal m^{-2} $year^{-1}$ (Table 6.4). Mean annual energy demand of *P. domesticus* populations was slightly higher in interior lowland valley locations and lower in foothill sites, and variation in energy demand was somewhat greater in maritime locations, less in valley situations. Maritime and foothill populations of *P. montanus* had similar annual energy requirements, on average, but populations occupying valley locations required nearly 10% more energy per year. For both species, foothill populations required a smaller proportion of the total annual energy demand during the breeding season than populations occupying valley sites; in *P. montanus* the proportion of annual energy demand required during the breeding season was also relatively low in maritime locations.

We have subjected these measures of annual energy demand and the proportion of the annual demand occurring during the breeding season to correlation and step-wise multiple regression analyses with several features of breeding biology within these regions. The results of these analyses are too inconclusive to detail here, but in general they suggest that a rather broad array of features contribute to variations in energy demands and that these features differ in their contributions between different regions, locations and years. The correlation between length of the breeding season and the proportion of the total annual energy demand required in the breeding season, noted above, holds for each region; for *P. montanus*, variations in hatching or fledging success also contribute to variation in breeding season energy demand, but in *P. domesticus* populations such secondary associations are less apparent. Variations in *P. montanus* energy demands during the breeding season thus appear to

231

Table 6.4. *Features of estimated* Passer *population energy demands averaged over sites within several general environmental 'zones' in Europe (Fig. 6.3)*

Region	Number of sites	Number of years	P. domesticus						Number of sites	Number of years	P. montanus					
			Annual energy demand (kcal m⁻² year⁻¹)			% Energy demand in breeding season					Annual energy demand (kcal m⁻² year⁻¹)			% Energy demand in breeding season		
			$\bar{X}$	S.D.	CV	$\bar{X}$	S.D.	CV			$\bar{X}$	S.D.	CV	$\bar{X}$	S.D.	CV
Maritime	4	12	1.99	0.28	14.0	51.1	8.2	16.1	7	11	1.50	0.18	12.2	47.5	10.7	22.6
Valley	3	18	2.04	0.20	9.8	55.7	5.4	9.7	6	26	1.67	0.23	13.6	55.1	8.1	14.8
Foothills	2	8	1.89	0.24	12.9	46.7	11.5	24.7	2	11	1.53	0.11	7.4	46.0	8.6	18.6
Total	9	38	1.99	0.24	11.9	52.3	8.5	16.2	15	48	1.60	0.21	13.1	51.3	9.7	18.9

be rather closely related to variations in the duration of the breeding season and in reproductive success, while *P. domesticus* population energy demands may be subject to a wider array of sources of variation. Other factors influence non-breeding energy demands, so that the closeness of these relationships is diluted when annual energy demand is considered. Seasonal and annual energy demands have direct ties to overall rates of food consumption (Fig. 6.1), and these analyses therefore suggest that food consumption (and potential impact) in these populations may vary in response to a complex of interacting factors affecting energy demands, quite apart from the variations resulting from features influencing the food selection process (e.g. Ellis *et al.*, 1976).

Food habits

The impact potential of a granivorous bird population is a function not only of its energy demands but also of the patterns of food selection exhibited by individuals. Information on food habits of *Passer* spp. has been obtained from several areas (see Grün, 1975), primarily through examination of stomach contents. Such analyses are sensitive to variations produced by sampling at different times of day (Grün, 1968), and diets may vary on quite a local scale with changing patterns of food availability. The dietary composition of *P. montanus*, for example, differs in areas of different soil type, primarily through the effects on species composition and abundance of weeds (Hammer, 1948; Kovacs, 1955; Simeonov, 1963; Pinowski & Wójcik, 1969).

Despite these variations, some general patterns in the diets of the two species do emerge, especially when we restrict consideration to major food categories. It is not our intention here to review studies of *Passer* diet composition (see Grün, 1975). A sampling of measures of diet composition from studies in Europe and North America (Table 6.5) suffices to demonstrate the major features. Seeds overwhelmingly predominate in the diets

Table 6.5. *Selected examples of dietary composition (%) of* Passer *species in several areas*

Location	Species	Spring[a]		Summer[b]		Autumn[c]		Winter[d]		Authority
		C[e]	W[f]	C	W	C	W	C	W	
Hungary	*P. domesticus*	58	21	47	47	53	46	57	42	Kovacs (1955)
	P. montanus	13	36	5	69	5	94	30	64	Kovacs (1955)
Czechoslovakia	*P. domesticus*	17	5	33	7	65	25	65	35	Ašmera (1962)
	P. montanus	21	11	18	13	37	46	49	46	Ašmera (1962)
Denmark	*P. domesticus*	49	18	50	31	70	25	78	21	Hammer (1948)
Bulgaria	*P. domesticus*	36	55	11	50	38	57	63	37	Simeonov (1964)
	P. montanus	13	76	7	53	5	92	15	78	Simeonov (1963)
Poland (north	*P. domesticus*	—	—	68	27	52	31	94	6	Pinowska (1975)
and central)	*P. montanus*	—	—	22	77	0	99	0	99	Pinowski & Wójcik (1969)
Romania	*P. montanus*	13	76	7	53	10	85	15	78	Ion (1973)
GFR	*P. domesticus*	—	—	—	—	48	36	48	36	Keil (1970, 1973)
	P. montanus	—	—	—	—	10	79	10	79	Keil (1970, 1973)
England	*P. domesticus*	75	10	60	20	80	15	80	20	Southern (1945)
USA	*P. domesticus*	82	9	75	22	69	31	85	14	Kalmbach (1940)

[a] April to June. [b] July to September.
[c] October to December. [d] January to March.
[e] Cereal grain. [f] Weed seeds.

of both species at all times of years, although generally to a lesser extent in spring and summer than in autumn and winter. There are regional differences in diets, but in nearly all locations cereal grain constitutes the major component of the seed food of *P. domesticus*. There are exceptions; in southern Hungary, Rékási (1968 *a, b*) found a strong 'preference' for weed seeds in *P. domesticus*, especially in autumn. There, the diet contained four to seven times as many weed seeds as cereal grain, depending upon the habitat type. The diet of *P. montanus*, on the other hand, normally contains a large proportion of seeds of weeds or native vegetation. Reviewing the diets of both species, Grün (1975) found the most important plant foods to be grains such as wheat, rice, oats, millet and barley, and seeds of wild grasses such as *Setaria* sp., *Echinochloa crusgalli* (L.) and *Digitaria* spp. Some weed seeds, such as *Polygonum aviculare* L. and *Chenopodium album* L., may be of equal importance, and *Polygonum persicaria* L., *P. convolvulus* L., *Amaranthus retroflexus* L., *Stellaria media* (L.), *Urtica dioica* L. and *U. urens* L. may be locally important seeds. Fruits of cultivated plants other than grains are fed upon only occasionally, but locally, crops such as sunflowers, hemp and perhaps lucerne may be damaged. *P. domesticus* sometimes damages soft fruits such as grapes, cherries and apricots.

The diet of *P. domesticus* is generally more variable than that of *P.*

233

J. A. Wiens & M. I. Dyer

montanus. House sparrows utilize a variety of foods in urban habitats and eat fruit, buds and some insect groups (e.g. ants, gnats) to a greater extent than tree sparrows. *P. montanus*, on the other hand, eats more weed seeds and appears to prefer smaller grain (Grün, 1975). Turček (1968) has interpreted these dietary differences in terms of the overall ecology and activity levels of the species. *P. domesticus* undertakes slight to moderate movements and occupies habitats with a diversity of food types. As a result it can subsist on foods of relatively low energy value. *P. montanus* is smaller, very mobile in activity, and more selective in its feeding, requiring foods of relatively high energy value from the woodland–cultivated ecotones which it inhabits.

Food consumption

To generate estimates of the patterns and magnitudes of food consumption by populations of *Passer* from the model projections of energy demand, it is necessary to combine this information on dietary composition at various times of year with measures of the calorific value of the various prey types (kcal g dry weight^{-1}). Measures of the calorific values of prey are available for a variety of foods (e.g. Golley, 1961; Kendeigh & West, 1965; Cummins & Wuycheck, 1971), but information on the dietary habits of *Passer* populations is much more limited, so that we are unable to specify the diet of each of the local populations listed in Table 6.1. For *P. domesticus* we have obtained measures of diet composition which may be applied to populations in northern and central Poland (Pinowska, 1975); southern Poland and Czechoslovakia (Ašmera, 1962); GFR, GDR and Finland (Hammer, 1948); England (Southern, 1945); and the USA (Kalmbach, 1940). Information on *P. montanus* diets is available for locations in northern and central Poland (Pinowski & Wójcik, 1969), southern Poland (Ašmera, 1962), Romania (Simeonov, 1963), and England, the Netherlands and the GDR (Hammer, 1948; Grün, 1964 a, b; T. Brugge, personal communication). In the following discussion we confine our comments to selected locations which represent each of these dietary 'regions'.

Average annual total biomass consumption by *P. domesticus* populations (Table 6.6) ranged from 287 kg dry weight km^{-2} year^{-1} in Missouri, USA, to around 400 kg dry weight km^{-2} year^{-1} at Yördenstorf, GDR and Dziekanów Leśny, Poland. Cereal grains of various types dominated the diets in all locations. Consumption of animal prey (chiefly insects) varied from 9 kg km^{-2} year^{-1} (Missouri) to 96 kg km^{-2} year^{-1} (Kraków, Poland). Consumption of cereal grain was greatest among the populations of northern and central Poland, less in the GDR, and still less in southern Poland and in Finland. *P. montanus* populations generally consumed less

234

Table 6.6. *Annual food consumption of standardized* Passer *populations at selected sites (kg dry weight km^{-2} year^{-1}). Values are means, with standard deviations in parentheses*

Locality	Number of years	Cereal grain	Weed seeds	Insects	Diet data source
		P. domesticus			
Finland					
Kangasala	2	187 (11.7)	80 (8.3)	48 (6.8)	Hammer (1948)
GDR					
Yördenstorf	1	237	111	55	Hammer (1948)
Steckby	0	—	—	—	—
Netherlands					
Oosterhout	0	—	—	—	—
Poland					
Gdynia	1	282	60	13	Pinowska (1975)
Dziekanów Leśny	14	310 (16.9)	73 (7.5)	16 (1.9)	Pinowska (personal communication)
Kraków	7	194 (16.6)	50 (2.3)	96 (14.7)	Ašmera (1962)
Romania					
Iaşi	0	—	—	—	—
USA					
Missouri	1	217	61	9	Kalmbach (1940)
		P. montanus			
Finland					
Kangasala	0	—	—	—	—
GDR					
Yördenstorf	0	—	—	—	—
Steckby	1	148	77	48	Van Balen (personal communication)
Netherlands					
Oosterhout	1	183	88	47	Van Balen (personal communication)
Poland					
Gdynia	1	10	236	4	Pinowski & Wójcik (1969)
Dziekanów Leśny	14	13 (0.3)	256 (8.4)	4 (0.4)	Pinowski & Wójcik (1969)
Kraków	7	99 (3.5)	78 (3.2)	90 (8.0)	Ašmera (1962)
Romania					
Iaşi	1	32	247	43	Simeonov (1963)
USA					
Missouri	0	—	—	—	—

total biomass per year, on the average, than *P. domesticus*. Average annual total biomass consumption ranged from 250 kg dry weight km^{-2} year^{-1} in northern Poland to 322 kg dry weight km^{-2} year^{-1} in Romania. In the Netherlands and the GDR, *P. montanus* consumption was dominated by cereal grains, while in southern Poland, cereal grains, weed seeds, and insects were consumed in roughly equal quantities over a year. Weed seeds accounted for the greatest share of food consumed by populations in

235

J. A. Wiens & M. I. Dyer

Table 6.7. *Breeding season food consumption, percentage annual total for each food type. Values are means, with standard deviations in parentheses*

Locality	Number of years	Cereal grain	Weed seeds	Insects
Finland		*P. domesticus*		
Kangasala	2	25.2 (1.06)	31.9 (0.92)	66.1 (6.01)
GDR				
Yördenstorf	1	35.5	43.0	79.4
Steckby	0	—	—	—
Netherlands				
Oosterhout	0	—	—	—
Poland				
Gdynia	1	24.8	51.1	32.8
Dziekanów Leśny	14	40.7 (5.32)	76.1 (7.65)	74.1 (15.46)
Kraków	7	26.3 (6.12)	22.0 (2.48)	71.4 (10.33)
Romania				
Iaşi	0	—	—	—
USA				
Missouri	1	45.9	56.6	60.9
Finland		*P. montanus*		
Kangasala	0	—	—	—
GDR				
Yördenstorf	0	—	—	—
Steckby	1	50.2	38.4	79.8
Netherlands				
Oosterhout	1	36.5	29.9	74.7
Poland				
Gdynia	1	43.7	27.1	54.8
Dziekanów Leśny	14	86.2 (11.95)	47.1 (2.57)	48.1 (9.59)
Kraków	7	19.5 (6.35)	25.7 (4.44)	69.6 (7.01)
Romania				
Iaşi	1	33.2	41.9	74.0
USA				
Missouri	0	—	—	—

central and northern Poland and in Romania. Consumption of insect prey ranged from 4 kg km^{-2} year^{-1} at Dziekanów Leśny and Gdynia to 90 kg km^{-2} year^{-1} at Kraków. In general, then, *P. domesticus* populations appear to consume rather large quantities of cereal grains over most of their European range (at least as represented by the locations considered here), while the food consumption patterns of *P. montanus* seem more variable and may perhaps be more closely attuned to local variations in food availability, but place greater emphasis on 'natural' food sources.

Consumption of foods should be expected to vary between years at a given location as a result of variations in total energy demands and in food availability patterns. We are unable to assess the importance of the latter, but the model estimations of energetics do permit consideration of annual variations in food consumption as they are related to energy demands. Coefficients of variation for annual consumption of cereal grain are lower than those for annual energy demand at both Dziekanów Leśny and Kraków, and weed seed variation at Kraków is roughly one-third that of energy demand; consumption of weed seeds and insects at Dziekanów Leśny and of insects at Kraków seems just as variable as energy demand.

The key to understanding these interlocality differences in variability of consumption rates of food is in the examination of the seasonal patterns of food consumption (Tables 6.6, 6.7). *P. domesticus* populations at all locations except Gdynia, Poland, consumed most of the insect prey taken during the breeding season, but for the other food types there were considerable differences between locations. Populations at Kraków ate only 22% of the annual weed seed consumption during the breeding season, while birds at Dziekanów Leśny consumed 76% of the yearly total during that time. Populations at Gdynia, with the same specified dietary composition as the Dziekanów Leśny birds (Pinowska, 1975), consumed only 51% of the annual total during the breeding season. Similar differences characterize the consumption patterns of *P. montanus* populations. At Dziekanów Leśny and Gdynia, cereal grains comprised roughly 5% of the total annual food consumption, yet 86% of this total annual cereal grain consumption at Dziekanów Leśny occurred during the breeding season, while only 44% was eaten at that time by the Gdynia population. Cereal grains were the most important constituent of the annual consumption in the Netherlands and the GDR, although 50 to 63% of that total cereal grain consumption occurred during the non-breeding period.

These results reflect the imposition of temporal variations in dietary composition on varying energy demands, creating patterns which are distinctly local in nature. Model plottings of short-term variations in estimated food consumption rates (Fig. 6.6) reinforce this impression. At Yördenstorf, GDR and Kraków, for example, *P. domesticus* consumption of animal prey occurred over a good deal of the year, while the consumption patterns at Dziekanów Leśny and Gdynia were more restricted in time, and insect consumption by *P. montanus* at these two locations was more restricted still. Consumption of cereal grain by *P. domesticus* was spread more or less evenly throughout the year at Gdynia and Kraków, but exhibited distinct late summer elevations at Dziekanów Leśny and Yördenstorf.

238

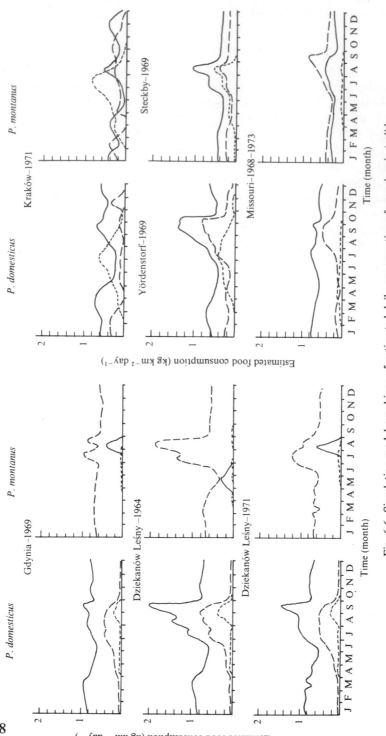

Fig. 6.6. Simulation model graphings of estimated daily consumption of cereal grain (solid line), weed seed (dashed line) and insect prey (dotted line) by *Passer* populations at selected locations and years.

Potential impact in ecosystems

Passer hispaniolensis *(Temminck)*

A population of the Spanish sparrow was studied in Kazakhstan, USSR, by Gavrilov (1962, 1963) over a three-year period. There, this species aggregates in large breeding colonies of several thousand individuals, radiating from these foci in daily foraging flights into neighboring steppe and cultivated areas where they may become major pests. The breeding season is relatively short, with females averaging 1.24 broods. During the period the birds occupy this area, a 'standardized' population of 100 adults plus their offspring requires an estimated 0.39 kcal m^{-2}, of which roughly one-fourth is taken by the young (Table 6.8). Cereal grains dominate the diet (Gavrilov, 1962), and the population consumes an estimated 43 kg km^{-2} during their stay in the area.

We have used the BIRD model to consider the energetics and food consumption patterns of several other granivorous bird species in an attempt to provide some perspective for the *Passer* simulation results (Table 6.8). These analyses are less intensive due to the lack of detailed studies of populations in various portions of the species' range. All of these species are seasonal migrants, and our analyses considered them only in breeding locations.

Agelaius phoeniceus *(L.)*

Red-winged blackbirds are abundant in many marsh, upland and agricultural regions throughout most of North America. In various regions of the USA and eastern Canada they are considered to be serious pests, largely owing to localized depredations of corn or rice crops or to the immense roosting aggregations which form in some areas of the wintering range (Meanley, 1971). Their populations have received especially intensive study in the Lake Erie Basin, where they have an agricultural impact upon corn production. We will describe a detailed model analysis of the dynamics of a specific population in this region shortly; the results summarized in Table 6.8 are for redwing populations in an area of the Lake Erie Basin of central Ohio (Dyer, Siniff, Curtis & Webb, 1973). The biotopes in the Lake Erie Basin range from old-field conditions and currently cultivated hay fields to traditional marshland areas consisting of cattail *Typha latifolia* L. Although no measurements of total production exist for the region, it is generally felt that production is highest in Ohio, somewhat lower in Michigan and lowest in southwest Ontario.

Redwings are considerably larger than the *Passer* spp., and this body size difference is reflected in their higher energy demands. During the time the birds are in this area, their diet consists principally of insects and corn (Hintz & Dyer, 1970). While the available foods have changed somewhat

239

Table 6.8. *Features of breeding biology and breeding season estimated energy demand and food consumption of standardized populations of several granivorous bird species*

Species	Locality	Reported breeding density (individuals km^{-2})	Days length of		Energy demand (kcal m^{-2} season^{-1})		Food consumption (kg km^{-2} season^{-1})					Adult body weight (g)
			Breeding season	Residency on site	Total	Percentage taken by offspring[a]	Insects	Cereal grain	Weed seeds	Wheat–oats	Corn	
P. hispaniolensis	Kazakhstan, USSR	2420	67	170	0.39	22.9	13	43	9	—	—	28.8
Agelaius phoeniceus	Central Ohio, USA	234.4	91	220	1.21	7.7	83	—	41	98	124	70 (♂) 47 (♀)
Quelea quelea	Chad	5000	40	82	0.38	16.7	6	8	31	—	—	19.0
Spiza americana	Kansas, USA	453	110	127	0.21	13.7	27	—	8[b]	—	—	27.0
Spiza americana	Oklahoma, USA	81	60	128	0.45	12.5	57	—	19[b]	—	—	27.0
Eremophila alpestris	South Dakota, USA	68.2	120	365	1.48	4.5	26	—	257[b]	—	—	31.0
Eremophila alpestris	Texas, USA	109.6	120	365	1.33	4.7	35	—	218[b]	—	—	31.0

[a] Nestlings+fledglings, during period of residency on site.
[b] All plant foods combined.

240

in this century, mainly because of intensive agricultural practices, there is little shift in the proportion of animal/plant diet ratios if one compares the present situation with that monitored before vast changes in the region (Beal, 1900).

Quelea quelea *(L.)*

The red-billed dioch ('quelea') forms breeding colonies, with individuals often numbering in the millions, and moves over large areas of sub-Sahara and southern Africa in patterns associated with seasonal rain fronts (Chapter 8, this volume). This brings the breeding period in a given area into conjunction with the development of ripening grain, and during this stage *Quelea* may assume importance as a major pest, especially in areas of millet agriculture. The single situation we considered in model analyses was a population at N'Djamena, Chad, studied by J. Jackson (personal communication); his observations were supplemented with information from the fine studies of Ward (1965 *a*, *b*, 1966, 1971 *a*; see also Katz, 1974). The birds occupy this area for only a short period, and therefore the total energy demands of a 'standardized' population of 100 birds plus offspring are relatively low (Table 6.8). Weed seeds are the major food of the birds during this period, and this is reflected in the overall consumption rates. The consumption of cereal grains is low for this standardized population, but in view of the vast number of birds which may concentrate their feeding in a localized area, the potential impact of an actual population is substantial.

Spiza americana *(Gmelin)*

Dickcissels are breeding occupants of abandoned agricultural fields, pastures and early successional stages of grasslands over much of the eastern Great Plains of the USA (Chapter 8, this volume). They winter in northern South America and Central America (Chapter 1, this volume), and during migrations may form large flocks which feed in agricultural areas (ffrench, 1967). We analyzed two populations: one near Manhattan, Kansas, studied by Zimmerman (1971) and one in northeastern Oklahoma, studied by Wiens (1974 *a*). The two populations differ in reproductive success, and these differences are seen in the total energy demands of the populations during their occupancy of the breeding areas (Table 6.8). The birds consume primarily insects during the breeding season, becoming highly granivorous only during migration and on the wintering grounds. Weed seeds and grass seeds account for most of the consumption of plant material during the breeding season.

241

Eremophila alpestris *(L.)*

Unlike the other species considered here, horned larks do not have significant impacts on agriculture through seed consumption, despite their frequently close association with agricultural habitats. We analyzed information from two populations studied as part of the IBP (USA) Grassland Biome Program (Wiens, 1973). At both sites, one in a mixed-grass prairie in western South Dakota and one in a shortgrass prairie in northwestern Texas, the species is present throughout the year, although there is apparently a complete exchange of individuals, breeding birds migrating south and being replaced in autumn and winter by migrants from the north. Total annual consumption of seeds is fairly great, but virtually all of this consumption is of weed seed and grass seed, even in agricultural areas where other economically important grains are potentially available. Since *Eremophila* is present at these locations throughout the year, their energetics and consumption may be compared with the resident *Passer* populations. Despite the larger body size of the larks, total annual energy demand is somewhat less than that of populations of either *P. domesticus* or *P. montanus*, probably because of the considerably greater reproductive output of the latter species.

Translating food consumption into impact estimates

An estimate of the amount of food of various types consumed by a population during its occupancy of an area provides an initial indication of the *potential* impact of that population. In order to translate this potential into estimates or predictions of actual, realized impact, however, additional information is required. The extent of the impact in either a natural or an agricultural system is related to the availability of the foods. In agricultural ecosystems it is often necessary also to attach an economic value to the loss of yield. Such economic calculations usually involve no more than multiplication of the units of the commodity lost by the monetary value per unit. Such economic calculations are incomplete, however, to the extent that the birds may have other negative or positive effects. If the granivorous bird populations consume pest insects during a small but critical period of the year, for example, these benefits may sometimes outweigh the crop losses to consumption. Collinge (1924–1927) suggested that in fruit growing districts in England, the benefits from insect 'control' by *P. domesticus* far outweighed the harm through seed consumption. while in areas of cereal grain agriculture the birds consumed larger quantities of wheat and the economic impact became negative. In other cases, cereal grain or corn plants suffering moderate consumption by *P. domesticus* or *Agelaius phoeniceus*, respectively, may exhibit com-

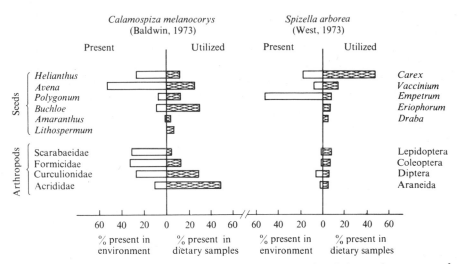

Fig. 6.7. Comparisons of consumption of various prey types in relation to the presence of those prey types in the environment for two granivorous bird species. Utilization of a prey type is measured by its percentage in the samples gathered directly from areas in which the birds forage.

pensatory growth, so that the weight of seeds or ears from 'damaged' plants may actually be greater than that from undamaged plants (Dawson, 1970; Dyer, 1975). Accessory effects of consumption such as this make accurate estimation of *net* economic costs of granivorous bird 'impact' difficult. Such features of granivorous bird impacts in agriculture are discussed in detail in the following chapter.

The dietary habits of a population, of course, have a major influence upon impact potential. The degree to which a population is selective in its feeding (i.e. consumes foods in proportions which deviate from the proportional availability of the foods in the environment) may determine its preferences for agriculturally valuable crops when they become available to the birds. The process of food selection by birds is affected by a wide array of variables (Ellis *et al.*, 1976), but under field conditions selection or 'preference' has most often been recorded as a simple ratio of the proportion of a food type in the diet to the proportion of that food in the environment. Baldwin (1973), for example, related the occurrence of prey taxa in dietary samples of lark buntings *Calamospiza melanocorys* Stejneger, taken in Colorado, USA, in May, to concurrent samples of insect and seed abundance in the prairie sod in which the birds fed. His results (Fig. 6.7) demonstrated a strong positive selection of grasshoppers (Acrididae) and of seeds of *Lithospermum*, *Amaranthus* and *Buchloe*, while the birds consumed seeds of *Helianthus* and *Avena* and ants (Formicidae) and scarab beetles (Scarabaeidae) less frequently than expected

243

from the occurrence of these prey taxa in the environment. West (1973) conducted a similar analysis of food 'selection' by American tree sparrows *Spizella arborea* (Wilson), during the breeding season in northern Manitoba, Canada. Here the birds used 22 of the 53 prey taxa available, appearing to prefer seeds of several *Carex* spp., *Draba*, *Vaccinium* and *Eriophorum* (Fig. 6.7). West's findings were in general agreement with the results of Willson's (1971) laboratory experiments on seed selection in this species (see also Willson & Harmeson, 1973). Pinowski, Tomek & Tomek (1973) likewise found a relatively close agreement between the seed preferences exhibited by *P. montanus* in laboratory tests and its food habits in the field.

The difficulty in evaluating food habits is compounded by daily and seasonal changes in both diets and standing crops. But even if detailed information is obtained, there may be substantial problems, since the abundance of prey types in the environment, as recorded through sampling procedures, may not coincide with the true 'availability' of these prey to the birds (Wiens, 1973). In areas of Africa, for example, *Quelea quelea* feed upon ripening grain of annual grasses (*Panicum*, *Echinochloa*) during the breeding season and continue to feed upon the seed fall from these plants covering the ground during the dry season. With the coming rainy season, the remaining seeds become inaccessible to the birds, forcing dietary shifts and contributing to substantial weight losses (Ward, 1965 *a*; Bortoli, 1969, 1974; Katz, 1974). When annual grains are not available the birds may turn to cultivated crops. Many granivorous species, including several species of *Passer*, *Quelea quelea* and *Agelaius phoeniceus*, exhibit marked preferences for grain or seeds at the 'milky' stage of ripening (Akhmedov, 1949; Ward, 1965 *a*; Dawson, 1970; Hintz & Dyer, 1970; Havlín, 1974), and grain that matures beyond that stage may be less accessible to the birds, even though its apparent 'availability' remains unchanged. In New Zealand, *P. domesticus* prefers to feed well above the ground level, and as a consequence grain ears which are fairly high suffer the greatest losses. Ears below 30 cm are left untouched (Dawson, 1970). In Poland, *P. montanus* feeds chiefly upon weed seeds during the winter, and the distribution and population dynamics of this species may be closely related to the actual availability (rather than abundance alone) of such foods. Large declines in sparrow abundance follow periods of deep snow in central Poland, where the birds are unable to subsist upon the seeds available in farmyards when the snowfall renders supplies in fields inaccessible. In submontane areas of southern Poland, where every year snow reaches a depth of 50 cm and persists for three to four months, tree sparrows are comparatively scarce, and their wintering distribution is closely tied to food provided by man (Pinowski, 1968). There are thus many factors which affect diet composition and consumption rates, and render simple correlations with food 'abundance' in the environment equivocal.

In assessing the impact of granivorous birds upon an agricultural crop, we are actually less concerned with the proportion of that crop that is actually 'available' to the birds than with that proportion that is available to man (i.e. harvestable). Under these conditions, we may assess impact by direct comparisons of avian consumption of a crop with agricultural yields. Unfortunately, there are few instances in which we know the population parameters which influence energy demand, the details of dietary composition of the birds, the true agricultural yields for specific local situations. Here, we consider two examples, more to demonstrate the approach to estimation of agricultural impact than to present precise values.

Pinowski (1973) has observed that *P. montanus* populations are neutral or even beneficial to farming in Poland, while *P. domesticus* may cause considerable damage to ripening crops of wheat, barley and sunflowers. In the Żuławy region of northern Poland the damage caused by *P. domesticus* is so great that winter barley is no longer an economic agricultural venture. At Dziekanów Leśny in central Poland, Pinowski & Wójcik (1968) measured the seed production of various weed seeds and of rye and oats in fields utilized by *P. montanus* flocks during 1967. Total annual seed production of the five commonest weed species eaten by the sparrows was estimated to be 12305 kg km^{-2}, of rye, 21314 kg km^{-2}, and of oats, 1178 kg km^{-2} (combined grain = 22492 kg km^{-2}). Estimates of annual consumption of weed seeds and cereal grain for a standardized population of 100 adults plus their young at Dziekanów Leśny for 1967 were 259 kg km^{-2} and 13 kg km^{-2}, respectively. *P. montanus* populations in this area normally number in excess of 500 individuals km^{-2}. Converting the model estimates to this density level, we calculate that the 'impact' of this species on weed seeds is 10.5% of the annual production, on cereal grain 0.3% of annual production. This of course represents an impact averaged over the entire year, and since the consumption of seeds by *P. montanus* shows a sharp seasonal peak (Fig. 6.6), the actual percentage impact during periods of high consumption may be substantially greater. The period of greatest consumption occurs in early autumn, coinciding with the production of a fresh seed crop by the weeds.

While it is not quite appropriate to apply values obtained for *P. montanus* to *P. domesticus* populations, since they forage in different areas, a comparison may nonetheless be instructive. Assuming a population of *P. domesticus* equivalent to that of *P. montanus*, we may convert the model estimates of weed seed and cereal grain consumption by a standardized population (81 and 323 kg km^{-2} year^{-1}, respectively) to values for a popu-

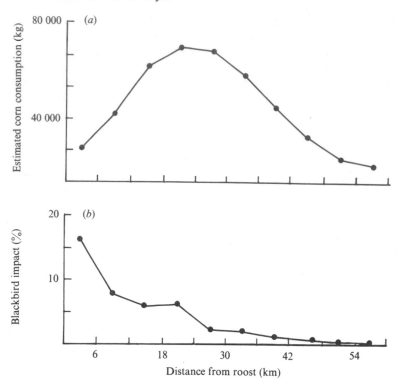

Fig. 6.8. Model-derived estimates of red-winged blackbird impact upon corn crops in the northeastern USA. In (*a*) the estimated consumption of corn by a post-breeding assemblage of blackbirds is projected as a function of increasing distance from the roost epicenter, within a 10° sector. (*b*) depicts the 'impact' (= consumption/yield × 100) of these blackbirds upon corn crops with increasing distance from the roost. See text (modified from Wiens & Dyer, 1975*a*).

lation of 500 individuals km⁻² and relate these to the same weed and cereal grain seed production measures. If *P. domesticus* were to feed upon these foods in these areas, projected 'impact' would be 3.3 % of the annual weed seed production and 7.2 % of the cereal grain production in 1967.

In another application of this modelling approach to estimating potential impact of granivorous birds, we considered the dynamics of a population of red-winged blackbirds during the late summer–early autumn post-breeding period in northern Ohio, USA (Wiens & Dyer, 1975 *a*). Here, adults and young-of-the-year from a large area gather in roosting assemblages in marshes along Lake Erie and disperse from the roosts in daily foraging flights into surrounding agricultural areas. It is during these foraging flights that the birds encounter vast, monocultured fields of maturing corn and have their greatest agricultural impact. Blackbirds in

246

this region are considered a major agricultural pest and are the object of intensive 'control' programs.

In our simulation, we considered the daily dispersal of roughly 232 000 birds from their roost in a 10° sector ranging out to 60 km (Fig. 6.8). Observations elsewhere led us to believe that the birds were distributed among the 10 concentric zones of this sector following a Poisson distribution. We then used the BIRD model to estimate energy demands and food consumption rates of the birds foraging in each of the zones from late July until the end of September, when the birds leave the marsh roost and begin their southward migration. Over this 68-day period, the birds consumed estimated totals of 244 437 kg insects, 470 139 kg corn, 662 396 kg 'detritus' in wheat–oat fields, and 169 560 kg weed seeds in the total area of the sector (314 km²). Insects and wheat–oat 'detritus' were the most important foods early in the post-breeding period, but when the corn crop matured there was a marked shift to this food, which continued until the crops were harvested. Since the birds were distributed unequally among the zones, consumption was uneven. The spatial patterns of corn consumption (Fig. 6.8 a) indicate that consumption was relatively low close to the roost and in outlying areas; 35 % of the estimated corn consumption was in the area 18 to 30 km from the roost, 65 % in the 12 to 36 km zones. Consumption was low in areas close to the roost partly because of the relatively small area involved but also because only about 15 % of the birds forage there. Farmers occupying areas close to this roost (which is a 'traditional' assembly area) have altered their agricultural practices to raise crops which are relatively unattractive to the birds, such as soybeans, thereby forcing the birds to forage farther from the roost.

Using information on crop acreages and yields we can estimate the amount of corn present and at least *potentially* available to the birds in each zone of the sector. Blackbird 'impact', estimated by calculating consumption as a percentage of yield, is greatest in areas close to the roost (despite the responses of the farmers); it remains relatively even at 6 to 8 % of the yield 6 to 24 km from the roost, and then declines to levels of less than 1 % of the yield beyond 42 km (Fig. 6.8 b). In general, these model-derived impact estimates agree rather closely with direct *post facto* field assessments of percent loss (US Fish and Wildlife Service, unpublished data).

Our exercise with blackbirds also serves to illustrate the importance of spatial distributions of consumption and impact. Other granivorous birds, especially those which feed in flocks, generally do not distribute their feeding at all evenly over the area occupied by the population. Flocks may concentrate their feeding in a few fields where the grain is accessible or at the proper stage of softness. As a result of this extreme patchiness in feeding dispersion, fewer than 5% of the fields in a region may bear 95% of the overall damage (Dyer, 1967). Flock feeding and the consequent economic losses are similarly of a local nature in areas subject to quelea foraging. In this species, the degree to which feeding is concentrated in agricultural areas is related to the availability of suitable areas of annual grasses, and there may thus be substantial local variation in the agricultural impacts of this species (Bortoli, 1974). Both *P. domesticus* and *P. montanus* prefer to forage around the edges of fields, with *P. montanus* restricting its movements to open areas more than *P. domesticus*. These behavior patterns result in a concentration of feeding activity about the periphery of fields (Akhmedov, 1949; Dawson, 1970; Havlín, 1974), and small fields may therefore suffer greater damage than large fields. Other examples could be cited, but it is apparent that damage or impact is *local* in its expression, and model estimates (or field assessments) derived from large-scale regional studies should be applied to local situations with caution.

Energetics and food consumption of granivorous bird communities

The foregoing analyses have dealt with populations of single species in various situations. But in the real world, populations of different bird species, with differing predilections toward granivory, occur together in combinations which vary in time and space. Perhaps the most conspicuous of these species combinations are the immense assemblages of blackbirds (Icteridae) and starlings which form about roosting locations in southern USA during the winter (Meanley, 1971). But granivorous species may constitute a functional group or 'guild' (Root, 1967) in many ecosystem types and may thus become 'important' not only through consumption of economically valuable commodities but also through some of the other dimensions detailed at the outset of this chapter. Here we will consider some features of total assemblages or 'communities' of granivorous birds in several agricultural and natural ecosystem types in North America.

Granivorous birds in Illinois agricultural habitats (biotopes)

In 1907–1909, Forbes & Gross conducted a comprehensive avifaunal survey of Illinois in which they systematically counted the number of individuals in bird populations during summer and winter in all major habitat types in northern, central and southern sections of the state. In 1957–1958, Graber & Graber (1963) repeated the survey, using the same methods in much the same areas. Their overall aim was to document avifaunal changes occurring in the state over the 50-year period and to associate these with changing land-use patterns. Their study provides a

248

valuable and detailed base which we may use to explore some features of assemblages of granivorous birds in different agricultural situations and in different sections of the state at the times of the two surveys.

The surveys encountered a great many species overall, but in order to make model analyses manageable, we chose to restrict our attention to 24 species which contributed significantly to the total densities in at least some habitat types. These species could be characterized broadly as granivorous, in that more than 75% of their annual diet was normally composed of seeds. We then used the summer and winter density values for these species provided by Graber & Graber (1963), together with information on temperature and photoperiod regimes for the three sections of the state and generalized information on migratory phenology, to conduct BIRD model simulations of energy demands and food consumption of the combination of species occurring in each census. We lacked detailed information on reproductive intensity and success, such as was available for the *Passer* analyses, and therefore excluded reproduction from our analyses; the model estimations are therefore conservative. General information on the dietary habits of the species, in terms of the proportions of plant and animal material in the diet at various times of year, was obtained from Martin, Zim & Nelson (1961).

Community attributes

We conducted model analyses of granivorous assemblages in small grain cropland (chiefly wheat and/or oats), fallow fields undergoing secondary succession, mixed hay fields, corn croplands and pastures (Table 6.9). Ten to nineteen granivorous species occurred in these agricultural types during the year, with the greatest number of species generally present in pastures and the lowest number in wheat–oat fields. The number of granivore species was remarkably stable in corn fields, both over regions of the state and between the 1909 and 1958 censuses. In general, the number of granivorous species varied little over the 50-year period in most habitats, with the exception of hay fields in northern Illinois, which experienced a considerable decrease in granivorous species. However, the species which numerically dominated the avifaunas differed between habitats, regions and years. We must briefly summarize the major patterns for the different agricultural types, since they provide a perspective for viewing the energy demand and food consumption patterns to be discussed below.

Granivores constituted a high percentage of all birds present in northern Illinois wheat–oat fields, and most of these individuals were summer residents. In the southern section of the state, there was a large increase in total densities of granivores over the 50-year period and an accompanying shift to predominance of summer residents. This was largely due to increases in populations of summering red-winged blackbirds and

Table 6.9. *Community attributes of granivorous species in the avifaunas of various agricultural habitats in Illinois (from Graber & Graber, 1963)*

Agricultural type	Zone	Year	Granivorous species present (no.)	Avifauna granivorous[a] (%)	Density of granivorous species (individuals km^{-2})	Granivorous individuals present (%) Summer	Winter	Predominant species[b]
Wheat–oats	North	1909	14	90.5	398	70	30	Lapland longspur (45)
		1958	10	95.5	420	70	23	Red-winged blackbird (51) Horned lark (20)
	Central	1909	13	78.7	238	39	61	Horned lark (30)
		1958	11	65.5	393	49	51	Starling (33) House sparrow (20)
	South	1909	11	73.4	300	31	69	Bobwhite (33) Meadowlark (28)
		1958	11	70.6	503	91	9	Red-winged blackbird (55)
Fallow	North	1958	10	76.0	712	81	19	Horned lark (36)
	Central	1958	11	79.5	845	84	16	Dickcissel (32)
	South	1909	16	48.1	348	71	29	Eastern meadowlark (35)
		1958	15	64.6	967	34	66	Bobwhite (33)
Hay	North	1909	19	83.5	509	97	3	Bobolink (26) House sparrow (22)
		1958	12	78.5	1340	93	7	Red-winged blackbird (41)
	Central	1909	12	78.2	509	97	3	Eastern meadowlark (26) Dickcissel (24)
		1958	10	93.3	2453	80	20	Red-winged blackbird (25) Dickcissel (21)
	South	1909	12	76.5	482	59	41	Eastern meadowlark (50)
		1958	10	78.2	941	72	28	Eastern meadowlark (37) Red-winged blackbird (32)
Corn	North	1909	13	71.7	131	92	8	Common grackle (25)
		1958	14	76.5	214	90	10	Horned lark (55)
	Central	1909	14	65.0	143	65	35	Common grackle (19) American tree sparrow (19)
		1958	14	89.4	263	31	69	Horned lark (41) Lapland longspur (39)
	South	1909	15	37.2	357	25	75	Dark-eyed junco (33) American tree sparrow (28)
		1958	14	25.5	434	30	70	Dark-eyed junco (54)
Pasture	North	1909	18	68.9	510	71	29	House sparrow (20)
		1958	17	78.4	566	88	12	Red-winged blackbird (23)
	Central	1909	18	68.0	553	67	33	American goldfinch (16) House sparrow (41)
		1958	18	78.2	555	84	16	Eastern meadowlark (20)
	South	1909	15	48.4	334	71	29	Eastern meadowlark (33)
		1958	14	50.4	611	64	36	Dark-eyed junco (31) Eastern meadowlark (23)

[a] Total individuals, summed over entire year.
[b] Proportion of total density in parentheses; only species contributing $\geqslant 20\%$ of total individuals are listed.

grackles, which also increased significantly over this period in the northern parts of Illinois.

Fallow fields were not plentiful enough in the northern and central portions of Illinois to justify a census in 1909. In the southern section, there was nearly a three-fold increase in total density of granivores over the 50-year span, accompanied by an increase in the proportion of the total avifauna contributed by granivores and a shift from dominance by breeding species to wintering forms. In northern and central sections of the state, summering individuals predominated and a somewhat greater proportion of all individuals present were granivorous. The data suggest that fallow fields in the southern part of the state may be important wintering habitat for granivorous birds.

Granivorous birds accounted for 76 to 93% of all individuals present in hay fields in Illinois, and in the northern and central portions of the state most of these individuals were present only during the summer. Total granivore densities more than doubled in all sections between 1909 and 1958, and this increase seems largely associated with tremendous increases in the abundance of red-winged blackbirds in this habitat.

While species numbers remained stable in corn fields over Illinois, total densities increased in all sections over the 50-year survey span. In the northern section of the state, granivores numerically accounted for roughly three-fourths of all birds present, most occurring during the summer. These attributes were reversed in the southern areas, where most granivores were present only in the winter and granivores as a group constituted only one-quarter of the individuals present. Total densities of granivores were generally lower than in the other agricultural types.

Granivorous birds contributed roughly 70% of all individuals present in northern and central Illinois pastures and around 50% in southern areas. Total densities of granivores were very similar in northern and central sections and were stable, but in the southern section, densities almost doubled between 1909 and 1958. In the southern area, roughly one-third of the birds were present in the winter only.

Several general trends are apparent from these data. Total densities of granivores are frequently greater in the southern part of the state, which also receives greater use as wintering habitat. Granivore densities increased markedly in different habitat types in several sections of the state over the 50-year period, and in several instances a substantial share of this increase was by red-winged blackbirds. Granivorous species appear to predominate numerically in the avifaunas of wheat–oat fields and hay fields to the greatest degree, but fallow fields and hay fields support larger granivore populations, especially in recent times.

Granivore energy demands

Estimations of the collective energy demands of all granivores present in these habitats and regions, derived from BIRD model analyses, ranged from 0.96 kcal m^{-2} year^{-1} (northern Illinois, corn, 1909) to 15.21 kcal m^{-2} year^{-1} (central Illinois, hay fields, 1958) (Fig. 6.9). Total annual energy demands were greatest in hay fields and fallow fields in the 1958 surveys and lowest in corn fields. Variation in energy demands over the 50-year

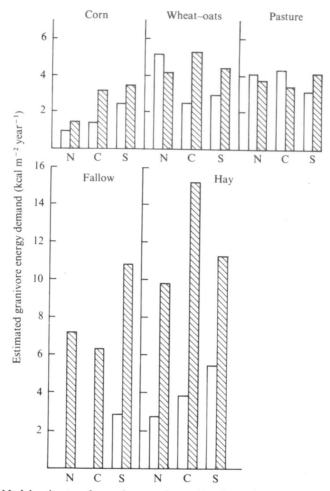

Fig. 6.9. Model estimates of annual energy demand by the total 'community' of granivorous birds occupying various agricultural habitats in northern (N), central (C) and southern (S) Illinois, 1909 (clear) and 1958 (hatched).

period was quite low in corn and pasture habitats and was expressed in moderate increases in recent times in wheat–oat fields in the central and southern sections of the state. Fifty-year changes were great in hay fields in all regions of the state and in fallow fields in the southern sections (early surveys from the central and northern sections were not available). It is in these situations that blackbird densities increased so dramatically between 1909 and 1958, and it is tempting to infer a causal relation between these observations. While a substantial share of the increase in 'community' energy demand may indeed be due to the presence of more

Table 6.10. *Ratios of estimated annual consumption of seed: animal prey by granivorous birds in Illinois agricultural habitats*

Agricultural type	North		Central		South		Overall average[a]
	1909	1958	1909	1958	1909	1958	
Corn	2.20	2.19	3.45	2.78	6.46	1.76	3.14 (1.73)
Wheat–oats	3.82	3.15	3.06	1.68	2.18	1.97	2.64 (0.83)
Pasture	2.28	1.15	2.17	1.62	1.34	1.07	1.61 (0.52)
Fallow	—	3.11	—	3.50	1.02	4.34	2.99 (1.41)
Hay field	1.30	1.56	1.02	1.70	0.78	1.38	1.29 (0.34)
Regional averages[a]	2.40 (1.05)	2.23 (0.90)	2.43 (1.08)	2.26 (0.85)	2.36 (2.35)	2.10 (1.30)	

[a] Standard deviation in parentheses.

blackbirds, other granivorous populations also underwent changes during this time period, so causal relationships are equivocal.

Regionally, the energy demands of granivorous birds as a group varied little in pasture and wheat–oat habitats and exhibited a slight increase toward the south in corn fields. Central Illinois hay-field populations required considerably more energy than those in other sections of the state in 1958, while in 1909 there was a general increase in demands of hay-field assemblages from north to south.

Regional food consumption patterns

The energy demands of these groups of granivorous species were translated into food consumption rates following the same procedures applied in the *Passer* population analyses. While our analysis was restricted to those species which we arbitrarily designated as granivorous, there was some variation among these species in their degree of granivory, and another way of examining the extent of granivory in the various agricultural situations in Illinois is through calculation of ratios of model-estimated rates of seed consumption to animal prey consumption (Table 6.10). In general, the highest ratios (greatest proportional seed consumption) were recorded in corn fields, fallow fields, and wheat–oat areas, with ratios in pastures and hay fields markedly lower. Ratios generally tended to be higher in the northern and central sections of the state than in the south, but there was considerable variability among habitat types. Ratios in most

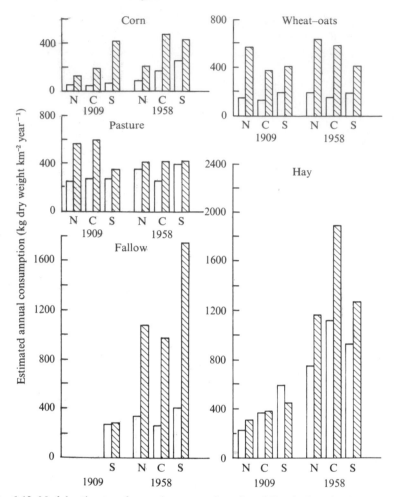

Fig. 6.10. Model estimates of annual consumption of seed (hatched) and animal (clear) prey types by the total 'community' of granivorous birds occupying various agricultural habitats in northern (N), central (C) and southern (S) Illinois, 1909 and 1958.

situations remained relatively unchanged between 1909 and 1958, although in the fallow fields of southern Illinois there were substantial increases in the degree of granivory over the 50-year interval.

The patterns of estimated annual consumption of seed and animal prey types (kg dry weight km^{-2} year^{-1}) are depicted in Fig. 6.10. In general, they parallel the overall patterns of energy demands, as should be expected. In hay fields, seed consumption rates were greater in all areas in 1958 than in 1909. Annual seed consumption was relatively low in corn fields in all situations, in hay fields in 1909 and in pastures in 1958. Seed

254

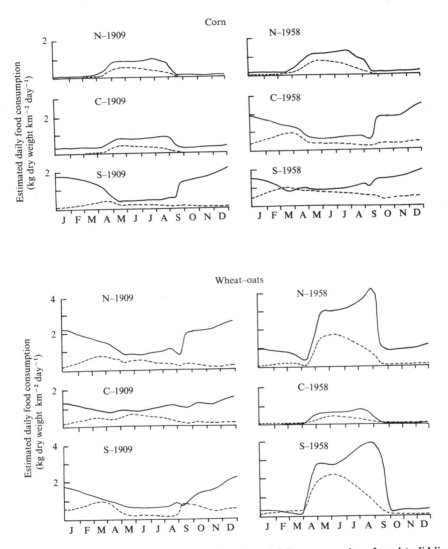

Fig. 6.11. Simulation model graphings of estimated daily consumption of seed (solid line) and animal (dashed line) prey types by the 'community' of granivorous birds occupying several agricultural habitats in northern (N), central (C) and southern (S) Illinois, 1909 and 1958.

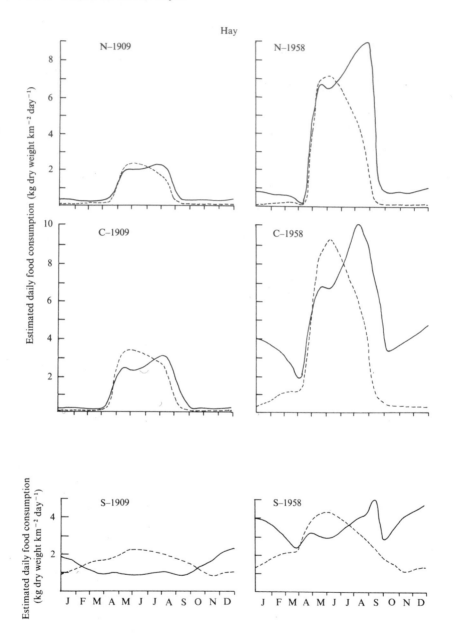

Fig. 6.11 (*cont.*). Explanation on p. 255.

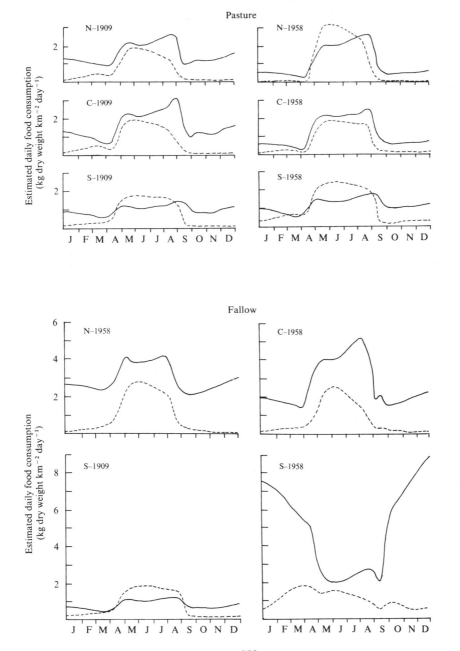

Fig. 6.11 (*cont.*). Explanation on p. 255.

consumption rates were greatest in hay fields in 1958, especially in the central section of the state, and in southern Illinois fallow fields in 1958.

As with the analyses of *Passer* model output, we can consider the seasonal distribution of food consumption among these granivore assemblages, although in this case summer increases in consumption are not associated with reproduction but with seasonal changes in the abundance and/or species composition of granivores. Model projections of seed and animal consumption rates are shown in Fig. 6.11. In wheat–oat habitats, consumption rates of plant and animal prey exhibited relatively little seasonal variation in all sections of the state in 1909, but 50 years later summer consumption was substantially greater than in winter, at least in the northern and southern areas. Seed consumption was greatest in early autumn, and insect consumption greatest in late spring and early summer. Food consumption in pastures was greater during the summer in all areas, and while on an annual basis seed consumption always exceeded insect consumption, there were several periods during which insect consumption was greater than that of seeds. Seasonal food consumption was greater in summer than winter in northern Illinois corn fields, but in other sections there was relatively little seasonal variation in consumption rates. Consumption of both seeds and insects was concentrated into summer months in fallow fields in northern and central Illinois, but in the southern section of the state winter consumption greatly exceeded that during the summer, at least in 1958. This, of course, is consistent with the heavy use of this area as wintering habitat by granivores, noted above. In hay fields, consumption patterns in the southern part of the state were relatively stable seasonally, especially for seeds. Elsewhere, however, nearly all of the annual consumption was confined to the summer, with insect consumption peaking early in the summer and seed consumption toward late summer and autumn. The seasonal peaks during 1958, when total consumption was very high in this agricultural type, were especially well defined.

Statewide consumption patterns

Graber & Graber (1963) obtained rather general approximations of the total area of each agricultural type over the entire state during 1907–1909 and 1957–1958. While these values are not precise, they provide a basis for extending the model results to estimate total annual consumption of insects and seeds in each of the habitat types over the entire state (Table 6.11). The area in agricultural use for the five types we considered decreased by over 40% between 1909 and 1958, largely as a result of decreases in area of pasture, wheat–oats and corn. During the same period, total granivore abundance increased markedly in hay fields, slightly in corn and fallow fields and decreased substantially in pastures. These changing

258

Table 6.11. *Estimated energy demands and food consumption rates (dry weight) of granivorous birds in several agricultural habitats in Illinois, 1909, 1958*

Agricultural type	Year	Total area (km²)	Average energy demand (kcal m⁻² year⁻¹)	Total granivore density (individual ×10⁶)	Animal foods Average (kg km⁻² year⁻¹)	Total over state (metric tonnes year⁻¹)	Seeds Average (kg km⁻² year⁻¹)	Total over state (metric tonnes year⁻¹)
Corn	1909	42087	1.596	8.9	60	2525	246	10353
	1958	33994	2.725	10.3	169	5744	368	12510
Wheat–oats	1909	26304	3.520	8.2	152	3998	448	11784
	1958	17401	4.618	7.6	215	3741	596	10371
Pasture	1909	24686	3.879	11.5	262	6467	504	12441
	1958	8094	3.791	4.7	334	2703	415	3359
Fallow	1909	6070	2.875	2.1	277	1681	284	1723
	1958	4047	8.105	3.4	345	1396	1268	5132
Hay field	1909	10100ᵃ	4.010	5.1	399	4030	382	3858
	1958	10100ᵃ	12.064	15.9	929	9383	1445	14595
Total	1909	109265	—	—	—	18701	—	40159
	1958	62726	—	—	—	25967	—	45967

ᵃ Estimated.

259

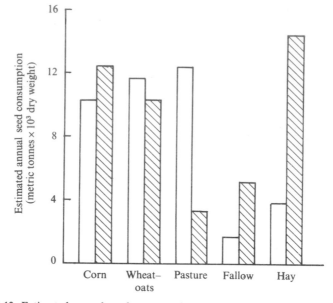

Fig. 6.12. Estimated annual seed consumption by granivorous birds over the total areas of several major agricultural types in Illinois, 1909 (clear) and 1958 (hatched). Values are statewide totals for the entire area planted in the indicated crop. Areas (km²) of each agricultural type are given in Table 6.11.

patterns of land use and avian abundance contribute to the variations in estimated statewide seed consumption (Fig. 6.12). In 1909, projected seed consumption was high and relatively similar (*ca* 10 000–12 500 metric tonnes year⁻¹) in pasture, wheat–oat fields and corn, and relatively low in hay fields and fallow fields. Fifty years later, overall seed consumption was relatively unchanged in corn fields (despite the considerable reduction in area under cultivation), lower in wheat–oat fields and pastures, somewhat higher in fallow fields, and had increased by over 250% in hay fields, despite the lack of any real change in the area of hay fields in Illinois. In all five agricultural types combined, total annual statewide seed consumption increased slightly over the 50-year interval, despite the substantial decrease in area devoted to these crops as the cultivation of soybeans increased.

If reliable information on regional or statewide levels of agricultural production in these habitat types was available for the two time periods, we could attempt a crude assessment of potential impact by comparison of these consumption rates with production values. Unfortunately, the information which is available does not really fit into this sort of analysis.

The bird populations which inhabit relatively undisturbed native eco-systems also exhibit varying degrees of granivory, and some analyses using this same simulation model conducted under the auspices of the IBP (USA) Grassland Biome Program permit a general assessment of the degree of granivory expressed in six grassland–shrub–steppe locations in central and western USA (Wiens, 1973, 1974 *a*, *b*; Wiens & Dyer, 1975 *b*) and in six coniferous forest stands located along an elevational–exposure gradient in the Cascade Mountains of western Oregon (Wiens & Nussbaum, 1975). The analyses were restricted to the populations present during the breeding season (a total of 190 days in the coniferous forests, 150 days in the grasslands) and include the costs associated with reproduction and recruitment of young.

Breeding avifaunas in grasslands are composed of a relatively small set of species (Table 6.12) (Wiens, 1973), and total breeding densities are low, in the order of 150 to 350 individuals km^{-2}. Estimated energy demands of these avifaunas varied considerably during the three years of study at each location, encompassing a range of 0.89 to 2.92 kcal m^{-2} $season^{-1}$; mean values were lowest in the arid Palouse shrub–steppe of eastern Washington, in lightly grazed shortgrass prairies in northwestern Texas, and in heavily grazed mixed-grass prairies in western South Dakota. The data show a tendency for energy demands of the avifauna to increase with increasing precipitation and primary production rates, but the highest breeding season energy demands were those of the bird populations in a heavily grazed Texas shortgrass prairie which had low precipitation and production rates. Overall, the energy flow through breeding bird populations in North American grasslands appears to be rather variable, within limits, and more sensitive to the intensity of grazing pressure applied to a habitat than to differences in habitat types.

Seeds constituted 4 to 24% of the estimated food consumption of the birds over these grassland sites, with the lowest proportions of granivory occurring in the arid Palouse shrub–steppe and in the lightly grazed mixed-grass prairie, and the highest proportions in the heavily grazed mixed-grass prairie area roughly 1 km away. Our dietary information from these locations permits separation of seed consumption into grass seed and herbaceous vegetation ('forb') seed components (Table 6.12).

Total seed consumption rates by breeding avifaunas varied from 9.8 to 69.7 kg km^{-2} $season^{-1}$, with the populations in tallgrass and mixed-grass prairies consuming greater quantities of grass seeds, those in shortgrass locations consuming a greater proportion of forb seeds. In comparison with the single species population consumption estimates given in Table 6.8, we note that a standardized population of 100 horned larks in South

261

Table 6.12. *Community attributes, estimated energy demand and estimated seed consumption by total avian breeding communities in several grassland and coniferous forest habitats in North America*

Habitat type	Number of species	Total breeding density (individuals km⁻²)	Energy demand (kcal m⁻² season⁻¹)	Percent energy demand from seed foods	Seed consumption (kg dry weight km⁻² season⁻¹) Grass seeds	Forb seeds
Grasslands						
Tallgrass prairie, lightly grazed (Oklahoma)	4	304	1.72	17	57.2	0.2
Mixed-grass prairie, lightly grazed (South Dakota)	5	204	1.50	7	18.1	1.4
Mixed-grass prairie, heavily grazed (South Dakota)	3	154	1.21	24	44.5	19.1
Shortgrass prairie, lightly grazed (Texas)	3	168	1.18	15	9.2	19.9
Shortgrass prairie, heavily grazed (Texas)	3	343	2.03	19	25.1	44.6
Palouse shrub-steppe, ungrazed (Washington)	3	203	1.16	4	7.8	2.0
Coniferous forests						
Low elevation, dry, Douglas fir	12	1779	10.73	16	291.6	
Low elevation, mesic, western hemlock	12	1380	10.49	29	501.7	
Low elevation, moist, western hemlock	12	2619	16.65	16	455.0	
Intermediate elevation, mesic, western hemlock/Pacific silver fir	15	2887	20.76	25	861.7	
High elevation, mesic, Noble fir	7	1910	12.26	17	341.7	
High elevation, dry, Pacific silver fir/mountain hemlock	13	1229	12.15	43	863.3	

Dakota or Texas consumes a greater quantity of seeds during a year than the entire complex of species breeding in these locations (which includes horned larks) consumes during the breeding season alone. The standardized *Passer hispaniolensis* population consumed an estimated 52 kg km^{-2} seeds during its 170-day occupancy of a breeding locality in USSR.

Species numbers and densities are substantially greater during the breeding season in coniferous forests than in grasslands (Table 6.12). As a result, estimated energy demands of the avian community during the 190-day breeding season were an order of magnitude greater than those in the grassland sites. Community energy demands in the coniferous forest habitats were lowest in the low elevation, dry-to-mesic stands, substantially higher in the wet, low-elevation stand, and highest in a 'transitional' stand located at an intermediate elevation. The proportion of the total seasonal energy demand obtained from seed sources was generally only slightly higher than that characterizing the grassland avifaunas but increased dramatically in the dry, high-elevation conifer stand. This stand also supported the lowest proportion of full-year resident species of any of the conifer stands. Wiens & Nussbaum (1975) suggested that these patterns might reflect the effects of general deterioration of environmental conditions (especially insect prey availability) during the early portion of the breeding season with increasing elevation and/or dryness. Eventually, a point may be reached at which resident species find it difficult to exploit the system, due to this early spring 'bottleneck' in resource conditions, and highly opportunistic granivores may be able to utilize resources more effectively. Evening grosbeaks *Coccothraustes vespertinus* (Cooper) and red crossbills *Loxia curvirostra* L., two granivorous species known for their erruptive and opportunistic population movements, together accounted for 40% of the energy demands in the dry, high-elevation stand, but were absent in the other coniferous forest locations.

Consumption rates of seeds (primarily conifer tree seeds) during the breeding season in the coniferous forest ranged from roughly 300 kg km^{-2} in the dry, low-elevation stand to 860 kg km^{-2}, in both the intermediate elevation 'transitional' stand (in which energy flow was high but the proportion derived from seeds intermediate) and the dry, high-elevation stand (with 40% less energy flow but a considerably greater reliance upon seed food sources). Unfortunately, we have no information on seed production rates or standing crops in either the grassland locations or the coniferous forest stands, so assessment of even potential impact is not possible.

J. A. Wiens & M. I. Dyer

Overview

The model-based, indirect approach to estimating 'impact' we have followed here marshals information on population dynamics, individual bioenergetics and diet composition to project population energy demands and rates of food consumption, which must then be related to 'supply'. These projections are only general, and the overall utility of this approach depends upon the accuracy of the estimations and their applicability to specific situations. The preliminary assessments we have presented here have provided some insights in this regard and have revealed several areas in which our knowledge of granivorous bird–crop relations is shallow. These areas should be the focus of future studies.

The accuracy of a modelling approach is largely a function of the model structure itself and the detail of the various input values. The BIRD model we employed was originally developed to provide general assessments of population energetics for ecosystem-level analyses (Wiens & Innis, 1974). The structure does not permit consideration of events occurring within a day, and while the simulation calculations are made on a daily basis, close interpretation of day-to-day changes in model outputs is probably unjustified as well. Further, several features of population dynamics, phenology, or bioenergetics are treated in the model structure using rather general, non-specific functions. This generality was intentional in the model development but necessarily carries with it some sacrifice of detail. Thus, an approach based upon a simulation model such as BIRD may provide reasonable estimates of energy demands over time periods of modest duration (e.g. weeks, months, seasons) but cannot be applied to short-term situations in which highly detailed results are required.

The accuracy of model predictions is even more strongly influenced by the input variables. The BIRD model requires specification of some 40 input variables, and while information on some may be readily obtainable, others are more demanding. We have indicated some constraints which have been imposed by the availability of accurate input data in the above discussions, but two features emerge as being especially important to the model analyses and deserving of closer study in field situations. A major determinant of population energy demands is population size in the area under consideration. Valid density estimates are not easily obtained, and the problems of density determination are especially acute for populations with highly aggregated dispersions. Unfortunately, such dispersion patterns characterize many granivorous bird species, especially during the times in which their potential agricultural impact is most severe, and we are frequently left with only very general *ad hoc* approximations of densities with which to work. Further, converting the model-derived energy demand estimates into consumption rates requires information on

264

the dietary habits of a population, and here also our knowledge is usually inadequate to permit detailed and accurate estimations. The diets of most species have been studied as if they were species attributes, and while virtually all careful studies of food habits have indicated important variations in time and space, such information is available for few populations, most of them insectivores. If we are to construct an approach which will allow prediction of impact on the basis of projected potential, we must know how the diet of a local population is composed and how it varies.

We have noted previously that the distribution of foraging individuals in a locality is often neither random nor uniform, but strongly concentrated in space. Since model projections of energy demand or food consumption are made on a per unit area basis, these projections must be adjusted to portray the actual utilization of a unit area by the birds. This is a function of the social organization and foraging behavior of the birds and of the types and interspersion of habitats characterizing the area. Application of the general estimations produced by model analysis to a local situation requires that we know where feeding occurs, for how long, and what factors contribute to variations in these patterns.

A modelling approach provides estimates of impact *potential*, as measured by the consumption rates of various foods during various periods of an annual cycle. Given that these estimates are accurate and are, in fact, applicable to a specified situation, it still may be difficult to convert them into estimates of *actual* or *realized* impact. This requires that we relate consumption to 'availability' for each of the food types of interest. Quite apart from the problems associated with determining what is actually available to the birds, there exist relatively few measures of the abundance or standing crop of food types other than those of direct agricultural importance. Ideally, we should determine not only the supply of each of the major food types present in the areas in which the birds actually feed at several points in time, but also the renewal rates of the prey resource pools.

It is apparent that we have only begun to understand 'impact' through a model-based approach. There are substantial gaps in critical areas of knowledge, but by identifying at least some of these our approach may serve to focus attention on some neglected areas of field studies of granivorous bird populations. Ultimately, of course, it must be determined whether a given level of 'impact' is important in any of the dimensions outlined at the outset of this chapter. If 'importance' is gauged only in economic terms, then we must have much better sources of information to establish the full set of economic associations, including marketing, cost–benefit ratios and a host of other agricultural inputs. But if 'importance' includes both economics and aesthetics, as we suggested earlier, then we must learn to make decisions using both quantitative and abstract parameters in order to determine the true nature of impact. We believe that somehow impact must be considered in a broader context than that of the simple monetary loss apparent in food consumption. Only when a complete assessment of the contributing factors has been made can decisions regarding the most appropriate strategies of managing granivorous bird populations be made, as detailed in the following chapter.

This endeavour would not have been possible without the enthusiastic support and gracious assistance of many students studying granivorous birds throughout the world. In particular, the following people provided unpublished information, assisted in literature syntheses, contributed to development and application of the BIRD model and/or participated in conceptual development, and to all we are personally grateful: T. Anderson, F. Balát, J. H. van Balen, C. H. Blem, R. Boyd, P. Clausing, J. B. Cragg, D. G. Dawson, V. R. Dol'nik, C. Ebling, J. Jackson, V. Jovanovic, C. Kaatz, W. Keil, R. B. Mason, J. A. L. Mertens, L. Mistry, R. M. Naik, C. North, H. Oelke, B. Pinowska, J. Pinowski, P. Rassi, J. Rékási, P. Ward, J. Weiner, A. Wendtland, G. C. West, M. Wieloch, R. L. Will and J. L. Zimmerman. Financial support was provided by the USA National Science Foundation through Grant No. GB-42741 to J. A. Wiens and through Grant No. GB-42700 to M. I. Dyer, and through Grants GB-41233X, BMS73-02027 AO2, and DEB73-02027 AO3 for the support services of the Natural Resource Ecology Laboratory, Colorado State University, and by the Smithsonian Institution which played a pivotal role in supporting international exchanges. This is contribution No. 51 of the Behavioral Ecology Laboratory, Oregon State University.

7.   Management of pest situations

M. I. DYER & P. WARD

It may seem surprising, at first glance, to find that of the hundreds of granivorous bird species in the world, many of them specialized for a diet of grass seeds, only a small number have so far become serious pests of man's cereal crops which are, after all, merely cultivated grass seeds. The reason for the relative paucity of bird pests is simple: there are few situations where granivorous birds can rely on an abundant supply of cereal seeds throughout the year (Ward, 1965 a; Wiens, 1973). Between periods of superabundance, when the crops are ripening and until the gleanings have disappeared, are interposed months when the birds must rely solely on wild seeds for their survival. Unlike many invertebrate pests, birds have no non-feeding stage in the life cycle in which to pass the between-harvest periods, and unlike rodents, their limited powers of reproduction prohibit a rapid increase in numbers within a single crop-growing season. Even in those relatively small areas in the humid tropics where rice production is virtually continuous, populations of pest species of birds do not reach plague proportions. Fortunately, even such a limitless supply of the birds' main energy food is not all that is needed for rapid population growth. At certain times, particularly for breeding, even the specialized seed-eaters require more protein than a pure cereal diet can provide (Jones & Ward, 1976). They obtain this from insects, on which their feeding is limited. Such is the present situation, but the move to introduce protein-enriched varieties into these same regions might one day alter the picture dramatically.

If, as we say, granivorous birds are generally precluded from increasing their population size in relation to the expansion of cereal production, how can any of them become serious pests? The answer seems to lie in the mobility of certain species and their habit of concentrating in large numbers within relatively small parts of their overall feeding area. The impact of such pests is not felt as a loss of production over a region or country (indeed, such loss may be very small in relation to production), but rather as heavy, occasionally catastrophic, damage to the crop of a particular farmer or small community.

These local concentrations of birds occur as a consequence of their flock-feeding habits, and it is noteworthy that the major pest species are those in which this type of foraging behavior, along with the associated formation of large communal roosts, is particularly well developed (e.g. *Agelaius*, *Quelea*, *Passer hispaniolensis*, *Psittacula krameri*). Further exacerbating the situation is the fact that many of them perform seasonal

267

Table 7.1. *Listing of the most common major bird pest species throughout the world that require management attention at various times*

Common name of pest species	Scientific name	Area
1 Red-winged blackbird (redwing)	*Agelaius phoeniceus* (L.)	North America
2 Common grackle	*Quiscalus quiscula* (L.)	North America
3 Brown-headed cowbird	*Molothrus ater* (Boddaert)	North America
4 European starling	*Sturnus vulgaris* L.	North America, Europe, North Africa
5 Eared dove	*Zenaidura auriculata* (Des Murs)	South America
6 Parakeets	*Aratinga* spp.	South America
7 Dickcissel	*Spiza americana* (Gmelin)	South America
8 Red-billed quelea	*Quelea quelea* (L.)	Africa
9 Red-headed quelea	*Q. erythrops* (Hartlaub)	Africa
10 Glossy starling	*Lamprotornis* spp.	Africa
11 Village weaver	*Ploceus cucullatus* (Müller)	Africa
12 Wood pigeon	*Columba palumbus* L.	Europe
13 Crow	*Corvus* spp.	Europe, North America
14 House sparrow	*Passer domesticus* (L.)	Europe, Temperate Asia, North America
15 European tree sparrow	*P. montanus* (L.)	Europe, Temperate Asia
16 Spanish sparrow	*P. hispaniolensis* (Temminck)	Temperate Asia, North Africa
17 Rose-ringed parakeet	*Psittacula krameri* (Scopoli)	Tropical Asia
18 Baya weaver	*Ploceus phillipinus* (L.)	Tropical Asia
19 Galah	*Kakatoe roseicapilla* (Vieillot)	Australia

migrations whereby large concentrations pass through or temporarily reside in relatively small areas. The concentration of birds within these regions, due to their foraging behavior, may then produce the spectacular clouds of birds which the luckless farmer may describe, with understandable exaggeration, as 'darkening the sky'.

Most of the major widespread pest species have been identified (Table 7.1). In addition to these well-known pests, many other species are serious pests at certain times within the restricted geographical areas they inhabit and these have not yet all been identified. Also, there are numerous minor pest species which may occasionally assume serious pest status following an unprecedented concentration in a small area. More commonly such situations result from mixed flocks comprising several species which individually are of little agricultural importance.

Only a few of the major pest species and hardly any of the minor ones have received serious study. Throughout this chapter, therefore, we must consider mainly *Passer, Quelea, Agelaius*, and to a certain extent, *Columba*, though the principles outlined should apply more generally.

Present attitudes

Clearly, no single management strategy can be applied to all pest problems involving birds. Indeed, there are some situations where none of the currently available strategies seems applicable. Yet it is almost a rule in many countries that an authority, confronted with a new or novel bird-pest situation, plunges into lethal control without adequate consideration of alternative ways of tackling the problem.

Compared with insect or rodent pests, birds must seem easy targets, partly because they are large, diurnal and generally easy to see, and partly because there is a general impression that bird life is 'vulnerable'. The story of the passenger pigeon *Ectopistes migratorius* (L.) of North America, which was reduced by indiscriminate shooting and habitat destruction from one of the world's most abundant bird species to total extinction in only decades, is well known. Thus lethal control is often adopted as the obvious strategy; discussion begins only with the choice of tactics: shooting, trapping, poisoning and other lethal methods. The expectation is that with only a modest effort, the birds will soon be so reduced in number they will no longer pose a problem. But such programs frequently escalate as the birds prove more intractable than expected. As more and more birds are accounted for, but damage levels remain unchanged, the arsenal becomes larger, more widely deployed and more sophisticated. If research is undertaken at all at this stage, the emphasis is on ways of making the population-reduction strategy work. Questioning the validity of the strategy itself is frowned upon. The result is that a program becomes self-perpetuating and is likely doomed to failure. Such is a classic case of 'taking on the impossible' in the sense of Hardin (1968).

Even when alternative strategies are proposed, either because population reduction has eventually proved too elusive a goal after a long and increasingly costly program or where it cannot be considered because of public antipathy, they are all too often involved with the birds themselves. Phrased simply, the aim is 'to get rid of the birds'. So, scaring is an obvious alternative to killing, and research is encouraged into clever mechanical devices or nowadays into the use of fright-inducing chemicals and other deterrents. A variant on this theme, the breeding of 'bird-proof' crop varieties, is also favored.

Yet, as we emphasize later, for the scaring strategy to be effective, there must be a place to which the birds can be diverted; this must not be simply the field of a neighbor who does not possess the deterrent. If, indeed, there is a plentiful alternative food supply for the birds in the form of wild seeds or gleanings from an already harvested crop, the decision to pursue research on scaring tactics is soundly based. But more often than not the strategy is chosen without any knowledge of the overall food situation of the birds.

We are not pretending that armed with a little basic knowledge of a pest species' ecology, reasoning ability alone can provide the ideal strategy for minimizing damage in a particular situation; the problems we are dealing with are too complex, too variable and too changeable for that. But even with a modest research effort it should be possible to rule out those strategies which have little hope of success or are environmentally undesirable.

Dubious economics

In addition to the headlong rush into avian management which results in questionable strategies, there is the question about economic justification for management in the first place. It is not always readily apparent when a basic cost–return economic threshold is reached during a presumed pest attack (Headley, 1972). The problem is twofold: (1) inability or failure to identify economic losses and (2) failure to account for benefits as well as losses in the crop–bird-pest association so that cost–benefit ratios to society can be described.

Perhaps the most dubious part of economics has to do with rather consistent overemphasis of crop loss or gross exaggeration of monetary values involved. For instance, in the late 1960s an estimate of $15 000 000 annual corn loss was made by officials in the state of Ohio, USA, in an attempt to justify introduction of a federal research and control program against red-winged blackbirds (Anonymous, 1967). This estimate was not based on fact but on guesses; subsequent field work established a value less than $1 000 000 (unpublished manuscript, US Fish and Wildlife Service, Richard Winters). A similar condition currently exists for the southeastern USA where unsubstantiated and exaggerated loss figures are used to exert political pressures to justify massive lethal control on blackbird and starling populations.

The non-critical lay segment of society has always assumed that a direct relationship exists between depredating birds, amount of energy or money spent on control, and reduction in depredation losses. We do not know of any instance where such a relationship has been demonstrated (Murton & Jones, 1973). There is an incorrect notion that direct action on bird populations will result in proportionately reduced depredation levels. Much of our present control efforts end up as simply costly, wasteful endeavors. By knowing cost–benefit ratios it should be possible to save much money and sometimes to avoid unnecessary environmental damage by ruling out strategies which have little hope of success. Again, it should be recognized that under some circumstances there is no ecological solution for alleviating pest problems using traditional approaches. Basing his remarks mainly on the wood pigeon problem in England where his researchers proved the ineffectiveness of a government-sponsored control program based on shooting, Murton (1968: 168) stated, '...knowing what not to do can save a lot of unnecessary expenditure of time and money'.

In order to know what the full cost–benefit ratio is for the bird–pest–agriculture problem, one must know the full range of processes in the environment. For instance, it is now known for the blackbird–corn association that the process of eating by the birds does not always cause a decrease in corn production and may even cause an increase (Dyer, 1973, 1975). A similar result has been examined for a house sparrow–wheat and oat association in New Zealand (Dawson, 1970), though the results were not as clearly shown. Quite possibly, many bird attacks on developing heads of cultivated crops of the grass family result in compensatory growth of the crop and thus more total crop production in the long run. Certainly this possibility has not been put into even the crudest forms of cost–benefit equations that have dictated avian control programs.

Overemphasis of the pest species

As is probably the case with pest situations in general, once the bird causing damage in a particular situation has been identified, all attention becomes firmly fixed on that species and the community picture is ignored. Remedial measures are concerned only with the pest and its control.

We feel that some redress is needed. Any problems posed by birds are due to some change in the agro-ecology of an area, not necessarily to a change in the birds' behavior. Though it is usually held that the pests have been attracted into an area by the crops, this is generally not the case. Rather, the crops have been introduced into an area already carrying a large population of granivorous birds for which, at times, the crops may provide a substitute for the wild seeds eaten exclusively before. Thus there is no evidence that the populations of *Agelaius* and *Quelea* have increased as a result of the expansion of cereal growing in North America and Africa, respectively.

With the community in mind, damage avoidance by crop substitution or changes in crop phenology ought to be contemplated. But because all attention is fixed on the pest, only those strategies which involve doing something to the pest (killing them, scaring them away, etc.) are considered.

A second danger brought about by ignoring the community picture is that the bird pest is not seen as only one of a whole range of granivorous

271

animals, such as other bird species, rodents and insects, but as a unique threat that must be eliminated. However, if eliminated, dangers of re-placement of the target pest by another (and possibly worse) pest species cannot be anticipated in the absence of the wider view. As an example of this short-sighted outlook, in the USA great attention is being paid to reducing the numbers of *Agelaius* by roost destruction, but little thought is being given to the effect such reduction might have on populations of other granivorous birds, and in particular on that of the starling which shows considerable dietary overlap with *Agelaius*, *Quiscalus* and *Molothrus* (Dyer, unpublished data).

Current practices in bird-pest management

Population regulation

One of the most basic and most frequently used modes of managing avian populations is through lethal control. It is at the same time the most controversial. Birds in agricultural crops can range from a few individuals (Murton, 1972) to extremely dense masses (Ward, 1973; Meanley, 1971; Dyer, 1968). With small populations there is the possibility of eliminating marauding birds with a variety of lethal agents, probably most effectively through shooting and limited use of poison baits. But with large numbers in huge flocks, or where there is constant replacement of relatively large numbers of individuals, the difficulty of lethal control increases immeasurably and the effectiveness drops dramatically.

Several current control programs involve the destruction of very large numbers of birds. The most intensive is that directed against the quelea in Africa, the annual toll being in the order of hundreds of millions (Ward, 1972) most of which are killed by organo-phosphorus poison sprays. In Kazakhstan, USSR, and also in North Africa, millions of Spanish sparrows have been destroyed in recent years (Gavrilov, 1962; Bortoli, 1973). The blackbird-control effort conducted by the US Fish and Wildlife Service in the USA is aimed at the permanent removal of several tens of millions of birds from the population. Smaller, but increasing, bird destruction campaigns are being waged against a variety of other pest species throughout the world.

It is not difficult to understand that the removal of such large numbers of birds, as well as the means employed to achieve this, such as aerial spraying of vertebrate poisons, poison baits, explosives and flame-throwers, must have considerable impact on the ecosystem. Yet in few places is this impact being adequately assessed, and in many cases it is not being considered at all.

There are instances where intensive measures taken to eliminate locally noxious birds have been successful. Mainly these have been in situations where the population size is not too large and can be localized, i.e. there are few potentials for immigration (house sparrow, Summers-Smith, 1963; starling and red-winged blackbird, Besser, DeGrazio & Guarino, 1968; cowbirds, Shake & Mattsson, 1975). Usually such bird–agriculture associations are easy to analyze. Numbers of birds before and after treatment can be determined, the crop being consumed can be easily sampled for damage, and there are usually minimal, if any, environmental side-effects.

Aside from the problems of determining whether lethal removal is justified economically, there is always the question of whether the strategy has any effect on the population itself. Murton, Westwood & Isaacson (1974) gave an excellent account of the problems encountered for the wood pigeon. With only sparse or sporadic lethal control measures being applied, there is the distinct possibility that this removal simply replaces mortality that would have occurred anyway and is not additive, as would be necessary for the strategy to achieve its intended goal. Whether lethal removal is directed at small localized populations or at large, regionally distributed populations, there is a threshold level of bird removal that must be achieved before any measurable benefit can be realized. Thus it seems to us that the problem at the outset, whether attacked on local or regional bases, is similar for many circumstances, but what is not the same is the relative magnitude of the problem. It is one thing to direct concentrated effort toward the small local population and quite another to put out lethal control programs against a regional or continental population. Also, it is much easier to assess the progress of local efforts than of larger-based programs. To our knowledge, there is no method for estimating the effect of large-scale lethal control on a population, simply because it is too difficult to census the populations.

An alternative method for measuring efficacy is to measure production of the crop presumably being protected to determine whether control actions are bringing about sought-after results. If, with a lethal removal program, there seems to be less bird damage and more crop being produced, the control program is likely to be continued. But since the mechanisms concerning the effects on the bird population are so little understood, any reduction in bird damage coincident with a lethal control program is seen as evidence of its success regardless of the reason for the reduction. In other words, the right result comes about for all the wrong reasons! While this is not a completely unwelcome situation, it does have its dangers. Inevitably, circumstances will change and at some point in time the seemingly successful program is suddenly of no value. One can easily imagine that instead of stopping the program involving lethal control, there would be pressures to increase the effort, which would be counter-productive.

Such are the difficulties with lethal control at present. Data are simply not available to make strong and valid decisions regarding the effects of lethal control on various populations of granivorous birds.

Because of environmental problems and criticism levelled by various individuals or groups throughout the world, severe constraints are placed on lethal control, even when otherwise justified and effective. Most of the complaints come from conservationists and humane societies. In such events, control of birth rate is generally first considered as a potential replacement for lethal control. Methods of sterilizing free-living animal populations have not been intensively investigated, aside from insects, but there seems to be little immediate potential for birds. Many compounds are known that will inhibit male and female reproductive functions in both mammals and birds, but except for two chemosterilants these will not be reviewed in detail here.

In North America, chemosterilants have been reported for the redwing and the starling. Davis (1962) showed that the chemical triethylmelamine (TEM) causes sterility in male and female starlings. Other chemicals inhibiting either gonad development or development of embryos or causing death in nestlings have been reported for pigeons and quail under experimental conditions (Elder, 1964; Wentworth, 1968; Wentworth, Hendricks & Sturtevant, 1968). The problems with field use of these chemicals lie with guaranteeing the action of the chemical (i.e. will it produce the same reactions in wild birds as in laboratory-held birds?), in the delivery of the material to the wild population and in lack of species-specificity. Both species-specificity and feeding behaviour problems pose almost insurmountable blocks. Similar difficulties exist for other gametocides (Murton, Thearle & Thompson, 1972).

Another subject that must be considered is deciding how much reproductive inhibition is necessary to reduce the population substantially. Savidge (1974) gives some indication of the degree of birth inhibition necessary to insure certain levels of population reduction. As was noted in Chapter 3 of this volume, few long-term studies exist showing the natural fluctuations of numbers for populations seemingly in 'equilibrium'. For red-winged blackbirds it seems that at least a three- or four-year period must be allowed in order to assess population change (Dyer, Siniff, Curtis & Webb, 1973; Dyer, unpublished data). Also, if these formerly stable populations were reduced over a several-year span, it would often be very difficult to assess any reduction, mostly because of the possibility of immigration into locally treated areas. Furthermore, if there were effective materials to use on the females, there is always a very real danger of reducing the population to levels lower than desired, since any constant degree of control is difficult to establish. Thus, an extensive program would run the very real risk of treating the conditions so drastically that other unforeseen problems could arise.

Murton *et al.* (1972), working on the pigeons *Columba livia* Gmelin in England, point out some of the problems one may expect to encounter in an operational program. In Manchester, England, feral pigeons were being effectively given reproductive inhibitors, but there were few, if any, changes over a long period of time in reproductive rates within the population. The explanation was that despite the fact nearly all pigeons were capable of breeding, only one-third were actively producing eggs and young. Consequently the chemosterilant was not necessarily affecting the reproductively active segment of the population. The experiments failed to produce any major reduction in the population also because of toxicity problems with the candidate chemosterilant, lack of breeding synchrony, and immigration into the area. One might expect such 'homeostatic' mechanisms to surface in any species, including those granivorous birds of concern here in this book. Even though research into chemosterilants should be carried out in depth to assess the potential for various granivorous birds, in view of the Manchester studies the effectiveness of this technique in agricultural ecosystems is open to question.

Habitat, or biotope, manipulation has long been espoused as the proper way to manage wildlife species (Leopold, 1939). Unfortunately, in most cases little has been done to alleviate problems with depredating granivorous birds through managing their habitat. To be really effective, managers of granivorous bird pests must concentrate either on breeding habitat or on wintering or post-breeding roost sites when dealing with large gregarious and mobile populations. Often, management programs which enhance a harvestable waterfowl species create excellent habitat for breeding or roosting populations of granivores (e.g. blackbirds and starlings), so it is often not possible to have one without the other.

For the most part, birds scattered uniformly over large areas create few problems, but when communal roosts form, problems develop. Recent attempts to eradicate roost sites or to break up roosting assemblies of blackbirds by thinning tree plantations in the southeastern USA are apparently quite successful in that birds are pushed out of places where they are not desired. The concentrations are made smaller and less dense so that the impact per unit area is substantially reduced (though, of course, the potential for total grain loss over the wider area is not altered).

Manipulation of the food supply is another way by which habitat modification can be employed to reduce depredation by granivores. Wiens & Dyer (1975 *a*) noted that because of land-use changes in recent decades there has been a major change in the diet of blackbirds compared with the diets known at the turn of the century. The proportion of insects and weed seeds is currently much lower, the difference in the corn growing areas being made up by several agricultural food crops. Wiens & Dyer suggested looking critically at overzealous management of insect and weed seed populations in areas traditionally occupied by blackbirds, with the idea that if allowed to, birds will return to their formerly preferred food items.

Scaring techniques

If it is realized in a particular situation that it is not possible to regulate the pest population in order to achieve relief from depredation and if it is not possible to exclude them completely from crops mechanically, recourse is made to repellents. Attempts to provide repellent effects have been with man for long periods of time (Murton & Jones, 1973), thus we consider that the use of a repellent in a cultivated crop is not a science but rather a long-practiced art which depends mainly upon aversive behavior in the bird that is responding to a stress. There are simply too many unknown conditions as yet concerning the granivorous bird–agricultural food complex to be able to state with certainty what one might expect from the use of a repellent at any given time.

Modern adaptation of scaring techniques still must utilize the basic receptor senses of birds, namely those of sight, hearing and taste. Any agent affecting these senses and subsequently putting sufficient stress on the bird to keep it away from a food source can be classified as a repellent. As we will see, there are not many agents or materials that maintain a high degree of efficacy.

For the most part there is no basic science behind the development of frightening agents that act as visual repellents, since many items used have been developed on the basis of what someone thinks might be frightening to a bird. These materials include various shiny objects strung across a food crop, silhouettes of raptors, both hawks and owls, hung above the food crop, and a variety of human effigies. To some degree these scaring devices function initially, but they usually lose their effectiveness because the birds habituate rather quickly.

To some degree, auditory repellents are more effective than visual ones. Birds readily perceive loud explosions created by shotguns and carbide exploders (bangers) and apparently find them noxious. However, with shotguns, it is not merely the noise that is effective but the reinforcement aspects of a human behind the sound. The various rattles constructed from pots, bamboo, tin cans, etc. strung across fields in many parts of the world likewise combine noise with the visual effect of the human manipulating the strings. Carbide exploders often will function well; however, birds learn to ignore these explosions if there is no reinforcement with shooting.

The use of recorded distress and alarm calls of various species deserves special attention. With good instrumentation it is possible to record accurately various distress and alarm calls and to play these back to populations or flocks of birds at a later time. These devices show considerable promise under certain circumstances, but the setting and conditions are crucial. For instance, the method is not practicable for the protection of large areas of food crop. The main difficulty is distance and power of the units (Dyer, 1968). If this tactic is employed, one of the most useful developments is the highly directional portable unit, which can be carried to affected areas, such as that used on vans and automobiles at airports to drive waterfowl and gulls from runways to reduce air traffic hazards.

For some granivorous birds it is not necessary to obtain the species-specific distress or alarm call for this method to work. Unreported work with blackbirds and starlings in the USA has shown that recorded distress calls of a yellow-headed blackbird *Xanthocephalus xanthocephalus* (Bonaparte) were quite useful in prompting the flight of red-winged blackbird, grackle, cowbird, and starling flocks (US Fish and Wildlife Service personnel, personal communication).

Another type of auditory repellent has been reported, but there seems to be considerable controversy about its usefulness. It is essentially a random-noise generator with elaborate electronics that produces a variety of undulating and piercing high-pitched frequencies which are broadcast in agricultural fields (Boudreau, 1975). In view of the expense and the questions about reliability, it seems that there is not too much to recommend it as a method of deterring birds from food crops.

There is a relatively large literature concerned with gustatory repellents (Kare, 1965), and much controversy about the usefulness of this approach. It seems reasonably certain that birds in general, in contrast with mammals, are not very receptive to subtle and ostensibly noxious tastes. Thus the use of gustatory repellents has a rather questionable future (Hill, 1972; Rogers, 1974). This is especially true if one must depend upon them for large-scale field applications. Nonetheless, there are reports of chemicals used in specific conditions where the repellency has been proven. Stickley & Guarino (1972) tested a carbamate material applied to seeds of corn in the USA. This treatment was judged efficient in keeping grackles and other blackbirds away from newly planted corn fields where sprout pulling was a major problem. This material has also been used on various fruit crops in the central USA (under some conditions it is also used as an insecticide). Although the reports are somewhat equivocal, under certain test conditions frugivorous birds have apparently been repelled from highly vulnerable fruit crops (Dolbeer, Ingram & Stickley, 1973).

Other than these few experiments, there are few data about efficient gustatory repellents. Some materials are known from laboratory experiments (Rogers, 1974), but the materials are so toxic that it would seem improbable that these or others like them could be used on a commercial scale in field conditions.

Another type of repellent relies upon the combined effects elicited when various materials are used or when chemicals have been developed which produce multiple stress reactions. The first such substance to give this reaction, and now the best known and most widely used, was reported by Goodhue & Baumgartner (1965). The chemical 4-aminopyridine (4-AP; trade name Avitrol) is extremely toxic (Schafer, Brunton & Cunningham, 1973) and only a few milligrams given to a 50–100 g bird is sufficient for an effective response, but its physiological function and the rate at which it is metabolized in the treated bird are as yet unknown (Schafer, Brunton & Lockyer, 1974). The treated bird emits distress calls and at the same time gives a highly visible, aerial distress display. Neighboring birds in the flock, or even in neighboring flocks, have been reported to respond to this stress display and react by fleeing from the area (DeGrazio *et al.*, 1971). However, it is difficult to predict what will happen since under some circumstances there is great attraction created by the treatment, and mobbing of the treated and displaying bird has been observed. This 'repellent' approach was used by DeGrazio *et al.* (1971) who stripped back corn ears in the field and painted a liquid solution onto the ear. Blackbirds feeding on the ear ingested the material, and thereby produced 'repellent behavior'. This cumbersome technique was later modified by placing chemically-treated small corn baits in fields where the blackbirds were feeding. Only a few birds had to feed on the treated baits to drive the flock from the area (Stickley *et al.*, 1972, 1976). At present, this approach and the specific chemical are being tried for various field crops in North America, and it is also being considered for other areas in the world, including Africa and South America, where granivorous birds are pests in field crops.

One of the more discouraging outcomes of analyses of bird response to repellents is that there is a very narrow range of conditions where they are useful (Dyer, in preparation). In conditions where low levels of damage exist it is probably not economical to use repellents, and in situations where alternative food resources are limited and competition is high, repellents are ignored by the birds. If these predictions are borne out by further scrutiny, then the utility of repellents will unfortunately be severely limited, especially in ripening grain fields, as noted by Dolbeer *et al.* (1976). However, to date, information about the use of Avitrol and carbamate materials on planted seeds seems promising (Guarino & Forbes, 1970; Besser, 1973). Present limitations to widespread use of repellents have to do with inability to answer the following questions. (1) What is a basic treatment level and how often should additional treatments be applied to keep the tactic economically feasible? (2) How do bird flocks feed? (3) What is the relationship between the ability of a repellent in keeping birds from agricultural crops and the amount of available food in the system? (4) What is the rate of feeding on the treated crop once birds learn to feed on the crop despite the fact that there is heavy treatment (it is known that this occurs (Stickley *et al.*, 1976))? (5) What are the cost–benefit ratios realized for immediate crop protection where the farmer is trying to achieve a profit or for long-term associations where there are many ecosystem interactions to account for? Until these questions are addressed, we feel that use of repellents, especially Avitrol, is little warranted.

Damage evasion

If agricultural crops that are attractive to granivorous birds are placed inside the birds' normal breeding range or within a traditional migratory path, there is a reasonably high probability that these birds will become pests (see Chapter 8, this volume). The problem is further aggravated by the phenomenon of mutual exclusion, that is, development of extensive areas of agricultural crops which reduces the area of native foods in which the birds can forage. The birds then face the alternatives of feeding in nearby regions that are not planted in crops, leaving the area altogether, or eating the new crop. The choice primarily depends upon the adaptability of the granivore, but serious crop loss can occur. In such situations it is up to the farmer to change agricultural practices, since the birds, although being opportunistic, do not have the same potential for adaptability.

If changes in crop phenology, that is, planting or harvest times of the food crop being threatened by birds, can be made, it is perhaps the easiest way to avoid bird problems. For instance, in the Lake Erie region of the USA and Canada where redwings feed on corn, fast-maturing varieties of corn, if planted early, are not susceptible to bird damage since the grain hardens before the birds aggregate in August. Unfortunately, these varieties give a reduced yield, but with nearby highly predictable roosting centers the grower ought to accept this. The reduction in yield is less than would be the probable damage caused by redwings to higher-yielding varieties that mature later. In the long run the producer stands to harvest more grain without having to spend time, energy and money on bird control. Despite the fact that this strategy was recommended in the late 1960s, little has been done to implement it in the Lake Erie basin. In the wintering grounds of blackbirds in the southeastern USA, there are sometimes problems with sprout pulling of winter wheat crops. However, when farmers make certain their crops are planted early in the autumn to produce a large growth before the blackbird flocks appear, there are few or no problems. In northern Botswana, native farmers have learned that by planting millet crops early in the wet season the harvest can be gathered before the hordes of newly independent quelea invade the fields. Any farmer who fails to plant early stands to lose the entire crop, no matter how intense efforts are to scare the birds away.

If it is not feasible to alter crop phenology, then it may be possible to try crop substitution or diversification, especially in monocultural areas. In the USA and Canada it is possible to rotate soybeans, sorghum and small cereal grains with the bird-susceptible corn crops. One farm in southwestern Ontario, Canada, stands out as a model in this respect. Cultivation of corn became difficult because of the proximity of redwing roosts. Rather than continue to fight a losing battle of protecting the corn crop, the owners also planted vegetable root crops, soybeans and sorghum and maintained a feedlot with approximately 5000 livestock animals to which the bird-damaged corn and sorghum were fed as ensilage. Neighbors subjected to the same corn depredations did not diversify or substitute crops and consequently did not fare as well.

We are not able to say whether, in general, monocultures of cereal crops are generally more or less vulnerable to bird damage than are similar areas carrying a diversity of crops, but in some areas there seems to be an increase. For instance, in Ohio, redwings display quite different foraging patterns in areas of high crop diversity than in areas where corn is grown in monoculture. In the monocultural areas birds gathered in larger concentrations, flew shorter distances to feed, and did significantly greater (that is, more concentrated) damage than in nearby areas where crops were more diversified and where the birds tended to scatter widely in small flocks (C. P. Stone, Jr, personal communication).

As noted earlier, much of our agricultural cropland has replaced traditional bird feeding habitats, and there is little to suggest that present populations are any higher now than they were before the crops were introduced. Indeed, in the middle 1700s, Joseph-Gaspard Chaussegros de Léry noted in his travel journals through the Detroit region that 'The inhabitants of the place say that it is not possible to settle the newcomers in a village on account of the large number of starlings (*sic* blackbirds) that eat the grain' (Lajeunesse, 1960: 47). Similar statements can be made about quelea in Africa. Large numbers of quelea have occurred in suitable parts of the continent for centuries, and they were probably engaged in the same migratory and feeding activities as after the advent of agriculture (Ward, 1971*a*, 1973). It therefore seems that in order to cope with large populations of bird pests, agricultural practices should change to allow the birds to fit into the ecosystem and not aim at their destruction. In order for this to be done, there must be provision for alternative food supplies. Wiens & Dyer (1975*a*) suggest that cultural practices which limit availability of insect and native seed production may seriously enhance conditions for avian depredation. If these food sources were available, there would be less pressure placed on vulnerable crops. The existence of alternative food supplies may also aid scaring practices. For example, it appears that in the successful clearing of blackbirds from corn fields in South Dakota, USA, by the use of Avitrol, the presence of alternative wild foods in the area was a key factor (DeGrazio *et al.*, 1971). Birds were observed to be flying over the treated corn fields to open pastures and prairies beyond for their foraging. Had not these alternative feeding sites existed, it is probable the birds would have continued to feed in the corn fields, despite the presence of repellents. A similar treatment in Ohio, an area with more limited alternative food resources, was much less successful (Stickley *et al.*, 1976). This observation tends to substantiate our hypothesis that alternative food supplies for birds are a viable and necessary management practice which can help reduce depredations.

We feel that, for the most part, changes in agricultural practices could solve the majority of bird-pest problems, nearly always to the benefit of the local and regional economy and always to the benefit of the ecosystem, in that birds could be retained as predators in the system and it would be unnecessary to add poisons aimed at bird control.

Control strategies

Thus far, our comments have been largely critical of present practices. Now we wish to suggest how management ought to set about the task of formulating a course of action, following reports of serious damage by birds. Our main concern is that first consideration must be given to those strategies which involve minimal ecological impact. Only if these are clearly unworkable and if the seriousness of the situation justifies it, should more disruptive strategies subsequently be contemplated. In our example scheme (Fig. 7.1) the strategies have been ranked, as one progresses through the choices, according to the degree of ecological disturbance. In general, the more desirable strategies from an ecological viewpoint are also likely to be the cheapest, if not in the short term then in the long run. Thus, economic considerations reinforce ecological considerations, and, where the latter are muted, serve as a forceful substitute. The last strategy in our list, total population control through destruction of birds, should be contemplated only after all other possible strategies have been shown to be inadequate. In normal pest–crop situations it would be expensive, often hazardous, and almost certainly would produce unforeseen changes in the ecosystem, including replacement by other species which may become pests of equal or even more serious proportions. There is an exceptional situation, however, where total population reduction to the point of complete extermination would be the initial choice. This is where

Hypothetical
impact on
environment
due to
action programs

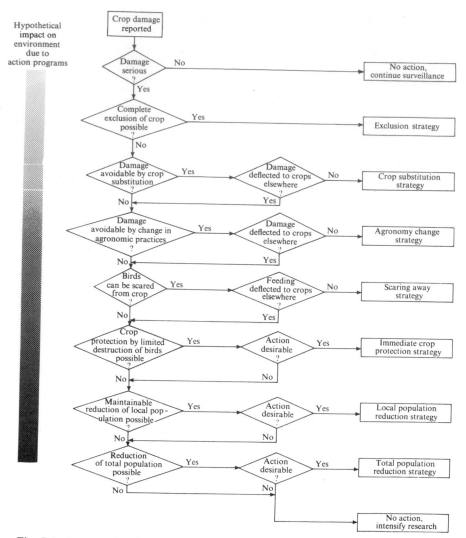

Fig. 7.1. An example of a systems flow diagram which permits the logical choice of an appropriate strategy for management of a bird-pest problem with minimal ecological disturbance.

a newly-introduced species, thought likely to become a pest, is still confined to a small area around the point of introduction. Ironically, in such situations where extermination can only be of benefit, it is rarely contemplated until too late, whereas in an established ecosystem involving native species, it is readily embarked upon!

The scheme in its present form is not applicable to all pest situations. In some cases ranking the candidate strategies according to ecological disturbance might produce a rather different order. It is intended merely as an example of the logical step-by-step procedure which management should follow and document before implementing any pest control action.

It may seem that we are overemphasizing a very simplistic notion that any management group would automatically develop of its own accord. But the need for emphasis is shown by the present situation with regard to quelea control in Africa, red-winged blackbird control in the USA, and probably much of *Passer* and *Sturnus* control around the world.

Quelea control began around 1950 in several countries in eastern and southern Africa and gradually spread so that now there are control teams operating in some 18 countries. None of these control programs has been based on carefully reasoned argument. If there was the excuse in the 1950s that lack of biological information precluded this procedure, this cannot be invoked now when so much is known about quelea biology. Furthermore, the amount of basic information needed for strategy choice is not overwhelming (see later) and could easily have been gathered before 1960. Since there was no coordinated research program, individual researchers investigated certain aspects of quelea biology in depth, resulting in other topics essential to strategy choice being neglected. In place of careful choice of strategy a selection of tactics was devised: destruction of communal roosts and breeding colonies by fire, explosives and poison sprays. These tactics have been refined continually over the last two decades, but the strategy implemented by these tactics is rarely, if ever spelled out. Presumably, some groups are operating the strategy of local population reduction while others have total population reduction as their unstated goal. Even now much of the research effort is being dissipated into tactics which presuppose strategies which have not been logically chosen and which will probably be quite unworkable. Thus, in the vast United Nations' quelea research program, research is being fostered on pathogens (presupposing 'total population reduction' strategy) and on repellents, including fright-inducing agents, bird-proof grains and acoustic devices (all presupposing a 'scaring-away' strategy). Yet it would be difficult to argue that the strategies in question would be chosen by any knowledgeable management organization as a means of reducing damage in view of what is already well known about quelea ecology.

The red-winged blackbird story is hardly more encouraging. Tactics similar to those used against quelea have been embarked upon without any prior decision-making processes such as those suggested in Fig. 7.1. Recent emphasis is on the widespread use of repellents in grain crops during the growing season and roost destruction on the wintering grounds. But at no time have the reasons for working within the 'scaring-away' and 'local population reduction' strategies been spelled out.

M. I. Dyer & P. Ward

Basic research requirements

Clearly the process of strategy choice can only be implemented after certain basic information is obtained. The nature and precision of this information will differ from case to case and in some instances may be procurable from the literature. This is more likely for information on climate and agricultural practices than on the pest species' biology, and even where it exists, considerable time is needed to locate it.

Where new research is required it cannot be undertaken quickly. Most ecologists would ideally like to know everything about the species, its position in the ecosystem and the animal–plant relationships, before advising management. Yet management is generally under great pressure to begin action immediately after damage has been reported. Obviously, compromise is needed. We feel that at least two years is a necessary period for research into the essential topics and at the same time is the longest delay that management can defend in its confrontation with farmers, politicians and others affected directly by the avian damage. This period is selected on the understanding that a high quality and well-led team of scientists can be found for the task and be adequately funded. Within such a brief period there is clearly no time for work on topics not required by the strategy-choice process. In a particular situation, one or more of the strategies listed in Fig. 7.1 may be decisively ruled out even before research is begun on the basis of existing information; in such a case any research pertaining to the choice of that strategy would be superfluous.

In making a plea for broadly-based research, followed by a rational choice of strategy, we are not suggesting that nothing should be done about a pest until a lengthy research program has been completed. If only for political reasons, a long-delayed start is untenable. We do suggest strongly that nothing be done until at least a one-year program has been carried out so that when the alternative strategies are considered there is at least some knowledge of, for example, the species' distribution, its diet throughout the year, the availability of its natural foods, and its movements. Clearly, it is not possible to work out a migratory pattern in a short time, but it is possible to decide whether the problem birds are sedentary or migratory. And in case it is thought that such elementary knowledge is already available in textbooks we would point out that in many parts of the world there is, as yet, no information on the seasonal movements of a large proportion of the avifauna. This is true even for many of the main pest species: e.g. the village weaver *Ploceus cucullatus* (Müller), red-headed quelea and glossy starling in Africa, and the rose-ringed parakeet and other *Psittacula* spp. in India. Similarly, there is scant information on the diet of these species and even in North America

284

there are too few dietary studies available throughout the year on black-birds and starlings.

We are omitting specifics about community studies for the sole reason that apart from general observations the ecologists conducting the initial research program could not reasonably be expected to carry out detailed research on the many other granivorous animals associated with the main pest. We would like to stress, however, that in the ongoing research program, community investigations merit considerable effort. Information from this approach becomes a valuable input for simulation models of the type used by Wiens & Dyer (Chapter 6, this volume). There, they combined data from sources cited in Chapters 3 and 5 to define impact. Future programs dealing with specific populations of granivorous birds could do likewise.

In the following sections we give our opinion of what the initial research should aim to provide in the way of broad information. Obviously, research will not, and should not, cease after such a superficial program. Each of the topics will, if the seriousness of damage warrants it, grow into a long and involved research program. We have omitted those lines of research which prejudge the issue. For instance, work on auditory responses and 'chemical scarecrows' which presuppose the choice of scaring as a viable strategy, or studies of pathogens, narcotizing agents, chemosterilants and poisons which presuppose population regulation as the chosen strategy, are untenable for most parts of the world at this time. However, once the correct strategy has been decided upon, certain of the topics omitted, because they concern tactics, will merit inclusion.

Identification of pest species

Generally speaking, when a bird-pest situation is reported in North America, Europe, or any other technologically advanced region, the identity of the pest species is either known to the farmers themselves or can be quickly and reliably established by reference to local ornithologists or museums. This is not the case, however, in many of the developing nations, where considerable effort may be required before the identity of the pests can be determined.

There are two separate problems involved. The first is concerned with deciding which of the several different kinds of birds commonly seen in the fields are actually consuming the grain, the second with obtaining specimens and getting them identified.

Sometimes the first is hardly a problem in that only one kind of bird is present, or the birds doing damage are easily distinguished (e.g. parakeets or icterids). But there are other situations where 20 or more species of small passerines are feeding in the field, not all of them granivores (for there may be abundant numbers of insects on or under the cultivated plants). Indeed, it may even be that not all the granivores are doing damage. In an African millet field, for example, a copious understory of wild grasses and herbs attracts a wide variety of waxbills, weavers and other seed-eaters; some of these may well be taking millet grains from the ripening heads, as well as fallen seeds and seeds of wild plants, but not all of them. It is not uncommon to see large clouds of quelea feeding in a millet or sorghum field, to the great consternation of the farmer, only to find by shooting into the flocks and opening the birds' gizzards that they are feeding almost exclusively on wild grass seeds. On one occasion, in northern Tanzania, the gizzards of the 'marauding' queleas were all packed with army-worm caterpillars! Red-winged blackbirds have also been reported to consume large numbers of corn-ear worms (*Heliothus* sp.) during periods when they were implicated in corn depredation (Mott & Stone, 1973). A further difficulty is that rodent damage to the crop, done at night, is frequently accredited to the birds seen in the crop by day.

Even when, by careful observation and sampling, the species actually taking grain is known, the task remains of putting each into perspective. In 1959, in a small area of millet fields, near Zaria in northern Nigeria, 31 species were commonly seen to be feeding on the ripening heads of millet *Pennisetum*: ten weavers and two sparrows (Ploceidae), six waxbills (Estrildidae), two serins (Fringillidae), one bunting (Emberizidae), one lark (Alaudidae), one bulbul (Pycnonotidae), two starlings (Sturnidae), two parrots (Psittacidae), three doves (Columbidae), a hornbill (Bucerotidae) and many other species were seen on occasion (Ward, unpublished data). This bewildering variety of pests, on a single crop, is typical of the situation in large areas of the dry tropics where the total granivore avifauna (many of them Graminae-seed specialists) is rich. DeGrazio & Besser (1970) describe similar situations in sorghum fields in South America where many species of fringillids, icterids, psittacids and columbids were implicated. Permutations vary from region to region, but throughout the dry tropics there is a great variety of actual or potential pest species. In the wet tropics, on islands and at higher latitudes, where the granivore avifauna is vastly simpler, identification of the pests and distinguishing which are most serious is obviously a much simpler task.

Given a complex situation, such as one of those described above, how does one decide which are the more serious pests? Unfortunately, there is no easy way to do this accurately within a season or two. Probably the best overall picture is obtained by making spot checks in a large number of fields scattered through the area with counts of the numbers of each species seen feeding in each during, say, the hour after dawn. This simple technique allows the species (in a particular area, on a particular crop and in a particular year) to be ranked. However, it is not permissable then to

extrapolate to other areas, crops, or years. For if there is any general-
ization that can be made about granivorous bird problems the world over,
it is that they are usually very local and sporadic.

Distribution of pest species

In a preliminary distribution study, three facets deserve attention. These
are: first, an overall picture of the species' total distribution; second, a
delimitation of areas where large concentrations occur either seasonally
or throughout the year; and third, any division of the species total range
into areas occupied by separable populations, whether or not these have
been formally named as sub-species.

Generally speaking, the total range can be determined, with sufficient
precision for a quick appreciation of the problem, from standard ornithological works, and further detail may be obtained from accounts of local
avifaunas. However, in less well-studied regions of the world, published
information is often incomplete or out of date, and broad ground surveys
will be needed.

Information on the density of a species over its range, and in particular,
on the location of concentrations, is not generally available anywhere.
Obviously it is impossible to locate these in the course of an initial survey,
but a reasonable picture can often be obtained by mapping areas of
preferred habitat (where detailed vegetation maps exist). In their absence,
probable concentration areas may be plotted by reference to topographical
and climatic data. For example, most of the concentrations of quelea in
Africa are now known from detailed surveys. But many of them could
have been predicted on the basis of early information that the species was
confined mainly to regions with between 200 and 800 mm annual
precipitation, and within this zone the main concentrations occur where
there is widespread seasonal flooding. Thus a map showing wide river
terraces and seasonal swamp areas within the 200–800 isohyet zone would
have reflected the actual pattern as it is now known.

Of course, these indirect methods can give only a rough working map;
detailed and repeated surveys are needed before seasonal and annual
changes and shifts of the concentrations can be appreciated. To us, the
need for a trustworthy distribution map is an obvious preliminary to
discussion of any strategy, and especially if it involves national coordin-
ation or international cooperation. Yet the need does not seem to be
widely appreciated. For instance, we know of no distribution maps accurate and detailed enough for our purpose for such widespread granivorous
pests as the rose-ringed parakeet of tropical and sub-tropical Asia, or the
golden sparrow *Passer luteus* (Lichtênstein) and village weaver of Africa
(although in the case of the last two, Hall & Moreau (1970) have mapped

in their atlas of African passerines localities from which museum skins have come).

The case of quelea illustrates further the low priority given to bio-geographical studies of this kind. At the First International Quelea Symposium (held in Dakar in 1955) a resolution to compile a map showing the species distribution throughout Africa was passed by the delegates. Yet at the Third Quelea Symposium, five years later, there was still no map; international cooperation was discussed without any clear picture being available to show which countries needed to group together. Incidentally, the first detailed map showing the total distribution of this important pest species was not published until 17 years after the need for such a map was first expressed!

For the North American scene, only recently has such distributional information become available for the redwing (Stone, Mott, Besser & DeGrazio, 1972). From their reports it is possible to get a relative idea about the degree of corn damage in the USA on a state-by-state basis, although the exact levels of damage remain in doubt.

The division of the pest species' distribution area into smaller units occupied by distinct geographic populations, where there are such divisions, is a refinement of the initial mapping exercise that can reveal much of interest to the control strategist. For instance, where adjacent regions are separated by a geographical or ecological barrier, there is a possibility that population reduction in one area (should this be the chosen strategy) will not be negated by constant re-invasion from the other area. On the other hand, where no such separation exists, or where there are two populations which intergrade clinally, re-invasion is much more likely.

Distribution and phenology of damage

Before any decision can be made on how to combat a particular problem, a rough picture of the distribution of damage in the region or country must be known. But even this level of knowledge is very difficult to obtain. Reports by farmers, unsolicited or in response to questionnaires, are frequently misleading or inaccurate; at worst they are complete fabrications. Even in agriculturally advanced areas results of questionnaires are difficult to interpret. Thus, after a survey of damage to soft fruit in New Zealand, Dawson & Bull (1970) commented on 'the small return of questionnaires' (about 16%) and how 'the occasionally false identification of birds and of the damage attributable to each kind, severely limited the usefulness of this survey'.

There seems to be no alternative at present to mapping all reports of damage in some reliability-weighted manner and supplementing this by wide-ranging surveys, partly to obtain information from areas not reported from and partly to check the truth of the reports received.

There is a great temptation to obtain accurate damage estimates, which can be translated into total depredation quantities, at the beginning of a research program. But, because of the laborious techniques required, only a few fields are surveyed, and even then no great statistical accuracy can be claimed, as noted by Stone *et al.* (1972) and Granett, Trout, Messersmith & Stockdale (1974). There is considerable utility in considering the use of simulation models to aid with impact assessment (Wiens & Dyer, 1975 *a*). In Chapter 6, the general approach is spelled out and there are estimates derived using this method that can help place a perspective on depredation levels caused by birds. At a later stage, when the results can be seen in perspective, detailed field work is of great value; the danger of starting out with such studies is that the figures (probably obtained from an area selected because of its high damage levels) are so multiplied as to produce exaggerated figures for losses over a whole region or country. When translated into cash terms they provide apparently complete justification for expensive, but often useless, control campaigns. This was done in an early period of the control problem with the redwing in Ohio.

Thus initially, even though large variation is encountered, it is of greater value to obtain a larger number of samplings of fields chosen at random over a wide area, with careful attention being given to statistical design, than to intensively sample a few fields. Such spot checks may only permit classification on a subjective basis (e.g. none, slight, heavy, or catastrophic damage), but they can provide an overall view of the damage situation. Also, when superimposed on maps of crop distribution they may reveal a previously unsuspected pattern. Again, careful attention should be paid to statistical design in order to avoid the problems posed by Granett *et al.* (1974) so as to obtain adequate sample sizes. The main problem that must be kept in perspective is that of spatial scale; a sampling program for answering questions in an isolated area as distinct from that addressing regional proportions.

Life history studies

To call a one- or two-year study of the annual cycle of a bird 'population dynamics' would be pretentious, though this is what a simple life history study ultimately grows into. Within a single year it is possible in most cases to discover: how an individual bird's time is partitioned between breeding, moulting and doing neither of these; how many broods are reared; and the average number of young raised to independence.

With this information, provided it is based on adequate samples, one may reasonably predict the natural annual mortality rate, by assuming the reproductive rate to be equal to the mortality rate (see Chapter 3, this volume). Obviously, such simple reasoning does not allow any distinction between the mortality rate of juvenile birds and the generally lower rate for adults. Still, it can provide a good indication of how high the destruction rate would need to be for a population control strategy to be effective. It has been pointed out repeatedly that the individuals destroyed in a typical bird control program merely constitute a part, often a small part, of the 'doomed excess' (Murton, 1972; Ward, 1972). This excess is the difference between the size of the population after it has been swelled by breeding during the most favorable period of the year and the number which can survive through the least favorable period when food is limiting. To achieve a real change in the size of the population, a control unit must destroy more individuals than that which constitutes the excess which will die anyway in the course of natural population regulation. Control planners are often incredulous when told that they must destroy 60–80% of the post-breeding population (a level noted by Savidge, 1974) before the next breeding season begins, to achieve a small net gain. However, achievement of such levels of mortality can also create additional factors to consider before program managers embark on such an operation. Dyer, Pinowski & Pinowska (Chapter 3, this volume) have noted that in many places of the world there have been large lethal control programs. They further noted that little or nothing is known to date about the effects on mortality or natality levels in the populations being controlled. Fretwell (1972 a) suggested that with increases in post-breeding mortality overwinter survival of the remainder of the population may be increased and/or breeding populations will encounter reduced levels of competition. Regardless of the situation, the result is increased potential for production the following year. In order to substantiate such high percentages, at least crude values for the species' reproductive rate must be available when the control strategy is formulated.

Despite the wealth of information now available to show that granivorous bird populations are not regulated by predators, and theoretical explanations of why they cannot be (Krebs, 1972), research on 'natural enemies' is often included in preliminary work on new pest species. Indeed, such work may be given high priority since many agriculturalists still accept old ideas of the role of predatory birds and mammals. As we see it there is no good reason for including research on predation in any initial research program of a granivorous pest, although at a later stage, when the population dynamics studies are being developed, predation will merit investigation.

Diet and feeding behavior

In this section and the next we have the most important components of the initial research program. For without an understanding of how damage to crops relates to the food situation of the birds over the whole year, meaningful discussion of control strategy is not possible. In most species the food supply dictates the size and fluctuations in size of the population. In at least an ultimate sense, it also determines such major features as migration and breeding seasonality, and not infrequently the distribution and intensity of damage to crops are governed by the availability of wild foods (see, for example, Ward, 1965 *a*).

One may safely generalize that the information on a species' diet that can be gleaned from general works on the birds of a region will be totally inadequate for the purpose at hand. At best, it provides a list of some of the food items the species is known to eat. The reason is that even a simple dietary study requires the sampling of large numbers of individuals, taken over a wide area, regularly over at least 12 months. The actual method for assessing the importance of the various food items is relatively straightforward, particularly if the undigested contents of the crop (which many of the pest granivores possess) are analyzed. The labor involved is very great. With frequent and large samples of birds, each of which may contain thousands of seeds, a great deal of patient sorting and counting is required. The temptation is to reduce the number of samples, or their size, to make the analysis more manageable. But instead, it is far better if large samples are obtained (and since we are dealing with pest species there should be no dificulty in getting these), and the mass of food removed from each large sample, carefully mixed and subsampled until an amount that can be sorted is obtained. The method used for quelea is described in Ward (1971 *b*).

Because we are dealing with granivorous birds, it should not be imagined that the seed component of the diet is all-important. In most cases the seeds do provide the bulk of the energy supply throughout the year, and for long periods they may furnish all essential nutrients. However, at certain times animal food is included in the diet and, although it may be in relatively small quantities, it can be of paramount importance. Its importance lies in the amino acid content, and simply to convert the insect food into calories and add this quantity to that derived from the seed food is to lose sight of its significance. Since the protein requirements of males and females may differ considerably, particularly during breeding periods, large differences in the need for animal food may exist between the sexes. Thus there is merit in treating the sexes separately in dietary studies, at least until the periods of dietary divergence are known.

Even during long periods of exclusive granivory, changes in the seed

291

composition in the diet may occur which can provide a valuable clue to the reason for the species propensity to attack cereal crops at certain times and not others (Ward, 1965 a). Knowing the reasons for a bird's interest in a crop may provide little comfort for the farmer but is valuable in that it permits a measure of the potential damage (Hayne, 1946; Dyer, 1967). More importantly, it can indicate whether scaring is a viable strategy to attempt. If it is discovered, for example, that the birds change from wild foods to crops only when their wild food is in short supply, the chances of scaring them off the crops are slight, as we noted earlier. If, on the other hand, crops are being raided when there is abundant alternative food, it is reasonable to assume that the birds simply prefer to feed on the cultivated grain. In this case it may well be possible to deflect the birds by making the crop, or feeding in the area of crops, less attractive by adopting one of the many tactics within the 'scaring-away' strategy, depending upon the strength of preference displayed by the birds.

The study of feeding behavior would normally be conducted in parallel with the dietary investigation, but only the study of birds feeding in the wild can provide the required information. Cage studies can be no substitute and must be seen as only a useful auxiliary technique, simply because the same cues utilized in the wild to prompt feeding are not present in laboratory conditions. Within this important subject certain topics stand out as essentials. First, there is the actual feeding technique employed by the birds under different conditions and the changes which take place with the seasons in the importance of the various techniques within a species' repertoire. Second, there is the deployment of individuals into flock units within a feeding area, and the variation in this according to geography, weather and season.

The techniques used by granivores to obtain the seeds on which they subsist are extremely varied. A single species may even be able to use several quite different methods. Thus, during the wet season when feeding on green seeds, quelea balances on grass or cereal stems while extracting seeds from the panicle. During the dry season when fallen seeds on the ground are the main food, the same species will pick these off the surface as long as seed food is plentiful but late in the dry season will dig into loosely textured ground for seeds trapped in the surface layers. When feeding on insects, quelea employs a further set of feeding techniques. Presumably, the various *Passer* spp. behave similarly; little is published about the feeding behavior of *Agelaius*, however, a large amount is known about physical adaptations associated with food gathering (Beecher, 1951).

The deployment of birds within a feeding area may be according to a system of territories, or flocks with no fixed composition may exploit the food supply over a large area. Many species of granivores change from the territorial system in the breeding season to a flocking system in other months, and it is generally when they are in flocks that damage to crops occurs. The study of flocking behavior is, therefore, of particular value. In particular, the control strategist must know how far the flocks move each day from the communal roost where the night is spent. Knowledge of roosting centers and roosting behavior is essential (Ward & Zahavi, 1973). For when the immediate crop protection strategy is discussed one must know from how far outside the crop perimeter the flocks are likely to come each day to consume crops within it. Some hypothetical considerations stemming from a modelling study on redwings (Wiens & Dyer, 1975 *a*) and starlings (Hamilton, Gilbert, Heppner & Planck, 1967) have been made, but there are few field studies reported following up these ideas.

Knowing the diet of only the main pest species provides but a partial answer. In order to have an understanding of how resources are partitioned during the year, dietary overlap between species ought to be known, especially for the periods during which crop damage might ensue. For the most part, such overlap is known only for the breeding season (Orians & Horn, 1969). For instance, in the southern USA much concern has been expressed about four species (redwings, grackles, cowbirds and starlings). Preliminary data about their diet collected during 1974 during preparation of an Environmental Impact Statement indicated a surprisingly high degree of overlap. Such information is valuable since it predicts that these flocking and communal roosting species all share the same food resources. If any one or a combination of species is removed, as has been attempted by massive lethal control, the remaining species could benefit from lack of competition and their winter mortality levels could be reduced dramatically. The most likely candidate benefiting from such a condition is the starling, a species that is unlikely to be affected by massive control measures because starlings predominately inhabit areas that cannot be treated. An hypothetical outcome would be to make any avian depredations problem worse than it is now.

Availability of natural and cultivated foods

Knowledge of a species' diet under specific conditions is of limited value unless it can be related to the availability of the various food items. Here we must emphasize that the quantity of a food item available to a species within a certain area may differ greatly from the amount of the food actually present. For example, we may consider a number of granivorous species feeding on grass seeds in an African grassland area in the dry season, and, for the purpose of the example, we will assume that they are all eating the same seed types. A number of species, such as the grey sparrow *Passer griseus* (Vieillot), cordon bleu, *Uraeginthus bengalis* (L.)

293

and fire finch *Logonosticta senegala* (L.) confine their feeding to shaded areas or areas of open grassland close to cover and water. *Quelea quelea* and the golden sparrow also feed in these places but, in addition, may range tens of kilometers away from cover and water to feed on the open plains. Thus the proportion of the seed stock present on the ground which is available to the quelea and the golden sparrow is far greater than that available to the other species. The techniques used by different species to find seeds also affect their availability. In a mixed group of estrildines feeding on a small patch of ground one may see combassous and whydahs (*Vidua* spp.) digging with their feet to obtain seeds in the soil, whereas cordon bleus wander about picking only the seeds lying on the surface.

It is extremely difficult to measure food availability quantitatively, but by observation of the birds' feeding behavior coupled with a knowledge of what they are eating (from gut content analyses) a general understanding of the situation becomes possible. Any marked local variation and seasonal changes in the availability may then be appreciated.

In the case of cultivated crops the availability is relatively easy to assess. Often the growing season is the same over a large region, being governed by spring and summer or the wet season. However, in warm, humid regions standing crops, of rice for instance, may be available throughout the year; and in warm, dry regions availability patterns may be modified greatly by irrigation or by planting cereals on the terraces and shores of receding rivers and lakes.

One might well wonder how the cursory knowledge of wild food availability that can be obtained by such a simple study can be of help to the control strategist. Perhaps the following example will show its utility. In the early years of research on quelea, it was assumed that the birds were positively attracted to the crops that they damaged on such a large scale. But it was later suggested on the basis of dietary information that, on the contrary, they turned to cultivated cereals only when and where they could not find sufficient wild grass seed (Ward, 1965 a). However, while this hypothesis could be substantiated in certain situations where there was heavy damage, in other situations it appeared that queleas were attacking crops in areas where wild grass seed was abundant. For example, Morel (1968) pointed out that the damage to rice in Senegal occurred at the end of the wet season when wild grass seed was more abundant than at any other time of the year. Similarly, total losses of millet, over a restricted area, were found to coincide with the time of maximum grass seed abundance in Botswana and elsewhere. Finally it was realized that although grass seed was present in large quantities at the end of the wet season, it lay on the ground beneath a tangle of dead grass stems (ultimately to be removed by termites and fire) beneath which queleas would not penetrate (Ward, 1971 a). Thus though the seed was abundant, available seed was not, and the damage to crops at this time did indeed reflect the birds' shortage of wild food.

Diurnal patterns of movement

Most of the grain pests are flock feeders, at least during the non-breeding season, and it is generally while they are in flocks that they cause heavy damage. As is typical of such flock feeders, they pass the night in well-protected communal roosts where birds assemble after feeding over a wide area during the day. In addition, there may be a number of 'day roosts' or more properly, 'secondary roosts' (Ward & Zahavi, 1973) within the feeding area around a night roost. These are attended in the middle of the day by sections of the population exploiting particular parts of the feeding area. Such day-roosting assemblies are very conspicuous in the dry tropics where they tend to be situated near water and to be occupied continuously through the hot hours of the day. But they occur also with temperate zone granivores, such as the house sparrow and icterids, where they feed in large flocks in farmland.

Feeding periods during the day must also be known. Hintz & Dyer (1970) showed that diurnal patterns exist for the red-winged blackbird. The patterns were different for males and females, with the females having nearly twice as many feeding periods as the males.

The distances flown between the communal roost and feeding areas vary greatly between species and intraspecifically according to season and overall food situation. In general, where birds are feeding in large flocks they tend to fly long distances to large communal roosts. Conversely, where they are feeding in small flocks they usually attend nearby small roosts.

Needless to say, communal roosts provide good targets in any strategy based on destruction of birds, although before destroying the roosting assemblages one must know to what extent the target species forms mixed roosts with non-pest (even protected) species, the kind of vegetation favored, and proximity to water and to human population centers (the latter in order to prevent undesirable impact as a direct result of control actions).

Seasonal movements and migrations

By this we mean migrations and irregular displacements which go under the vague name of 'nomadism'. In Europe, North America and northern Asia, migrations of birds are well documented, but throughout the tropics and the Austral sub-tropics the study of bird migration is still in its infancy. There can be no doubt that a large number of species are migratory, at least among the birds living outside the rain forest, but few avian communities can yet be divided with confidence into sedentary and migratory species. In tropical Africa many granivorous birds are clearly performing seasonal migrations over relatively short distances, although only the species quelea has been studied in detail (Ward, 1971 *a*). In this species the seeds lying on the ground, that comprise the bulk of the diet throughout the dry season, germinate at the start of the rains. This deprivation of food is probably the ultimate cause of the birds' departure. Many other granivorous species are affected the same way, although others, like the icterids, perform seasonal movements at a leisurely pace, probably because food is abundant. Some, like quelea, migrate to areas where rain has been falling for two months and where they can find green seeds forming (as wild grains and cultivated crops). Others do the opposite, moving ahead of the rain front to places where germination has not yet begun (Ward & Jones, in preparation).

In many of the regions where migratory patterns of the pest species are still unknown, ringing (banding) is either not practical or, at least, cannot be expected to give quick results. Migratory species must, therefore, be picked out by amassing circumstantial evidence of regular absences or arrivals over a large area. At first this can only suggest that a particular species is migratory, although eventually it may be possible to work out the migratory pattern in this way. The method is simple, but the work involved is great. An assessment of the species' abundance in a delimited area is made at intervals of a month or less over a period of at least one year. The same is done at a number of other points in the species' distribution area. It is not, as might be imagined, possible to determine the comings and goings of a pest by asking farmers. In the main, farmers see the pest only when it is doing damage and will swear that the birds 'come at the beginning of the harvest time, and leave when it is ended'.

First and foremost, the control strategist needs to know whether a particular population of a pest is sedentary or not, for on this knowledge will rest the decision for or against the strategy of local population reduction. So if the only outcome of an initial investigation based on routine observations over a year or two is a positive statement that the pest species does perform seasonal movements over a wide area, this is of immense value to the strategist, even though the details of the movements may be long in coming.

The role of management

The administrators responsible for the management of bird-pest situations have a complex task to perform in order to satisfy their many customers (the agricultural community, general public, special interest groups such as naturalists and hunters, and the environmental lobby generally). The diverse distribution of granivore pests is such that these reflections on the overall problem take on different dimensions throughout the world. As a result decisions must be made in a spirit of compromise, rather than attempting to provide all that is desired by any one group, but identical decisions cannot be made in all countries.

Formulation of strategy and tactics

Management agencies are responsible for the careful choice of strategy best suited to a particular pest situation. And, as we have shown, there is a way of doing this which ensures that the strategy chosen is that which best combines efficacy with minimum ecological disturbance, safety and long-term economic benefit. Following the strategy choice, they must decide upon the tactic to be employed, using similar guidelines, and finally the actual technique to implement the tactic. In North America and Europe especially, there is a growing need to have such decision processes part of the public record.

While it is not our aim in this chapter to discuss tactics, we have given an example of tactics choice process in Fig. 7.2 using the strategy of local population reduction. In our example we have chosen first the least environmentally harmful tactic we can imagine, reduction by trapping and shooting, and have progressed through reduction by birth control (using nest destruction and methods of chemosterilization) to reduction by use of poisons. Clearly, the research requirements for tactical choices will differ substantially from those we have listed in connection with strategy choice, being more directly concerned with practical-control matters.

Establishment of research priorities

Management must ensure that the information required for strategy-, tactic- and eventually technique-selection will be available when the three stages in the decision-making process (which may follow one another quickly or at long intervals, depending upon the urgency of the situation) are undertaken. Generally speaking, part of the information will be available from published literature or from agricultural, meteorological and other sources, but much will have to be obtained by the research and development (R & D) teams answerable to management. The requirements will differ from case to case, and we have indicated earlier how basic research topics should be selected so that strategy-selection can proceed without undue delay. Similarly, in deciding the nature of R & D input for the subsequent tactic- and technique-selections, topics must be selected

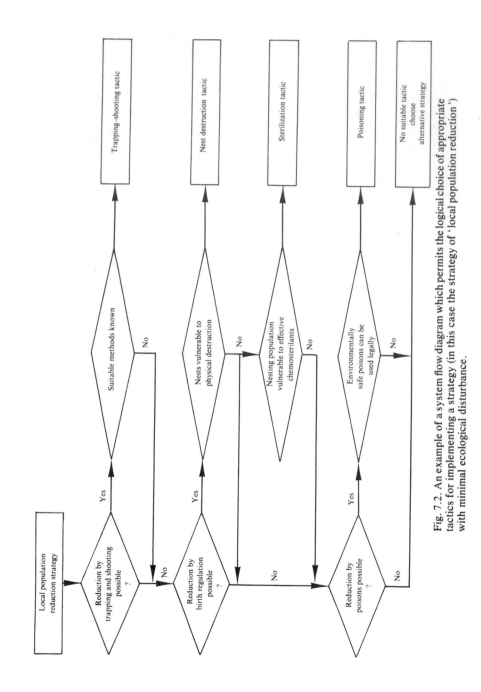

Fig. 7.2. An example of a system flow diagram which permits the logical choice of appropriate tactics for implementing a strategy (in this case the strategy of 'local population reduction') with minimal ecological disturbance.

298

with a view to answering specific questions expected to arise when the choices are made. Ensuring that the R & D input is available in appropriate sequence is perhaps the most arduous task of management. Such a task challenges the best that is available in systems integration theory and practice. Management teams must not get locked into any given strategy and should review their actions from time to time. Any given pest situation may be expected to alter as the birds themselves change in their distribution, density, or behavior; behavioral changes may even be induced by the actions taken against them. Equally liable to alter the picture are changes in agricultural practices, economic factors and public sentiment; and new information will be coming in from ongoing research following the basic work on which the initial strategy choice was dependent. Thus, what is the best strategy at a given time may not be so later; even where there is no need to change strategy, the tactics and techniques implementing it will merit periodic re-examination.

Management has the responsibility of explaining to all interested parties how a particular course of action was decided upon after careful consideration of alternatives during which they were guided by information carefully supplied by research commissioned specialties. This seemingly tiresome duty is not without benefit, for though such an explanatory statement will invariably attract some adverse comment from groups who feel their interests have been compromised or not given due weight, constructive views and technical information will also accrue. As noted before, in the USA it is now mandatory for an 'Environmental Impact Statement' to be prepared and widely circulated well before any action is taken that might be expected to cause major environmental changes.

Surveillance

Management must also ensure continuous surveillance. Pest situations require monitoring even where no action is contemplated, so that in the event of a minor problem starting to grow in importance appropriate research can be implemented early in anticipation of a call for action. A further responsibility of management, or of a designated body, is watching for the introduction of exotic species which are known to be pests in their country of origin or might be expected to become pests in their new home. Examples of introductions where the immigrant population is still small and confined to a small area are the crested mynah *Acridotheres cristatellus* (L.) from Asia, in Vancouver, Canada; the red-whiskered bulbul *Pycnonotus jocosus* (L.) and rose-ringed parakeet from the Old-World tropics, in Florida, USA (Owre, 1973; Carleton & Owre, 1975); and the little bishop *Euplectes orix* (L.), a close relative of *Quelea quelea*, from Africa, in Australia. It seems amazing that these nuclear populations are not only

299

tolerated but often afforded a measure of protection, when one recalls how the European starling has swept across North America and the African village weaver has become a major pest of rice on Hispaniola (from where it may eventually spread through the Caribbean and into the North American continent). New Zealand is in the unique position of having imported nearly all its pest species, but now, like Australia, rigidly controls the import of exotic species. Elsewhere in the world, and especially in the tropics, controls are very lax indeed.

At present, the nuisance value of newly introduced species frequently receives attention only when it is too late to check the spread, and the opportunity of a 'final solution' has been lost forever.

Conclusions

We feel that management of agricultural bird pests in many parts of the world is needed, but management is at a critical stage. In some areas nothing appears to be known about their impact, while in other areas damage is known to occur, but there is insufficient knowledge as to its extent. As a result, bad decisions are made, many of them political in nature. In some instances, there is not sufficient impact to justify major management attempts, especially for potentially environmentally damaging efforts involving massive lethal control. In other instances, there is demonstrable damage, but because of lack of knowledge or simple tolerance of the conditions, there is no action taken when there should be.

It seems to us that it is necessary to keep a very close watch on the various granivorous bird populations about which we have given short synopses. In a world with increasingly large demands for human foods, particularly high protein cereal grains, there is little reason to waste large quantities by neglecting proper management of granivorous birds. However, 'heavy hammer' techniques are equally out of place. Thus we encourage administrators of management programs to assess carefully what is needed for management in light of our systematic approach to the problem and in light of the information given elsewhere in this book about the biology of granivorous birds.

We are indebted to many scientists for critique of our ideas, especially those attending a workshop session at Fort Collins, Colorado, in October 1974: Drs J. A. Wiens, S. C. Kendeigh, J. Pinowski, R. F. Johnston and J. B. Cragg. Help with translation and interpretation of Russian literature has been given by Dr V. R. Dol'nik. Several have read earlier drafts, including many of the above mentioned. The work has been supported in part by a USA National Science Foundation Grant GB-42700, the support services of the Natural Resource Ecology Laboratory funded by NSF Grants GB-41233X, BMS73-02027 AO2 and DEB73-02027 AO3, and travel funds supplied by the Centre for Overseas Pest Research, London, PL-480 funds administered by the Smithsonian Institution, and the IBP (London).

8. Adaptive correlates of granivory in birds

J. A. WIENS & R. F. JOHNSTON

The success of granivorous birds in natural ecosystems is a consequence of distinctive complexes of adaptations or 'adaptive strategies' that have been molded by natural selection. While we may discern a number of distinctive adaptations enhancing the utilization of seeds as the primary food resource, in fact, these features are interwoven into integrated suites of characters – what Pitelka, Holmes & Maclean (1974) have termed *exploitation systems*. Crook (1963), for example, suggested that the ancestral stock of the present *Ploceus* finches 'became adapted to an open country environment through the evolution of a number of correlated characteristics of which seed eating, gregariousness, polygamy, colonial breeding and seasonal and sexual dimorphism are the most important'.

Such evolved characteristics may act as 'pre-adaptations' in appropriate circumstances, predisposing granivorous bird populations to achieve extraordinary success when agricultural ecosystems are made available to them. Such species may then achieve pest status or 'pesthood' when their utilization of these ecosystems conflicts with that of man. The propensity of *Quelea quelea* (L.) for feeding on wild grass seeds, for example, predisposes it to attack cultivated cereals (Ward, 1965a), and bullfinches *Pyrrhula pyrrhula* (L.) could perhaps have been expected to become a major pest in orchards, since in their natural habitat they take a greater proportion and variety of buds over a longer period of the year than other native birds and possess several behavioral and morphological adaptations which enhance bud consumption (Newton, 1972). There are, of course, constraints on pre-adaptations to agricultural systems; a species which is ideally equipped to utilize a particular agricultural crop may be prevented from establishing itself because certain critical requirements are lacking. Chaffinches *Fringilla coelebs* L., for example, could be expected to prosper from cereal grain production in many parts of their range, but the absence of high, elevated singing posts such as hedgerow trees (Murton, 1972) or suitable alternative food sources at critical times of the year (P. Ward, personal communication) may prevent them from occupying many areas.

Some of the features that predispose granivorous species to pesthood are apparent: species that form large feeding aggregations, that are opportunistic in their choice of foods, that can locate and rapidly exploit locally concentrated food sources and that are largely insensitive to conventional human control efforts would seem to be likely candidates.

301

J. A. Wiens & R. F. Johnston

Proper management of granivorous bird populations requires an understanding of these and other, subtler, adaptive dimensions of granivory. In this chapter we explore several dimensions of the adaptive strategies of granivorous bird populations. In essence, we are asking what attributes characterize successful granivores, why some species form close associations with man as pests or commensals, while others do not, and whether by considering adaptive complexes we may recognize some species as incipient pests or commensals, were man to alter their environmental setting slightly. Unlike previous chapters, which considered facets of the biology of granivorous birds within a factual or empirical framework, we adopt a more theoretical and intuitive approach. We do not intend to review all adaptive features of all granivorous birds but rather draw attention to several dominant features in the lives of some notably successful granivores. With Dyer & Ward (Chapter 7, this volume), we believe that understanding the relationships of granivorous birds in agricultural ecosystems and framing appropriate management strategies and tactics rely upon discerning the underlying exploitation systems of the birds.

Seeds form a major component of the diet of a rather broad array of avian taxa, although few families are primarily granivorous (Morse, 1975). Of these, the seed snipes (Thincoridae) are a small South American group, while sandgrouse (Pteroclidae) are widespread throughout the Old World, especially in arid regions. The weaver finches (Ploceidae), other finch groups (Emberizidae, Estrildidae, Fringillidae), and pigeons and doves (Columbidae) are large families which make substantial contributions to all continental avifaunas. The considerable success of the finch families and the relative scarcity of other primarily granivorous groups suggested to Morse the possibility that these families 'have preempted most available niches on continental regions'. Be that as it may, granivory is well developed in several other avian groups, notably larks (Alaudidae) and at least some members of the diverse New World blackbird family (Icteridae). Not all of these granivorous groups have received close study, but the studies of Crook (1962, 1964, 1965), Ward (1965a, b, 1966, 1971a) and Skead (1964) on ploceids, Orians (1960, 1961b) and others on icterids, Newton (1972) on finches, and the more general reviews of Lack (1966, 1968) provide the background for an exploration of the adaptive correlates of granivory in these groups. In particular, we wish to discern whether there are common trends to granivorous exploitation systems among forms of differing evolutionary lineages, occupying different bio-geographic zones.

The context of granivory: seeds as resources

In order to understand the various adaptations promoting successful granivory, we must first consider some characteristics of seeds and of the vegetational communities in which granivory probably developed to its fullest extent. Many of the adaptive dimensions of granivory are determined by the characteristics of seeds as prey items. Most seeds are small, are protected by a hard seed coat and, while frequently rich in calories and easily metabolized carbohydrates, are deficient in other essential nutrients, especially proteins. As a consequence, rather large quantities of seeds are required to satisfy the energetic and nutritional needs of individual birds. Special nutritional demands may be especially critical during the breeding season, when females must produce eggs and young must grow at suitable rates, and many granivores must feed upon insects at these times (Morse, 1975; Jones & Ward, 1976). Some species, notably several cardueline finches, are able to raise young rapidly on a seed diet (Newton, 1972), but in such forms special modifications of the digestive tract are necessary to enable the adults to transport large numbers of small seeds between the feeding grounds and the nest. Breaking the seed coat is not easy, and most granivores possess a conical bill with well-defined palatal grooves, a broad, powerful skull and large jaw muscles which enable them to husk seeds rapidly and efficiently (Kear, 1962; Ziswiler, 1965; Newton, 1967, 1972). Further breakage of the seed coat is facilitated by the extremely muscular gizzard and the large quantities of coarse grit.

These are features which relate to the exploitation of seeds as prey items, but the success of various granivore populations may rest upon adaptive responses to the distinctive patterns of seed availability in time and space. Seed production is not continually required for direct survival of plants, and patterns of the timing of seed production and the quantity and quality of seed produced may thus vary in response to the many selective factors influencing germination probabilities (Janzen, 1971; Harper & White, 1974). Seeds are normally produced at discrete times, between which the resource pool is not renewed, although much of the crop may remain present following production. During an annual cycle of seed production and depletion, two periods may be especially stressful to granivorous birds (Fig. 8.1). Following production the seeds remain on the plant for a time, then fall to the ground, below the vegetation cover. If the cover is dense (as in many grasslands) such fallen seeds may become unavailable to at least some species, such as quelea, until the vegetation cover becomes compacted or disappears. Another stressful period occurs at the onset of the growing season, when most of the seeds remaining in the resource pool germinate, and seeds become quite limited in availability until the new crop is produced (Ward, 1971a). At this time many granivores

303

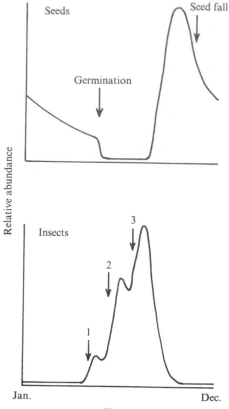

Fig. 8.1. Schematic diagram of the seasonal patterns of relative abundance of seeds and insects in a north-temperate location. Seed abundance diminishes during winter as seeds are consumed, and the supply disappears almost entirely when the remaining seeds germinate. In late summer or autumn, plants set seed which first is carried on the plant and then falls to the ground. Insect abundance is low throughout autumn, winter and early spring, as most individuals are hidden and/or in dormant stages. In spring abundance increases, often exhibiting 'waves' of abundance corresponding with the emergence of new generations (three shown here).

switch to feeding upon insects. The patterns of resource availability confronted by insectivorous species are quite different (Morse, 1971). Insect populations may undergo several generations in a single season but in most environments are quite restricted in seasonal availability (Fig. 8.1). The flush of insect abundance in many habitats during the growing season coupled with the frequent dearth of active insects in the non-growing season promotes characteristic adaptive responses among insectivorous birds, perhaps most notably seasonal migration and/or substantial plasticity in foraging techniques and food choice (Morse, 1971). Superimposed upon these differences in seasonal abundance are differences in yearly variation between insect and seed resource pools. While abundance levels of both food types vary substantially between years, seed production may be far less predictable than insect levels in many temperate or dry, tropical environments (Murton, 1972). This no doubt reflects the closer degree of determination of seed production by climatic variation, such as rainfall, and the irregularities are especially conspicuous among annual plants.

Circumstances favoring the evolution of large-scale granivory among birds are especially prevalent in pioneer or early seral stages of successions, grasslands and arid or semi-arid savannahs. The vegetation in such communities is subjected to widely varying environmental conditions, either through the temporal transiency of early successional stages or the unpredictable stresses imposed by climatic fluctuations (e.g. Noy-Meir, 1973; Wiens, 1974b). In especially unstable environments, annuals frequently dominate the vegetation. Plant populations in such environments are subjected to selection favoring high and rapid reproductive rates – what has been termed '*r* selection' (Hairston, Tinkle & Wilber, 1970; Pianka, 1972) – and a major portion of productivity is allocated to reproductive output rather than individual growth and maintenance (Gadgil & Solbrig, 1972; Hirshfield & Tinkle, 1975). When and where environmental conditions are favorable, vast quantities of seeds are produced. Since growing conditions vary substantially from place to place, the production of abundant· seed resources is usually localized and concentrated, and these locations are often unpredictable from year to year.

While many birds occupying woodland or forest habitats are at least partially granivorous at times, it is apparent that the most highly developed cases of granivory, and the most closely attuned exploitation systems, occur among birds exposed to the resource circumstances just described: large concentrations of small seeds produced at discrete intervals and exhibiting tremendous yearly variation in abundance at a given locality. As human civilizations developed, the more productive grasslands and savannahs were among the first biotopes to be converted to agricultural croplands, and the native granivorous birds were admirably pre-adapted to consume the crop seeds which replaced those of the natural vegetation or to form commensal associations with man (Johnston & Klitz, Chapter 2, this volume). As agriculture has developed, efforts have been made to maximize seed production by controlled breeding and hybridization programs. Crops are planted over large areas and are maintained in analogues of early successional stages more or less permanently. Many arid regions are now irrigated, increasing both the seed yield and the spatial and temporal predictability of the resource. Harvested grain is stored in massive concentrations close to human dwellings, fostering the

305

development of close commensalism. While harvesting removes a substantial portion of the seed crop, considerable seed 'detritus' is often left on the ground, and this, together with livestock feeding areas, may provide the food resources necessary to permit overwintering of large populations of birds in some areas. Thus many of the advances in seed crop practice have had the effect of increasing the suitability of grain crops for birds. Under these conditions, some species have become pests or commensals while others have not. We believe the differences are closely tied to differences in the strategies of granivory followed by the different species.

Components of granivorous adaptive strategies

We may identify several major threads of adaptation to exploitation of patchily distributed seed resources among granivorous birds: components of morphology, feeding behavior, distributional dynamics, reproduction and various features of social organization. While these act in concert in nature, we may initially consider them separately for convenience. Later, we will consider them together.

Morphological features

The most obvious morphological correlate of a seed diet is structure of the bill which in most granivores is large, relatively short and deep at the base. There are, of course, variations on this theme. Finches of grassland and savannah biotopes tend to have relatively blunt bills in comparison with woodland granivores (Lack, 1968), and species which husk large thick-shelled seeds, such as the hawfinch *Coccothraustes coccothraustes* (L.) of Europe or the evening grosbeak *C. vespertinus* (Cooper) of North America, have truly massive conical bills with well-developed palatal ridges and knobs (Sims, 1955). Species that include insects in their diets with some regularity frequently have longer and thinner bills than forms which are more higly granivorous; the diet of horned larks *Eremophila alpestris* (L.), for example, is composed almost entirely of seeds during the winter and of over one-third seeds in the breeding season, yet the long, narrow bill structure more closely resembles that of insectivores than most granivorous birds. Crossbills *Loxia* spp., of course, have a highly specialized mandibular structure enabling them to extract seeds from the hard, closed cones of conifers (Newton, 1972).

Bill size is usually assumed to be closely related to the sizes of prey items selected, and this relationship is the foundation of the many studies using bill dimensions as measures of trophic niches (e.g. Schoener, 1965; Van Valen, 1965; Willson, 1969; Rothstein, 1973). Several laboratory studies (Kear, 1962; Newton, 1967, 1972; Willson, 1971) have demonstrated that large-bill finches husk large seeds more efficiently and consume a wider range of seed sizes than small-billed forms and also exhibit an apparent preference for the larger seeds. Under more natural conditions, however, a wide variety of factors may exert proximate influences on seed selection (Willson & Harmeson, 1973), reducing the closeness of the bill size to seed size relationship. Quelea, for example, has a massive bill yet normally feeds on very small grass seeds; the bill may be large to permit exploitation of larger seeds when food is scarce (Ward, 1965b) or perhaps to minimize wear when picking seeds off hard ground. While differences in bill structure and size between species probably do have long-term correlations with differences in diet composition, it is an oversimplification to assume that these relationships hold under all conditions and at all times. Such, unfortunately, is an assumption of most current 'niche variation' theories.

Many of the more highly specialized granivorous birds have structural modifications of the digestive tract which enable them to store temporarily large quantities of seeds. Bullfinches, for example, have special gular pouches used for carrying food to the young, and many other cardueline finches carry masses of seeds in esophageal pouches (Newton, 1972). *Passer* spp. also have distensible crops, as does *Quelea*. The chief function of these structures seems to be storage of seed quantities for efficient transport. Murton, Isaacson & Westwood (1963) pointed out that the evolution of a storage organ is essential for a nidicolous species feeding on large quantities of food of relatively low nutritive value or needing to make long flights between feeding grounds and the nest to feed developing young. In quelea the well-developed crop enables the birds to gather all of their food during two feeding sessions roughly two hours in length during the cooler portions of the day, returning to roosts some distance away during the midday heat (Ward, 1965a). Beyond this, possession of a crop may allow individuals to eat rapidly, minimizing actual feeding time and thereby maximizing the time available for searching for the patchily distributed seed concentrations, and further enables an individual to take as much as possible from a localized food source before other individuals discover it and consume the remainder (P. Ward, in preparation).

Granivorous birds also differ in the locations in which they feed, and these are closely related to adaptations of leg structure and locomotion. Covariation of head and leg elements in house sparrows *Passer domesticus* (L.) is considerable (Johnston, 1976), and this probably occurs in other species. Seeds may be obtained from the ground, from standing vegetation, or from both. Among the British finches discussed by Newton (1972), the most agile species, which commonly feed from seed heads on the vegetation, are either light in weight or have very short, thick legs.

The species that feed best standing on the ground have relatively long tarsi, and the remaining species, feeding in both locations, are intermediate in leg length and behavior. At least some of the ground-feeding finches can walk or run. Newton suggests that this enables them to cover more ground and thus find more seeds more quickly than the carduelines that feed on vegetation, which can only hop.

Within-species as well as between-species variations in morphology may be associated with differences in biotope utilization patterns. *If* bill size is a reliable index of trophic niche, for example, considerable intraspecific size variation may indicate that individuals within the population are specializing upon different prey resources, while morphological uniformity may reflect a common resource specialization among all members of the population (Soulé & Stewart, 1970; Van Valen & Grant, 1970; Roughgarden, 1972; Hamilton & Johnston, 1977). The most readily apparent form of morphological variation within a population is sexual dimorphism. While many granivorous species are dimorphic for plumage features for different parts of the year, relatively few are dimorphic for characteristics such as body size or bill dimensions, and fewer still show geographic variation in degree of size dimorphism (as in house sparrows; Johnston & Selander, 1973a). In most cases, such size dimorphism is associated with polygynous mating systems (see below), and it seems likely that the primary selective forces favoring size divergence are associated with increasing the attractiveness of males to females. Still, such differences may have important ecological consequences. For house sparrows, Hamilton (1974) has demonstrated an inverse relationship of variance in body size (roughly, sexual dimorphism in body size) with degree of interspecific competition for seeds, noted above (Chapter 2). Ward (1965b) has suggested that the bill-size dimorphism in quelea may be important in times of food shortage, males then taking larger seeds than females. Similarly, Fretwell (1972b) has suggested that in the dickcissel *Spiza americana* (Gmelin), which is dimorphic for size, males on the wintering grounds may feed upon larger grain (e.g. sorghum) than females. He further proposes that the increasing availability of sorghum in wintering areas resulting from agricultural development may enhance male survival, contributing to a skewing of the breeding sex ratio in favor of males, perhaps to the point of females representing such a small proportion of the breeding population that populations are threatened with extinction.

Dietary opportunism

The relative flexibility or stereotypy of a species' diet may have major consequences upon its potential to attain pesthood. Intuitively, we would expect that opportunistic, generalist feeders should be more likely to

308

become agricultural pests than highly specialized forms, since they may exhibit strong functional and numerical responses (Holling, 1965) to the appearance of an abundant new food type (e.g. an agricultural grain) in their surroundings. Many granivorous species are to some degree opportunistic in their selection of seed types. Chaffinches in Britain, for example, feed almost entirely upon seeds outside the breeding season and opportunistically exploit a broad range of seed taxa (Newton, 1972). Their feeding is strongly influenced by agricultural practices, and where conditions are suitable they may eat cereal grains for up to two-thirds of the year, obtaining them from fields of freshly sown or ripening corn, from grain stubbles, or from around granaries. In general, however, the birds take grain only from the ground and not from developing or ripe seed heads, and their agricultural impact is thus quite limited. On the other hand, crossbills, which are highly restricted in the range of seed types consumed, respond dramatically to variation in the distribution and abundance of the preferred seed types but are largely insensitive to variation in the availability of other foods, such as cereal grains. Their pest potential is thus quite limited.

Seed consumption during the breeding season may be especially critical, both in determining the adaptive complexes of the bird species and in its potential effects upon developing grain crops. Many carduelines, for example, feed their young regurgitated seeds or a mixture of seeds and insects (Newton, 1972), while other highly granivorous species, such as quelea or the sociable weaver *Philetairus socius* (Latham), must feed their young insects or other animal prey during part or all of the nestling period (Ward, 1965a; Maclean, 1973e). Ward's studies of quelea provide a particularly detailed picture of seasonal dietary shifts in a granivore. In the Lake Chad region, quelea feeds upon seeds on the ground through the dry season, maintaining constant minimum body weight (Ward, 1965b, 1971a; Fig. 8.2). With the first signs of coming rains, the birds begin to eat heavily and gain weight. When the heavy rains begin in late June, the remaining seeds germinate, and the birds feed primarily upon termites (*Odontotermes*) which emerge briefly. Upon disappearance of the termites, most birds leave the area in a pre-breeding migration (see below). Those individuals which remain in the area turn to small insects, but this is a time when substantial food shortage causes stress and body weight declines; starvation may be frequent. However, as the vegetation becomes more lush, green seeds and larger insects become available, and quelea occupying such habitats regain weight in preparation for breeding. They breed at the end of the rainy season, weight again declining after the young are hatched because of the time and energy expended in feeding and carrying water to the young. The young gain independence, and the adults return to a diet of dry seeds on the ground. These seed supplies are

309

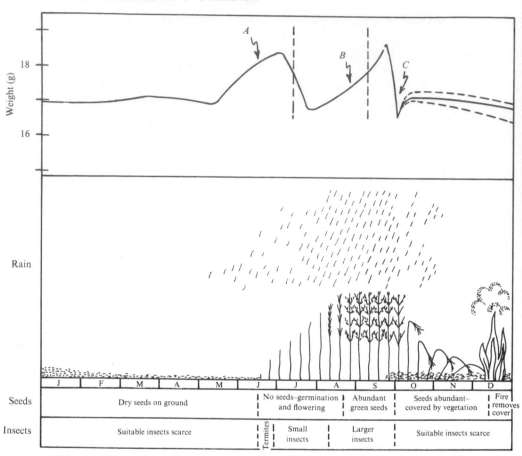

Fig. 8.2. Diagrammatic portrayal of seasonal changes in female body weight, rainfall and availability of grass seed and insects to quelea in a semi-arid area with one wet season in the year (e.g. the Lake Chad area). The initial weight increase (*A*) precedes the 'early-rains migration', when most birds leave a locality. The interval between the dashed lines denotes their absence from the area. Weight again increases (*B*) in preparation for breeding and return to the breeding quarters, and then drops rapidly (*C*) as the birds lay a clutch and initiate incubation. There follows a weight gain, but the final level may vary in relation to the availability of proper food supplies and the clutch size and drains placed upon individuals by the activities of rearing young. Modified from Ward (1965*b*, 1971*a*, personal communication).

initially covered by standing, dead vegetation, and the birds again are subjected to stress caused by food shortage. Only when fires or consumption of dry grass by herbivores removes the standing vegetation are the abundant seeds readily available for consumption.

The basic problem faced by a granivore-like quelea is when individuals should feed upon different prey types. Katz (1974) has used dynamic optimization techniques to approach this problem. The basic premise of

310

his optimization procedure is that selection should favor that foraging behavior that maximizes the overall time–energy benefits to individuals. Katz considered optimization of quelea feeding patterns with the long-range goal of minimizing total annual feeding time, and his model produced feeding patterns and a history of individual weight variation that bore close resemblance to the actual patterns documented by Ward (1971*a*, *b*). In addition, his model suggested that minimization of long-term foraging time might involve maximization of daily feeding at some times of year, especially during the lush rainy season, and minimization of daily feeding time at other times.

Katz has applied such long-range optimization models to another granivorous species, the dickcissel feeding on its wintering grounds (Katz & Fretwell, unpublished). The challenge dickcissels face is to survive for roughly the four months spent on the wintering grounds (Trinidad, in this case) and gain enough fat reserves to migrate north to the North American breeding grounds. During the winter period, the birds have access to two general food types. Food 1 consists of dry seeds (e.g. rice, weed seeds), which are eaten on the plants, above ground. Since these seeds are relatively large, individual energy demands may be satisfied by relatively little foraging time, but since the seeds have hard coats they require prolonged digestive pauses; during this time the birds may seek more sheltered areas, where they are less vulnerable to predation than when feeding in the open fields. Such seeds thus represent a low profit, low-risk food type. Food type 2 offers greater rewards, but at greater risks. These are small seeds and insects which are obtained on the ground. Such food is harder to find, and thus more time must be spent in foraging, during which the birds are exposed to predation. However, the food is soft and readily digestible, so more can be gathered during a day. Katz's simulation analysis suggests that the optimal strategy of dickcissels faced with a choice between such food types may be to 'prefer' type 1 at all times, because of its lower risk. However, the birds are unable to achieve the necessary pre-migratory weight gain on food 1 alone, so they must supplement the diet with food 2. The optimal pattern is to do this as late as possible during the wintering period, switching to large-scale consumption of food 2 to gain weight rapidly, rather than gradually increasing weight through the winter period. The model results show good agreement with the actual weight changes in wintering dickcissels documented by ffrench (1967). Interestingly, large males feeding in rich agricultural areas, upon sorghum, may be able to avoid completely the shift to food 2 and the attendant increase in mortality, because of the increased richness of food 1 supplies.

Opportunism in feeding behavior and food selection may also be affected by learning and the development of 'traditions'. An individual's experi-

311

ence with various food types may lead to the establishment of preferences which differ from those of other members of the population whose experience with food types has been different. Repeated encounters with specific food types may lead to the development of 'specific searching images' (Tinbergen, 1960; Gibb, 1962; Croze, 1970) for those food types, causing consumption to be disproportionately great in relation to their availability. Similar learning processes may lead individuals to forage preferentially in locations which have proven to be profitable in the recent past (Royama, 1970). Dyer (1967), for example, noted that once a flock of red-winged blackbirds *Agelaius phoeniceus* (L.), established a feeding site, there is a marked faithfulness to this area as long as the flock remains in the vicinity. The abilities of birds to learn the identity of profitable food types and/or profitable feeding patches have been tested experimentally, although most of the studies have been conducted on insectivorous tits (*Parus*). Alcock (1973), however, tested the learning potential of young red-winged blackbirds with respect to both locational cues (*where* to feed) and visual cues of food items (*what* to feed upon), and found that both abilities were well developed. This is clearly adaptive in an omnivorous opportunist which is likely to encounter food in a variety of complex biotopes, as redwings do. By learning where the productive patches are, the birds can concentrate their feeding there and thus avoid wasting time and energy in searching in less productive areas. And by learning the features of specific prey types, individuals may be less sensitive to other 'irrelevant' food stimuli in the selected patches and may exploit more efficiently the main foods present there.

Frequently, individuals occupying an area share common food preferences. Such similarities could arise because the individuals share the same resource base and thus have similar individual feeding experiences, or because they learn food preferences or foraging techniques from one another. Klopfer's (1961) experiments with such observational learning demonstrated that species differ in their abilities to profit by watching other feeding individuals. In particular, greenfinches *Carduelis chloris* (L.), which are normally rather solitary feeders in nature, were relatively uninfluenced by the learning experiences of other individuals, while the more social, flock-feeding great tits *Parus major* L. rapidly assimilated the feeding characteristics of individuals they observed. It seems likely that the potential for observational learning to be an important determinant of dietary preferences is associated with the tendency to feed in aggregations, and since many granivorous species feed in flocks over much of the year (see below), such learning may be widespread. Observational learning may produce local traditions in feeding, in which individuals in a given local population share common food preferences that differ from those of other populations of the species, even if the underlying food resource base is quite similar (Newton, 1972). Redpolls *Acanthis flammea* (L.) adopted the habit of eating fruitbuds in New Zealand, where they were introduced, and such feeding is now so widespread that in some areas (e.g. the South Island) the species is an agricultural pest. This feeding habit has only recently appeared in British redpolls (Newton, 1972). Such learned local traditions clearly complicate the determination of the pest status or potential of a species over various parts of its range.

Distributional opportunism

The distributional stability or instability of a species may have profound consequences for its potential agricultural impact and it susceptibility to various management tactics. Thus, a form that is capable of regional and/or seasonal shifts in distribution and abundance in response to changing food supplies may be more likely to respond to regional agricultural patterns than a species with more conservative distributional dynamics. Closely related species may differ in distributional stability. In Africa, for example, the distribution of *Ploceus cucullatus* (Müller) is strongly influenced by the availability of conditions suitable for colony location, and new sites are sought out before colony establishment; *P. nigerrimus* Vieillot is a resident of forest edges and exhibits much more stable distributional patterns (Crook, 1963; Collias & Collias, 1969). In semi-arid and arid regions of south-western Africa, sociable weavers are remarkably sedentary, individuals remaining at the colony location throughout the year. Inter-colony movements, even over distances of less than a kilometer, are quite rare (Maclean, 1973*a*). Among the five species of Australian *Corvus*, three are relatively large (ravens) and sedentary, adults remaining on territories throughout the year, while two are smaller (crows) and nomadic. The nomadic species spend most of the year roaming over large areas, exploiting seasonal food abundances, frequently in biotopes unsuitable for breeding. Species pairs of crows and ravens co-exist locally, and the nomadic crows have several features that may act to minimize direct competition with the larger residents (Rowley, 1973*a*, *b*; Rowley, Braithwaite & Chapman, 1973; Rowley & Vestjens, 1973).

Regional or seasonal shifts in the distribution and abundance of granivorous species are usually associated with variation in seed supplies. This relationship is perhaps most apparent among the crossbills, whose entire biology is closely geared to the production of seeds by conifers. Reviewing their annual cycle, Newton (1972) observed that the birds move between areas during the summer as seed crops are forming, leaving areas where the crops are poor, and settling in more productive areas. If the seed supply persists, the birds remain in one location and undergo breeding and moulting, moving again the next summer in search of other areas of high

313

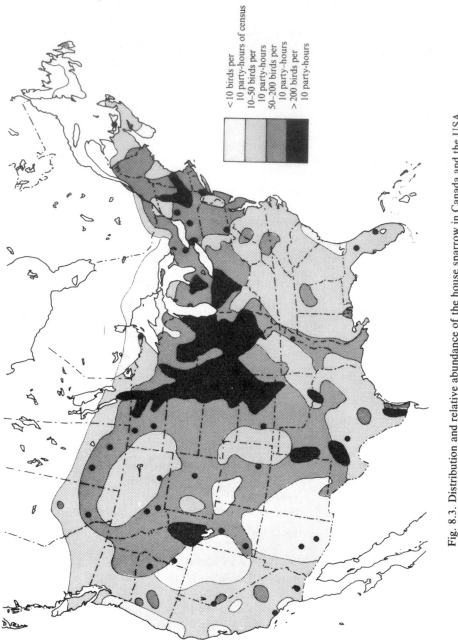

Fig. 8.3. Distribution and relative abundance of the house sparrow in Canada and the USA in December 1973 (data from Arbib & Hailbrun, 1974).

Legend:
< 10 birds per 10 party-hours of census
10–50 birds per 10 party-hours
50–200 birds per 10 party–hours
> 200 birds per 10 party-hours

seed production. The length of the breeding season varies in relation to the quality and persistence of food supplies. Crossbills may breed over three months each year in larch areas, up to seven months in pine, nine months in spruce, and even longer in mixed conifer forests where the cones of different species open at different times and thus provide the birds with a more stable food supply. Crossbill nesting phenology has become divorced from the photoperiodic control which is customarily thought to govern the breeding cycle of most passerines. Peter Ward has pointed out to us, however, that the breeding cycles of all passerines may be governed by nutritional state as well as photoperiod. Crossbills thus may attain the proper nutritional state for breeding at various times of year, while most 'typical' passerines are so induced to breed only in spring and summer. In any event, crossbills are highly adapted in both distributional opportunism and reproductive phenology to a spatially and temporally unpredictable but abundant seed supply.

The regional patterns of abundance of more sedentary granivores may also show close relationships to resource availability. A particularly graphic example is provided by the distribution of house sparrows in North America. This species was introduced to several localities in the eastern USA beginning in the 1850s and spread westward to cover virtually the entire country by 1915 (Robbins, 1973). While local or even regional population densities have not yet stabilized, especially in the more recently colonized areas of the west, the present distribution reveals an informative pattern. Fig. 8.3 plots the densities of house sparrows over North America (exclusive of Mexico) during the mid-winter. The two ecological dimensions of house sparrow commensalism with man are clearly indicated. On the one hand, major concentrations of house sparrow abundance during winter occur in areas of large-scale cereal grain production, such as the upper midwest, the eastern Great Plains and the Mississippi Valley. Even rather isolated areas of grain production, like western Texas, the Pecos River Valley and the irrigated areas north of Great Salt Lake in Utah, support high densities of wintering house sparrows. In these areas, winter food supplies, such as grain detritus in cultivated fields and scattered grain in livestock feedlots and farmyards, are presumably abundant. The other concentration of abundance is in urban centers, where the birds are able to overwinter in large numbers by scavenging the abundant human detritus and by utilizing buildings as shelter.

The adaptive aspects of the distributional dynamics of granivorous birds are especially evident in the patterns revealed by close studies of several notably opportunistic species. Ward's studies of quelea (1966, 1971a; Jones & Ward, 1976) demonstrate the precision with which this species distributionally tracks variation in its food supplies. Quelea are most numerous where their staple foods, seeds of annual grasses, are most plentiful. The most suitable areas are those in which the annual grasses growing during the wet season have been especially productive, such as river terraces, lacustrine plains, or shallow depressions. With the coming of rains, the seed supply becomes unavailable, and the birds feed temporarily upon emerging termites (Fig. 8.2). When the termites disappear the birds move out of the area completely. The seasonal rain fronts move across regions of Africa in bands, and Ward suggests that quelea undertakes an 'early-rains migration' by moving back across the front to regions which began to receive heavy rains roughly six weeks previously and where green seed is now becoming available. As conditions begin to deteriorate in these areas, one to two months later, the birds perform 'breeding migrations', following the rain front into areas where seed and large insects are now appearing. As the birds undertake these migratory movements in relation to the rain front, large detachments of individuals in breeding condition may suddenly appear in an area where conditions have become suitable, and establish huge breeding colonies. There is no fidelity to particular locations from one breeding season to the next, and the occupancy of a given area in a given year depends very much upon whether it has received heavy or light rains. Quelea exhibits 'itinerant breeding', in which individuals apparently stop to breed several times during the rainy season migrations, raising successive broods in colonies which may be hundreds of kilometers apart. In equatorial regions, the rain front moves first south and then back north across the equator, producing a pattern of successive quelea breeding episodes such as that depicted in Fig. 8.4.

Quelea occupancy of a given area during the breeding season is limited by the phenology of the vegetation. As the annual grass seeds ripen they fall to the ground and are temporarily unavailable under the herbage (Fig. 8.2). During this time the birds must turn to other foods. In areas under agricultural development, the crops of small grain cereals (e.g. rice, wheat, sorghum, millet) grow more slowly than the wild grasses and do not shed the seed when ripe. These food supplies are thus available at the end of the rainy season, when grass seed is not, and massive numbers of quelea may turn to such food sources, creating enormous damage. Past records indicate that immense numbers of quelea were present in the drier parts of central Africa before the recent increase in grain crops, and in both east and west Africa the birds still live primarily in uncultivated areas (Lack, 1966). It is unclear whether quelea numbers have increased with the advent of agriculture, although P. Ward (personal communication) thinks not.

In western North America, the tricolored blackbird *Agelaius tricolor* (Audubon) exhibits a distributional opportunism which parallels that of

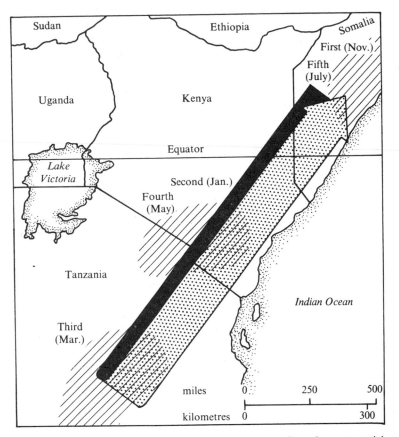

Fig. 8.4. The probable pattern of migration and itinerant breeding of an equatorial quelea population, showing the succession of areas becoming suitable for breeding as the rain front moves first south and then back north across the equator. How many of these breeding opportunities an individual exploits of course depends upon many factors. Breeding also occurs outside the shaded areas. From Ward (1971*a*).

quelea in many ways. Tricolors are largely restricted to the central valley of California, where they are nomadic and highly colonial. Like quelea, large numbers of individuals suddenly appear in an area and quickly breed where conditions are suitable. Fidelity of young or adults to breeding areas is absent, and colony location shifts extensively from year to year, the birds responding to areas in which food (especially insects) is plentiful (Orians, 1961*b*; DeHaven, Crase & Woronecki, 1975). As breeding areas become hot and dry in mid-summer, the birds depart to form large aggregations in flooded areas and marshes, feeding upon weed seeds, maturing rice and other grains and insects. If conditions are suitable in the autumn, some individuals may breed again (Orians, 1960; Payne, 1969),

317

in a manner equivalent to the itinerant breeding of quelea, although on a much reduced scale. Orians (1960) is of the opinion that such autumnal breeding is a recent innovation in tricolors, associated with the development of agricultural irrigation.

The distributional dynamics of dickcissels appear to involve opportunistic movements over considerable distances. In autumn the birds depart from the breeding grounds in grasslands of the central USA, migrating south through Mexico and Central America into northern South America, where populations winter in the llanos of Venezuela and Colombia. The llanos become quite dry by late December, however, and dickcissels move elsewhere, largely into more favorable areas in southern Central America or Trinidad (ffrench, 1967; Fretwell, 1972b). The departure of dickcissels from Trinidad is quite sudden and predictable, most individuals leaving within a week of mid-April. Apparently some birds move west back into Venezuela, where spring rains have increased food availability (ffrench, 1967), while others migrate directly north across the Gulf of Mexico to Texas and Louisiana (Fretwell, 1972b). Fretwell suggests, without much evidence, that birds which breed in Texas in April may move north to breed again in the central and northern prairies in June, following the northward progression of spring and abundant food supplies.

Dickcissel breeding densities and distributions are quite erratic from year to year (Emlen & Wiens, 1965; Wiens & Emlen, 1966; Sealy, 1971). In some areas these fluctuations in abundance may correspond with variation in local conditions, but often there are no apparent differences in biotope suitability or food supplies between locally 'high' and 'low' dickcissel years. Particularly perplexing is the breeding occurrence of this species on the east coast of the USA. There, it was a common breeding species during the first half of the nineteenth century, but thereafter it declined in abundance, disappearing from this eastern segment of its range by 1900. Hurley & Franks (1976) have attributed this range shift to changing agricultural practices and increasing urbanization in the east, but Fretwell (1972b) has proposed a relationship to events on the wintering grounds. About the time dickcissels began to disappear from the east coast, livestock grazing was markedly increased in the South American winter grounds. Winter dickcissels exhibit a strong preference for tall grass cover, and the extensive reduction in such habitat by heavy grazing may have substantially reduced populations, perhaps especially influencing those birds with more easterly wintering distributions. It seems likely that the erratic fluctuations in distribution in the present breeding range may also be associated with events influencing survival in wintering areas.

One other aspect of the distributional behavior of some granivorous birds deserves mention. Male dickcissels with established territories may periodically leave the territories in 'distant flight', especially early in the breeding season. This behavior is substantially more frequent in unmated males than in males with several mates (Schartz & Zimmerman, 1971). Bachelor bobolinks *Dolichonyx oryzivorus* (L.) exhibit similar behavior, departing from established territories for as much as a day at a time (Martin, 1971). Schartz & Zimmerman suggest that birds engaged in distant flight survey other possible breeding areas nearby. Both species occupy early seral stages of grassland and meadow succession, and suitable patches of habitat are thus somewhat ephemeral in time. Under such conditions it may well be to the advantage of a male with poor breeding prospects (no mates) in one location to search for other areas coming into the most suitable successional phase, while at the same time maintaining a base in a currently suitable situation.

Reproductive biology

The reproductive potential of granivorous birds has a profound effect upon the patterns of distribution and abundance of populations in natural biotopes and the potential of a population to assume pesthood. Populations with high reproductive output and rapid population turnover may be able to respond rapidly to environmental changes and exploit new resource opportunities (e.g. crops, seed flushes) through numerical responses (Holling, 1965), and such populations may also be able to recover numbers quickly following localized lethal control measures where they are serious pests. Populations with lower reproductive potential and slower population turnover rates, on the other hand, may be especially stable in suitable biotopes but are relatively sensitive to environmental alterations or fluctuations. These alternative reproductive modes are recognized in the definition of *r* strategist and *K* strategist patterns (Gadgil & Solbrig, 1972; Pianka, 1972; Pianka & Parker, 1975; Jones, 1976). Compared with many other taxa, birds are low-fecundity organisms (Williams, 1975), and therefore their potential to express rapid population growth rates (high *r*) with the attendant adaptations usually attributed to *r* strategists is quite limited. Still, many of the most important and spectacular granivorous bird populations inhabit relatively variable and unpredictable environments (deserts, savannahs, grasslands, etc.). Theory leads us to expect that under such conditions bird populations might express various traits which lead to relatively high reproductive output (e.g. large clutch size, multiple broods, rapid developmental rates, early sexual maturation), at least in comparison with bird populations occupying more stable and predictable environments. Can we in fact distinguish features of the reproductive biology of granivorous birds which enhance their exploitation of seed-rich habitat situations?

319

J. A. Wiens & R. F. Johnston

Timing and duration of breeding

In most birds occupying temperate localities, the initiation of breeding activity is associated with photoperiodic changes (Follett, 1973; Immelmann, 1973; Farner, 1975). Generally, changes in day length in these areas act as reliable proximate predictors of oncoming conditions suitable for producing eggs and rearing young. In tropical regions or arid and semi-arid environments, however, photoperiod may not be so closely associated with periods of maximum primary production and food supply, and this may be especially true for granivorous species whose food supply is so closely geared to vegetative phenology. In areas with highly seasonal or unpredictable rainfall, the development of breeding condition is geared to the onset of the rainy season. Thus quelea in sub-Sahara Africa begins to accumulate fat deposits and gain weight with the first indications of early rains, in preparation for the lean period when the remaining seeds germinate (Fig. 8.2) (Ward, 1965b) and sociable weavers in the Kalahari sandveld of South Africa rapidly initiate breeding following a certain minimal rainfall, regardless of the time of year (Maclean, 1973c). In Australia the larger resident *Corvus* spp. have regular breeding seasons which are presumably photoperiodically-coordinated, while the smaller, nomadic forms have quite variable breeding seasons, apparently responding to local climatic conditions (Rowley et al., 1973). Here, the resident species occupy territories throughout the year, thus insuring continuing access to food; the nomadic species are in a sense superimposed upon the distribution of the larger residents, opportunistically exploiting favorable feeding conditions when and where they occur.

These timing mechanisms bring individuals into readiness for breeding, but the actual initiation of clutches by individual females in at least some granivores may be more directly determined by food intake and nutritional condition. In quelea, Jones & Ward (1976) propose that successful formation of a breeding colony can occur only when sufficient protein food is available for the females to accumulate the reserves necessary to initiate egg production and sustain the rapid decline in body protein as eggs are formed. While seed foods supply sufficient calories for the birds, they do not provide enough protein for females to achieve the necessary levels, and breeding activity is thus closely tied to the availability of insect food, which is of course associated with the onset of the wet season. Autumnal breeding in tricolored blackbirds in California is tied to abundant insect prey, and spring breeding may also be initiated by the presence of adequate amounts of the proper sorts of food (Payne, 1969).

In the arid and semi-arid habitats in which granivory is most highly developed, the period of seed and insect abundance associated with wet seasons is often quite limited. Under such conditions, selection should

320

favor a high degree of efficiency and speed in breeding. Among the various African and Asian ploceids, the more highly insectivorous species, which are usually associated with forest fringes, have longer breeding seasons than the granivorous species occupying semi-arid grasslands and savannahs (Crook, 1963; Lack, 1968). The two *Corvus* spp. in Australia which are nomadic and respond opportunistically to localized food abundances have substantially shorter breeding periods than the sedentary ravens (Rowley, 1973*b*; Rowley & Vestjens, 1973). Where wet seasons are of variable duration, at least some granivores may be able to respond with breeding seasons of flexible length. In the Kalahari, for example, sociable weavers are resident at colony locations throughout the year, and the length of a given breeding period is governed by the time of year at which rain falls and the amount of rain, longer periods generally following greater rainfall amounts (Maclean, 1973*c*).

Breeding efficiency and speed in grassland and savannah granivores are promoted chiefly by two adaptations: breeding synchronization among adults and rapid development of young. Synchronization of breeding activities is especially close among populations of quelea (Ward, 1965*b*; 1971*a*; Jones & Ward, 1976), tricolored blackbirds (Orians, 1961*a*; Payne, 1969) and Spanish sparrows *Passer hispaniolensis* (Temminck) (Gavrilov, 1962, 1963). In these species, birds appear and establish colonies in an area quite suddenly, and virtually all of the females may begin incubation within one or two days. All of these species are highly colonial, breeding aggregations frequently numbering in the hundreds of thousands or millions of birds, and less aggregative granivores seem to lack such close synchronization. Red-winged blackbirds, for example, subdivide suitable habitat into large breeding territories and spend considerable time in territory establishment and defense and in courtship displays. There is thus individual variation in breeding phenology among the birds breeding in a given location (Orians, 1961*a*). Jones & Ward (1976) have suggested that in quelea the remarkable degree of colony synchronization may stem from the nutritional control of breeding activity. Individual females within a massive flock moving through regions on their early-rains or breeding migrations may be at different nutritional levels, reflecting their past and current feeding histories. At any given time, a portion of the females may exceed the nutritional threshold to initiate egg formation, and these birds then break away from the main body of the migration to seek out a suitable location, establish a breeding colony and begin nesting. Whatever the mechanism, breeding synchronization enhances the exploitation of the extremely abundant but ephemeral resources on which such vast colonies feed.

Efficient exploitation of such foods is also facilitated by a short breeding cycle and rapid development of the young. While growth rates are subject to nutritional, environmental and genotypic influences, there is rather remarkable uniformity among altricial land birds, and Ricklefs (1968, 1972) suggests that this is close to the maximum rate physiologically possible. Still, some granivorous species have extremely rapid breeding cycles. The nomadic, opportunistic Australian *Corvus* spp., for example, require about one to two weeks less for incubation and rearing nestlings than the larger resident species, and upon fledging the young are rapidly led from the breeding territories to form large feeding flocks (Rowley, 1973*b*). Rowley & Vestjens (1973) suggest that the accelerated breeding cycle acts to reduce competition with the sympatric resident species by minimizing the duration of territorial overlap. Quelea females begin laying eggs before the nest is complete, and the 9- to 11-day incubation period is among the shortest of all birds. Young remain in the nest for 10 days, during which time they accumulate fat reserves to ease food stress when they attain independence from the parents, some 19 days after hatching (Ward, 1965*b*). Sociable weavers, on the other hand, require 13–14 days for incubation, and the young remain in the nest for 21–24 days before fledging, after which they retain close association with the parents (Maclean, 1973*c*). The breeding cycle of tricolors is on the order of 20 days between the start of incubation and fledging, while red-winged blackbirds require roughly 30 days (Payne, 1969). Certainly, there is no general pattern for all granivores to have reduced breeding cycles and rapid development, although the more highly exploitive colonial species seem to. By contrast, however, most cardueline finches have prolonged nesting periods and relatively slow developmental rates, perhaps because the diet fed the young is composed largely of seeds (Newton, 1972).

Natality patterns

Despite all the discussion of clutch size in the literature, the annual fecundity of birds is poorly understood (Ricklefs, 1973). Clutch size is, of course, a major component of natality rates and has received by far the greatest attention, probably because of its relative ease of measurement. While latitudinal and maritime–continental gradients in clutch size exist for many species (Chapter 3, this volume), there is often substantial yearly variation at any single location (Pinowski, 1968, this volume). This variation mirrors the effects of many proximate influences on clutch size and tends to obscure broader comparisons. Overall, the clutch sizes of many granivorous birds seem about average for terrestrial passerines, and trends which hold for other species (e.g. smaller clutch size in tropical regions, larger in hole-nesting forms) are also found among granivores.

The effects of proximate environmental influences upon clutch size are particularly visible in species inhabiting arid or semi-arid regions. Maclean's (1973*c*) studies of sociable weavers showed that clutch sizes became significantly greater as the vegetation became greener and the food supply presumably increased, both results of increased rainfall. In quelea, females begin to lay eggs within four days of the establishment of a colony, and are just beginning to produce yolks when they arrive (Jones & Ward, 1976). Females apparently normally begin development of one yolk more than the number eventually ovulated, but under unusually favorable feeding conditions, all developed yolks may result in eggs. Jones & Ward postulate that all females begin egg production upon reaching a similar initial level of protein reserve and cease when they reach some lower threshold or begin to approach this too rapidly. Differences in clutch size thus arise from differences in daily rates of protein loss from the reserve, assuming that the supply of kilocalories is not limiting. When protein intake is high during laying, all developing follicles may produce eggs, leading to the exceptional clutches of five, six, or perhaps seven eggs. When daily protein intake is quite low egg production may be sustained only by the rapid exhaustion of the reserve protein, resulting in one or two egg clutches. Thus, female quelea may attempt to lay as many eggs as possible in a clutch, but limitations are imposed by the demands placed upon body protein reserves and the rate at which these can be supplemented from daily food intake. Under 'normal' feeding conditions, the average clutch size (around three) corresponds to the maximum number of young the parents can successfully rear (Ward, 1965*b*), but under exceptionally good conditions females have the capacity to produce larger broods successfully.

Variation in the number of broods which a female successfully produces during a breeding period has a far greater influence on natality potential than clutch size variations and provides most of the flexibility and environmental responsiveness in fecundity. In Poland, European tree sparrows *Passer montanus* (L.) usually have three broods per year, but variation in brood frequency is closely related to proximate climatic conditions (Pinowski, 1968, this volume). The proportion of old (i.e. at least two-year-old) females in the population also affects the frequency of third broods in the population, since old birds tend to begin nesting earlier in the spring and thus are more likely to have a third brood. Spanish sparrows in Kazakhstan have two broods among individuals breeding early, while later breeders in the same locality have only a single brood (Gavrilov, 1962, 1963). House sparrows frequently have four broods per year in Europe, and in good years some birds may have five. In south temperate (the southern USA) and tropical (India) regions, *P. domesticus* may produce even more successive broods. These differences in brood number have obvious effects upon annual rates of production of young. Where Pinowski (1968) conducted his studies in central Poland, *P. montanus* has a significantly greater clutch size than *P. domesticus* and

J. A. Wiens & R. F. Johnston

the number of broods is similar in the two species, therefore annual egg production is higher in *P. montanus* (see Appendixes 3.7, 3.8).

Few other granivorous species have received such intensive study during the breeding season, but general patterns in brood frequency are nonetheless apparent. The ability of a female to produce multiple broods is constrained by the time required for the eggs to hatch and the young to develop to independence and by the climatic and resource regimes of an area. *P. montanus* requires 36–38 days between successive broods, for example (Pinowski, 1968). Since reproduction is not initiated until weather conditions become suitable in the spring, and the birds must complete moulting following breeding in the autumn before conditions become limiting, the total time available for successive broods is restricted. Environmental limitations are much more severe in arid and semi-arid regions, where the availability of food for adults and offspring is closely determined by the occurrence of rainfall. Many of the ploceids occupying such habitats have but a single brood per season (Lack, 1968). Others exhibit greater flexibility. Sociable weaver breeding is closely attuned to the duration and the magnitude of rains, and when the rains are good the birds may produce up to four successive broods per breeding period (Maclean, 1973c). Quelea, tricolored blackbirds, and dickcissels normally are single-brooded in any given area, but under suitable conditions individual females may be able to produce two or more successive broods in different areas, through itinerant breeding (Payne, 1969; Ward, 1971a; Fretwell, 1972b). Many of the cardueline finches inhabiting temperate areas are multiple-brooded (Newton, 1972), and those which feed upon spatially concentrated but unpredictable and ephemeral seed supplies, such as crossbills and redpolls, may also exhibit itinerant breeding to some degree (Peiponen, 1957).

Mortality and survivorship

Relatively high fecundity is generally coupled with relatively high mortality and low survivorship. Such populations exhibit rapid turnover in membership, and have the potential to undergo rapid and dramatic fluctuations in size as proximate environmental factors influence natality or mortality components. As with natality, there are gradients in the magnitude and nature of mortality factors with latitude and habitat. Ricklefs' (1969a) analysis of avian nesting mortality revealed, for example, that mortality rates are generally greater among field- and marsh-nesting passerines than among tree-nesting species, especially those nesting in cavities. Within general taxonomic groups, mortality appears to be lower in arctic than in tropical regions, and in arid than in humid habitats within the tropics. In many tropical and temperate woodland species, predation and nest desertion are major sources of nesting mortality, while in habitats

324

with variable food supplies, such as grasslands, marshes and arid regions, starvation of young appears to be the dominant mortality agent. Many granivorous species may experience the latter conditions. If starvation is consistently a major mortality source, selection should be expected to counter its effects, either by reducing fecundity (fewer mouths to feed) or prolonging developmental time (less food required per day). Ricklefs suggests that neither of these options is normally available. Sacrificing fecundity essentially avoids the problem rather than adapting to it, and lengthening of developmental periods in many of these habitats is countered by the relatively short time during which adequate food supplies may be available. Cardueline finches in temperate areas occupy habitats with longer growing seasons, and they appear to have countered the starvation problem (which may be especially acute when the young are fed a seed diet) by prolonging developmental time (Newton, 1972).

Mortality of young during the breeding period is quite sensitive to proximate environmental conditions and thus varies between species, between populations or localities, or between years in the same locations. One of the most careful and intensive studies of these mortality variations in a granivorous species is that of Pinowski and his colleagues on the Polish *P. montanus* (Pinowski, 1968; Mackowicz, Pinowski & Wieloch, 1970; Pinowski & Wieloch, 1973; Pinowski, personal communication, reviewed in Chapter 3, this volume). Average nesting mortality (percentage of eggs which failed to produce fledglings) of second broods (usually the most successful) ranged from 17% at Nowy Targ to 53% at Gdańsk. Long-term information was available from Dziekanów Leśny (near Warsaw) (14 years) and Kraków (7 years). At Dziekanów Leśny, second-brood nesting mortality ranged from 28 to 54% ($\bar{x} = 38\%$); at Kraków, 19 to 81% ($\bar{x} = 51\%$). Mortality following fledging is heavy, roughly one-third of the fledglings not surviving the first month of independence. At fledging, young *P. montanus* have an average life expectancy of about six months, but for those individuals which survive through November, life expectancy increases to roughly a year. The greatest mortality is of young birds in winter. The heavy post-fledging mortality in *P. montanus* populations produces a layered age structure, individuals representing as many as five annual cohorts occurring in the population at any given time.

Information on mortality schedules in other granivorous species is far less complete, but indicates substantial variation, at least in the relatively easily measured components of nesting mortality. Hatching and fledging success is relatively high (*ca* 75–90%) in *P. domesticus*, and survivorship in colonies of quelea and *P. hispaniolensis* is often in excess of 90% (Gavrilov, 1962; Ward, 1965b). During Maclean's (1973c, d) studies of sociable weavers in the Kalahari, however, he recorded overall nesting success values of only 3.1–17.8%. Success was directly related to clutch size, but even in clutches of six eggs only one-third of the eggs produced fledged young. Predation, especially by snakes, appeared to be the major source of mortality. In North America, dickcissel nesting mortality is apparently related in a complex manner to brood parasitism by brown-headed cowbirds *Molothrus ater* (Boddaert) (Fretwell, 1972*b*; Fretwell, Francis & Shane, 1974). Parasitized nests are rarely successful in producing young, and nest parasitism is more frequent in northern prairie regions than in more southern parts of the range (Texas, southern Oklahoma). Dickcissels are polygynous, but in the northern parts of their range only one-half to one-third of the males are mated. Likewise, breeding densities are low in relation to southern areas, where most males have mates and many are polygynists. Fretwell postulates that cowbird densities are limited by factors other than the availability of dickcissel nests. In northern areas, where there are relatively few nests in a breeding population, cowbirds can parasitize them all, lowering nesting success to virtually zero. In the southern areas, on the other hand, there are many nests, the cowbirds cannot parasitize many (and may be actively harassed when near the breeding population by the larger number of dickcissels), and survival is greater.

Other breeding adaptations

In addition to these general elements of granivore reproduction, two additional features of the breeding biology of at least some species require comment. Some granivores introduce added flexibility into their reproductive output by practicing 'brood reduction', in which by initiating incubation before the clutch is completed, the young hatch on different days and are thus of different sizes. Should environmental conditions deteriorate after the clutch is laid, the smaller, later-hatching young may starve within a day or two after hatching, and thus the relative amount of food for the remaining nestlings is increased. On the other hand, if environmental conditions are suitable, all young may receive adequate food, and productivity is thereby increased. Quelea begins incubation of the clutch with the laying of the penultimate egg (Ward, 1965*b*), as do bobolinks (Martin, 1974), so one young customarily hatches a day later than its nest-mates. Martin has provided an especially detailed examination of the operation of brood reduction in relation to proximate food supplies and the polygynous mating system of bobolinks. In sociable weavers, incubation usually begins with the laying of the second egg of the clutch, or sometimes even upon laying the first egg (Maclean, 1973*c*). Clutch size varies from two to six ($\bar{x} = 3.5$), so the asynchronous hatching often produces young of a variety of ages and sizes in a single nest. The size differences persist for roughly two weeks, but by the time the birds leave the nest the differences among the surviving young have usually evened out, and fledging normally occurs in the same day. In other granivorous species, notably *Passer* spp., brood reduction appears to be uncommon (but see Seel, 1970).

Where populations are localized in distribution, sedentary throughout the year and socially closely knit, the potential exists for the development of cooperative breeding. Such systems have been most extensively documented in the studies of Brown (1974) and Woolfenden (1975) on New-World jays. Cooperative breeding reaches its fullest development in several species in which some young remain associated with the parental territory following fledging, defer breeding the following year, and assist in rearing the offspring produced by their parents that year. Among granivores, such a degree of cooperative breeding has been documented only in the sociable weaver (Maclean, 1973c). As soon as the chicks of a second or later brood hatch, the young of the previous brood(s) help their parents feed them. Since the adult weavers may initiate a new brood quickly after fledging young from the previous brood, young birds may be only 25–30 days old when they begin to feed the next brood. This happens with each successive brood, so that chicks of a fourth brood may be fed by as many as 11 birds (nine young plus the two parents). The parents thus are required to do much less work toward the end of a breeding period, when they are in worn plumage and perhaps exhausted from several months of virtually non-stop breeding activity. The food supply also dwindles toward the end of a breeding period, and the additional birds procuring food for the chicks undoubtedly have significant effects upon survival (which, as noted above, is low, despite the 'helpers'). During the incubation and nestling periods of second and later broods, the young of previous broods continue to roost in the nest chamber with their parents until too many birds accumulate. When there are as many as six young, the nesting chamber becomes crowded, and the fledglings begin to sleep in adjacent chambers in the colonial nest mass. Young remain closely associated with their parents through a given breeding period and, with few exceptions, remain at their home colonies to breed in later breeding periods. The possible genetic consequences of such high fidelity to localized colonies and the kin selection operating in the cooperative breeding system are intriguing but unexplored.

Social organization

Various elements of sociality among individuals are among the most important components of the adaptive systems of granivorous birds which predispose them to pesthood. Birds are most likely to become serious agricultural pests when large numbers of them are concentrated into a localized area, at which time traits that in a more dispersed population

327

might well pass unnoticed become noxious or damaging. A tendency to form social aggregations during the breeding and/or non-breeding seasons is thus of obvious consequence, especially as it is associated with granivory. Other social aspects of exploitation systems, such as the occurrence of monogamous or polygynous mating systems, may also be associated with granivory. Additional elements of social organization could also be explored, but here we will limit consideration to breeding aggregation (colonialism), feeding aggregation (flocking) and mating systems. These features of social organization are closely related to the environmental settings of populations, varying in their expression and development with dispersion and abundance of food supplies and the importance of predators as mortality agents during the breeding period (Crook, 1965; Lack, 1968; Pitelka *et al.*, 1974).

Colonial breeding

Many granivorous species tend to breed in colonial aggregations; indeed, the greatest concentrations of breeding individuals among terrestrial birds occur among seed-eaters. Colonies of quelea typically consist of over 100000 individuals and not infrequently contain over a million birds (Ward, 1965*b*; Lack, 1966). *Passer hispaniolensis* breeds in similarly dense colonies, which often are as large as 800000 pairs, and in one area of Kazakhstan 2600000 birds were estimated to be breeding within an area of 1 km² (Chapter 3, this volume). In North America, colonies of tricolored blackbirds frequently number in the hundreds of thousands of individuals, and the extinct passenger pigeon *Ectopistes migratorius* (L.) bred in vast aggregations of several million birds. These are extremes, of course, and more often the colonially-nesting granivores occur in groups of less than 10 to several hundred individuals (e.g. *Ploceus cucullatus*, Collias & Collias 1969; *Passer montanus*, Pinowski, 1968; *Spiza americana*, Zimmerman, 1971; Fretwell, 1972*b*; *Philetairus socius*, Maclean, 1973*a*; *Agelaius phoeniceus*, Orians, 1961*b*).

This variation in the development of colonialism is closely associated with environmental features. Crook (1965) and Lack (1968) discerned the following patterns. Solitary or territorial breeding appear to occur where the nesting sites are easily accessible to predators. By being cryptic and widely dispersed, the nests are protected, and the dispersion of nesting locations further permits individuals to feed near their nest sites on food resources that are largely unexploited by other individuals. Such a distribution of feeding individuals may be especially significant in areas where food is more or less uniformly distributed and/or is limited in supply; large aggregations of individuals in such situations could rapidly exhaust food supplies in the colony area and be forced to forage at prohibitive distances from nest sites. Where food occurs in patches of local abun-

328

dance, on the other hand, foraging in tight flocks is not only feasible but may enhance feeding efficiency of individuals (see below). The model developed by Horn (1968) demonstrates that if food is clumped and unpredictable, travel time to exploit these resources is minimized by individuals nesting colonially in the geometric center of all food patches, while an even dispersion of food favors individual territorialism.

Under these food conditions, then, aggregative breeding may be advantageous. Whether it develops may largely be determined by predation. If the preferred nesting areas are relatively accessible to predators, species feeding gregariously may become territorial or solitary about nest sites, but where nesting locations are relatively inaccessible, colonies may develop.

These patterns are revealed in the tabulations of breeding systems of avian sub-families developed by Lack (1968). Colonial nesting is found in all three sub-families of birds that feed aerially upon insects, while solitary or territorial nesting is typical of all of the roughly 40 groups that feed upon insects in other ways. Of the sub-families that consume mainly seeds, 83% include colonial species, but colonialism is expressed in only 17% of the fruit-eating groups and 12% of the taxa feeding generally upon both insects and vegetable matter. The correlations between breeding system and feeding preferences are even more apparent when examined within more closely-defined taxa. Thus, among the insectivorous forest-dwelling ploceid finches, 15 species nest solitarily, two breed in small groups and one is colonial; of the granivorous savannah species, none nests solitarily, while 16 breed in colonies (Crook, 1962; Lack, 1968). In the North American species of *Agelaius*, the blackbird *A. tricolor* is highly colonial, individuals defending only a small perimeter about the nest itself, while *A. phoeniceus* is highly territorial, vigorously defending rather large exclusive areas of breeding habitat. Orians' (1961*a*) studies revealed that tricolors compress breeding activities into a short period of time, presumably in association with the frequently limited duration of availability of the locally concentrated food, while redwings prolong courtship and breeding over a longer period. The tricolor system is thus more demanding of energy but less demanding of time than the territorial system of redwings. Tricolors therefore have a high rate of food gathering during the short breeding period and are more closely dependent upon abundant, easily available food supplies than redwings.

Among the finches, most fringillines rear their young on insects (even though the adults may remain largely granivorous), while most carduelines feed the young primarily protein-rich seeds of herbaceous vegetation. The fringillines defend large territories, spread relatively evenly through suitable habitat, and individuals obtain most of the required food within the territories. Carduelines, on the other hand, feed upon a resource which

329

is patchily distributed and locally concentrated. They normally breed in loose colonies, defending only small nest-centered territories, and forage away from the colony in flocks rather than singly (Newton, 1972). This combination of traits alone implies that cardueline finches may normally be closer to the 'threshold' of becoming pests, should conditions change somewhat, than fringillines.

Flock feeding

A tendency among individuals to feed in aggregations should be advantageous when food supplies are patchily distributed, occurring in local concentrations that are scattered and unstable through time (Newton, 1972; Wiens, 1976). Under such conditions, there is generally more food in any one location than can immediately be consumed or economically defended by the individual that finds it, so the disadvantages of permitting other birds access to the food supplies are slight. But gregarious feeding may confer distinct advantages. In addition to the possible roles of flocking in reducing the vulnerability of individuals to predation (Lack, 1968; Hamilton, 1971; Newton, 1972; Pulliam, 1973; Vine, 1973), flocking may enhance the ability of individuals to locate the scattered patches of abundant food, since each bird may respond to the presence of other feeding birds rather than independently searching over a large area. Such behavior is apparent in Ward's (1965*a*) description of feeding quelea. 'Should a flock settle to feed for any length of time it is joined by an ever-increasing number of other groups, both large and small, coming in from all directions. If, on the other hand, a flock makes a series of touch-downs without ever stopping on the ground for more than a few seconds – presumably because food is not being found in sufficient quantity – groups will detach from it and fly to join a more successful flock feeding some distance away. As a result of this behaviour a small group which has found a good feeding place may swell into an enormous flock in a matter of minutes, while a veritable swarm which is unlucky can dwindle to nothing with equal rapidity.' Such feeding by 'local enhancement' (Hinde, 1961) has been observed in a large number of granivorous species (e.g. Crook, 1965; Hamilton & Gilbert, 1969; Newton, 1972; Feare, Dunnett & Patterson, 1974).

One consequence of such a fluid feeding behavior is that flock size may be quite variable and responsive to local conditions. Stewart (1975) noted seasonal changes in the size and localization of roosting–feeding aggregations of starlings *Sturnus vulgaris* L. and several icterids in the eastern USA. In the Mohave Desert of California, Cody (1971) noted that mixed-species flocks of seed-eating finches increased in size as the winter progressed and food supply dwindled and became more patchily distributed. Quelea flocks become larger as the dry season advances (Ward, 1965*a*), and the changes in food distribution and abundance which underlie these shifts in flock size also appear to influence feeding tactics of individuals within the flocks. Early in the dry season, when seeds are plentiful over wide areas, individuals within the small flocks are spread out and forage more or less independently of one another. Later in the dry season, seeds become less abundant and more locally concentrated, and members of the now large flocks forage by 'roller feeding', in which the flock as a whole adopts a directional pattern of movement. Birds that stop to pick up seeds are rapidly by-passed by other flock members, and after a few such stops they find themselves in the wake of the advancing flock, where little food must remain. They then fly over the main body of the flock and take a position in front of the first rank of birds, where the chances of finding food are good. The flock as a whole thus has a constant overhead flow. It is possible that the turning rate of the flock as it advances is related to food density in a manner that optimizes the search of yet unscanned areas (Cody, 1971) and maximizes the return time to patches already visited (Cody, 1974), although convincing evidence for either is lacking.

During the non-breeding season, members of the flock usually gather in large communal roosts from which they radiate in daily flights. In a somewhat oversimplified manner, Hamilton and his colleagues (Hamilton, Gilbert, Heppner & Planck, 1967; Hamilton & Gilbert, 1969; Hamilton & Watt, 1970) have suggested that the patterns of dispersal of individuals from a roost epicenter are determined by individual assessments of the costs and benefits associated with flying various distances from the roost. An individual foraging close to the roost incurs a small energy cost in flying to the foraging area and spends less time getting there, but since the density of individuals in such areas is relatively high, intraspecific competition for the available resources may be intense. An individual travelling farther to less heavily exploited feeding areas may have more ready access to food as it forages, but expends more energy and time to get there, and thus must consume more food in a shorter period of time. Hamilton suggests that the observed dispersal patterns about a roost reflect the adoption of different but equivalent cost–benefit options by individuals. While there is little doubt that the patterns of abundance of individuals with increasing distance from a roost epicenter are related to individual energetics and food supplies (e.g. Wiens & Dyer, 1975a), actual documentation of the cost–benefit correlates has not been accomplished.

Occupation of communal roosts is usually associated with flock feeding in granivores, as in other birds. Finches that in winter cover large areas in search of patchily distributed food, such as greenfinches, linnets *Acanthis cannabina* (L.), chaffinches, or bramblings *Fringilla montifringilla* L., occupy large communal roosts, often relatively far from the feeding areas. Other species that forage individually within small areas, such as

331

bullfinches or hawfinches, roost solitarily or in small groups (Newton, 1972). In flock-feeding birds which forage by local enhancement, the roost may function as an 'information center', enabling individuals lacking knowledge of locations of patches of abundant food to profit from the searching abilities of other individuals in the social group (Ward, 1965*b*; Ward & Zahavi, 1973). Ward suggests that this information transfer occurs at the central gathering place (the roost or colony), presumably by marked differences in behavior between birds that have been successful in the previous day's foraging and those that have met with poor foraging returns. Successful individuals may in a sense 'advertise' their status, and unsuccessful birds may follow them out of the roost on foraging trips to good feeding areas the next day. Advertisement of the roost by conspicuous gathering displays and vocalizations may further function to attract large numbers of birds to the roost location, and since a larger group will be able to search over a larger area for food, the chances that all good feeding locations within the area about the roost will be discovered are increased. The level of advertising may also indicate the 'mood' of the roost: when the 'mood' is good (i.e. the majority of individuals behave as successful foragers), the area is clearly worth returning to the next day, but if the 'mood' is poor (most individuals behaving unsuccessfully) this may act as a stimulus for all birds to shift to a new roosting and feeding area before the food situation becomes critically poor. Ward was drawn to this idea from his observations of quelea roost dynamics; in addition, the studies of starling aggregations by Hamilton and his colleagues (Hamilton *et al.*, 1967; Hamilton & Gilbert, 1969) indicated that some individuals in fact followed others in morning foraging flights, although no information was available on the previous success of the leaders or followers. Hamilton suggested that 'in-transit' stops along flight paths permitted continuing evaluation of resource conditions closer to the roost, potentially reducing the distance flown and associated energy expenditures. The 'information center' concept may also apply to assemblages of dickcissels (ffrench, 1967) and some European finches (Newton, 1972), but documentation that an information transfer actually occurs is difficult.

Mating systems

In theory, selection should favor polygamy as the fundamental mating system, since males and females usually have unequal 'investments' in individual offspring (Trivers, 1972). A superabundance of food during the breeding season should further enhance the development of polygamy (especially polygyny), since it enables a single female to raise the offspring while the male partakes of additional matings. Orians, Verner and Willson (Verner, 1965; Verner & Willson, 1966; Orians, 1969) have extended this argument to consider the effects of local variation in food supply or

Table 8.1. *Food habits and mating system among ploceid finches according to pattern of spatial dispersion during the breeding season. From Crook (1962)*

Spatial dispersion	Food habits		Mating system	
	Insectivorous	Granivorous	Monogamous	Polygynous
Solitary pairs	19	0	16	3
Flocks: territories in 'neighborhoods' in breeding season	0	17	2	15
Flocks: colonial in breeding season	2	13	2	13

habitat 'quality'. Thus, if male territories differ markedly in quality, a female may realize greater success by pairing with a mated male on a prime territory and rearing the young alone than by forming a monogamous bond with a male on an inferior territory, even if the male assists. Environments exhibiting considerable patchiness in productivity and food distribution, such as marshes, grasslands, savannahs and the like, should provide the best setting for such conditions. Lack's (1968) survey of avian sub-families indicated that most (93%) are typically monogamous. Granivorous groups, such as carduelines, estrildids and most cardinalines, for example, are monogamous, as are most Passerinae. Polygyny is well developed, however, among some ploceid finches, and Crook's (1962) analysis has related its incidence to other elements of ploceid exploitation systems. Of the insectivorous forest and forest-edge species, 19 nest solitarily, two in small groups; 16 of the 19 solitary species are monogamous, as are both of the colonial species (Table 8.1). The granivorous ploceids inhabit savannahs, grasslands and marshes and breed in grouped territories or dense colonies; almost all are polygynous. Quelea and the sociable weaver differ from almost all other colonial savannah weaver finches in being monogamous (Ward, 1965b; Maclean, 1973c). For quelea, Ward suggested that the immense colony size places such severe demands upon the food resources of the surrounding area that both parents must be involved in active foraging during the daylight hours in order to provide sufficient food to the young, thereby precluding polygyny. Crook & Butterfield (1970), on the other hand, attribute the monogamous mating system in quelea to a roughly balanced sex ratio at the time of breeding. In many of the polygynous *Ploceus* spp., males do not breed until their second year, while females breed in the first year. In the relatively lush savannah habitats occupied by these species, annual mortality of the sexes is presumed to be roughly equal, so breeding colonies are occupied by a high proportion of females, favoring polygyny. In quelea, however, the more severe environment, coupled with male dominance in feeding encounters, effects a disproportionate mortality among females, skewing the sex ratio in favor of males, thus imposing monogamy. Fretwell (1972*b*) has suggested that differential mortality between sexes of dickcissels on the wintering grounds may produce sex ratios skewed in the direction of males in some areas, which have a similar effect upon mating systems. Unfortunately, all of these arguments are to some degree speculative, and careful field measurements are needed.

The incidence of monogamous and polygynous mating systems among the New World Icteridae parallels that among the ploceids in many respects, although they are generally less well studied. The insectivorous forms, such as *Icterus* spp., normally nest in well-defined solitary territories and are monogamous. *Xanthocephalus* and most *Agelaius* spp. are closely convergent to some *Euplectes* spp., breeding in grouped territories, feeding upon a mixture of seeds and insects and mating polygynously. *Agelaius tricolor* resembles quelea in many respects; while it is not entirely monogamous, it exhibits a substantial reduction in the degree of polygyny in relation to the closely related *A. phoeniceus*. This may be associated with the formation of immense breeding colonies which feed upon quite temporary food sources, conditions in which the assistance of the male is required to rear young successfully (Lack, 1968).

Strategies of pesthood: a synthesis

We have noted on several occasions that these various components of granivorous bird adaptation are interwoven into exploitation systems or 'strategies'. We may now return to our initial question: can we detect complexes of adaptations which characterize successful granivory, and do those complexes differ between 'pest' and 'non-pest' species? To clarify relationships among the various adaptive responses to granivory, we compiled a set of characteristics summarizing many of the ecological, behavioral and morphological attributes discussed above for 26 species of birds (Table 8.2). Several of the species are either synanthropic commensals or are granivorous pests, but the 'pest matrix' includes other non-commensal granivores that are not, or are only rarely, pests. It is apparent that certain species share suites of characters. Yet, if we depart from assessing clearly natural units, such as *Passer* or *Corvus*, the possible groupings become obscured in direct relation to the number of species we try to include. This suggested to us that crude enumeration of variables was not useful of itself, and we thus elected to consider them simultaneously.

Such consideration is through some multivariate statistical technique

334

Table 8.2. *The pest matrix: a tabulation of the occurrence (X) and absence (dash) of features among a variety of more or less granivorous species*

Species	Morphology: Large bill	Crop	Sexual size dimorphism	Dietary opportunism: Eats grass seeds	Seed opportunism	Animal food only in breeding season	Animal food at any time	Distributional opportunism: Migrant	Nomadic	Ecologically ecotonal	Reproductive biology: Single brooded	Edificial nesting	Nest site versatility	Rapid population turnover	Social organization: Feeding flocks outside breeding season	Feeding flocks during breeding	Associates in mixed flocks	Feeding flocks > 5000	Roosting centers	Colonial breeding	Polygynous
Tympanuchus cupido	—	X	X	X	X	X	—	—	—	—	X	—	—	—	X	X	—	—	X	—	X
Colinus virginianus	—	X	X	X	—	X	—	—	—	X	X	—	—	X	X	—	—	—	—	X	—
Columba livia	—	X	—	X	X	—	—	—	—	—	X	—	X	X	X	X	X	—	X	—	—
Eremophila alpestris	—	—	—	—	X	X	—	X	X	—	—	—	—	X	X	—	—	—	—	—	—
Corvus frugilegus	—	—	—	—	X	—	X	X	X	X	X	X	X	—	X	X	X	X	X	X	—
C. monedula	—	—	—	—	X	—	X	X	X	X	X	X	X	X	—	X	X	X	—	X	—
C. corone	—	—	—	—	X	—	X	X	X	X	X	X	X	—	X	—	X	—	X	—	—
Agelaius phoeniceus	X	—	X	—	X	—	X	X	X	X	—	—	—	X	—	X	—	X	X	X	X
A. tricolor	—	—	X	X	X	—	X	—	X	X	X	—	—	—	X	X	—	X	X	X	X
Dolichonyx oryzivorus	—	—	X	X	—	—	—	X	X	—	—	—	X	—	—	X	—	—	—	—	X
Passer domesticus domesticus	—	X	—	X	X	—	X	—	—	X	—	X	X	X	X	—	X	X	X	X	—
P. d. bactrianus	—	X	—	X	X	X	—	X	—	—	X	—	—	X	X	X	—	X	X	X	—
P. montanus	—	X	—	X	X	X	—	X	—	X	X	X	X	X	—	X	—	X	—	X	—
P. hispaniolensis	—	X	—	X	X	X	—	X	—	—	X	—	—	X	X	X	X	—	X	X	—
P. ammodendri	X	X	—	—	X	X	—	X	—	X	X	X	—	—	X	—	—	—	—	—	X
Quelea quelea	X	X	—	X	X	X	—	X	X	—	—	—	—	X	X	X	X	X	X	X	—
Petronia petronia	—	X	—	X	X	X	—	X	—	—	X	X	—	X	X	—	—	—	—	—	—
Spiza americana	X	—	X	—	X	—	X	X	X	X	—	—	—	X	—	X	—	—	X	X	X
Carduelis tristis	—	—	—	—	X	—	—	X	X	X	X	—	—	X	X	X	X	—	—	—	—
Acanthis flammea	—	X	—	X	X	X	—	X	X	—	—	—	—	X	X	—	X	—	—	X	—
Emberiza citrinella	—	X	—	X	—	X	—	X	X	—	—	—	—	X	X	—	X	X	X	—	—
Coccothraustes vespertinus	X	—	—	—	X	—	X	X	—	X	—	X	—	X	—	—	—	—	—	—	—
Calamospiza melanocorys	—	—	—	—	X	—	X	—	X	—	X	—	—	X	X	X	—	—	—	—	—
Amphispiza belli	—	—	—	—	X	—	X	X	X	—	—	X	—	—	—	—	—	—	—	—	—
Melospiza melodia	—	X	—	X	X	—	X	X	—	X	—	—	—	—	X	—	X	—	—	—	—
Calcarius lapponicus	—	X	—	X	—	—	X	X	—	—	X	—	—	—	X	—	X	X	—	—	—

that provides a suitable display. Principal Component (PC) analysis provides an easily interpreted view of the taxa and of the variables that cause them to assume certain positions in component hyperspace (Morrison, 1967; Blackith & Reyment, 1971). In our analysis, principal components of variation were computed from a correlation matrix generated from an input or raw data matrix of 26 OTUs (= species) and 21 variables (Table 8.2). The first three of the 21 possible axes or components of the hyperspace accounted for 46.7% of the total variance in the original variables (PC I = 18.2%, PC II = 15.1%, PC III = 13.4%). These three axes repre-

335

sent new linear combinations of the original 21 variables such that each includes several original variables. Each new axis is statistically independent of all others – they are wholly uncorrelated.

The positionings of the OTUs with respect to these three components are depicted in Figure 8.5. PC I contrasts a tendency toward single-broodedness at one extreme with a combination of a prediliction for seeds of grasses (Gramineae), possession of a crop and high population turnover at the other. PC II contrasts the effects of granivorous habits with tendencies toward opportunistic breeding, edificial nesting, nest site versatility and occupancy of ecologically ecotonal habitats. PC III summarizes variation in tendencies toward roosting in flocks, polygyny, marked sexual dimorphism in body size and flocking in large groups (more than 5000 individuals).

The three-dimensional plot summarizes nearly half of the information in the 21 variables, but relationships of the OTUs in the three-space inevitably are to some extent distorted from what they would be were all of the variation summarized in the three dimensions. To suggest in what ways distortion occurs within the present ordination, the minimum spanning tree from a matrix of taxonomic distances has been superimposed on the plot of OTUs. Whenever OTUs that are close to one another in the ordination space are linked with more distant OTUs by the distance net, it may be assumed that some distortion of the overall relationship has occurred. Thus, as an example that is especially clear, *P. montanus* and *Columba livia* Gmelin share close ordination proximity but are joined in the distance net only by the links with *P. domesticus domesticus*.

Axis II, PC II, provides the greatest separation of commensal and non-commensal species, although realization of such separation is best seen in conjunction with PC I. Species occupying the lower left part of the ordination space tend to nest site versatility and edificial (human) nesting, and tend also to occupy ecotonal or marginal ecologic conditions; they tend not to be nearly as completely granivorous in their diets as those species plotting in the upper right sector of the diagram. Although PC I does not separate commensals from non-commensals, the boundary between the two angles from the upper left diagonally to the lower right, taking a slight jog back to include *Agelaius tricolor* and *Spiza americana*, neatly separates species which are frequently agricultural pests and/or commensals from those which are rarely, if ever, pests or commensals (see Fig. 8.5*b*).

In this depiction the commensal or pest species are restricted to a fairly small part of the three-space. The non-commensal or non-pest species can be seen to have about half of the ordination space, and this proportion would undoubtedly be even greater if a broader array of such species (e.g. insectivorous woodland species or shorebirds) were to be included. This

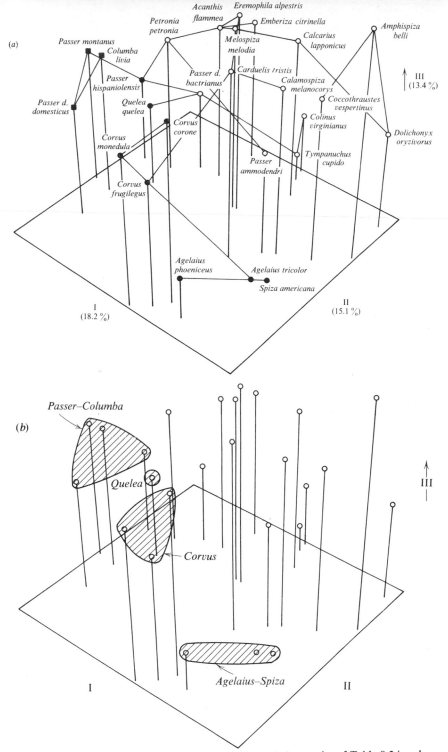

Fig. 8.5. (a) Three-dimensional plotting of the positions of the species of Table 8.2 in relation to principal components (PC) I, II and III (see text). Lines connecting species are minimum multivariate distance linkages. Solid circles = frequently a pest species; solid square = frequently a commensal with man; open circle = never or infrequently a pest and/or commensal. (b) The four major modes or species groupings discussed in the text.

implies that there are more ways of assuming a 'normal' or non-pest ecology than there are ways of becoming a pest. But perhaps this is only a truism, emphasizing that only certain ecological and phyletic lines have the potential for becoming commensal granivores or assuming true pesthood. We note these as follows:

(1) The *Passer–Columba* mode. The house sparrow, European tree sparrow, Spanish sparrow and rock dove share an intense social flocking and roosting tendency, sometimes in mixed-species assemblies; they all possess at least a functional crop; and they have potential for high rates of population turnover. They commonly nest in human edifices, exhibit considerable versatility in nest site selection, and occupy ecologically ecotonal areas. They are neither markedly dimorphic in size nor given to polygyny. *P. hispaniolensis* is the link species, joining with non-commensals *Petronia petronia* (L.) and *Passer domesticus bactrianus*. A fourth member of *Passer*, *P. ammodendri* (Gould), is well within the non-pest sector, but shows relationships to this group via a distance network connection with *Petronia*.

(2) The *Corvus* mode. The three species of crow (rook, carrion crow and jackdaw) that we considered all occupy extreme positions on PC II, being only sporadically granivorous but showing nest site versatility and ecotonal orientation. They fall to the right of the midline on PC I because they tend toward single-broodedness and lack a distinct crop. The link species is *C. frugilegus*, joining by minimum-distance linkage both *Agelaius tricolor*, another pest species, and *Carduelis tristis* (L.) (the American goldfinch), which is not a pest species.

(3) The *Agelaius–Spiza* mode. The two blackbirds and the dickcissel lack a distinct crop, are not persistently granivorous, are not versatile in choice of nest sites and are polygynous, bringing them to the right on PC I. They frequent ecotonal situations, tending to place them to the left on PC II. The group has only one distance network connection, through *A. tricolor* to *Corvus frugilegus* L.

(4) The *Quelea* mode. Quelea possesses a number of somewhat distinctive granivorous adaptations, only some of which were included in the variables entering into this analysis. Still, it remains rather distinct in the ordination, although its closest links, through *Passer domesticus bactrianus*, are to the *Passer* group. Quelea differs, however, in showing a substantially greater flocking tendency while lacking the propensity for edificial nesting and nest site versatility which characterizes the *Passer–Columba* group.

The remaining 15 species are here very roughly characterized as non-pest or non-commensal species, which is generally adequate. Of the 15, only the rock sparrow *Petronia*, the mid-continental migratory house sparrow *Passer d. bactrianus* and the American goldfinch *Carduelis tristis* show

338

distance links with any of the established pest species. On the whole, these non-pest species maintain this status, although they may have pesthood thrust upon them, as is the case with the redpoll *Acanthis flammea* in New Zealand (Newton, 1972) and the bobolink *Dolichonyx* in the south-central USA following the introduction of rice agriculture (Forbush, 1929; Meanley, 1971).

We suppose that species occupying component space near *Passer* spp. that are presently commensals should themselves be capable under the right environmental opportunities of adopting commensalism. Or, since quelea is also very close by, they could alternatively assume a granivorous-pest status. Species at the other end of PC I could, with a shift in a number of character-states important on both PC II and PC III, approach the *Agelaius–Spiza* mode; this possibility seems extremely unlikely, however, and it is largely through approaching the *Passer–Columba* or *Quelea* behavioral modes that additional commensals or pests might arise.

Beyond this, however, since the four commensal or pest granivore strategies represent four different adaptive peaks, there is a clear implication that other strategies, unused by any species we considered in the present ordination, exist in the diagram of Fig. 8.5. If so, the evolution of commensal granivory is not closed with the set of bird species treated here, and given that humans continue doing as they have in providing granivorous food sources, we should expect to see the evolution of additional commensal or pest species.

What species might be involved and what commensal or pesthood modes would be involved invites speculation, only part of which is of very low prior probability. For instance, some species or races of *Passer* that are not yet known to be commensals, like *P. ammodendri*, might well move into commensalism if other commensals are absent from some place or uncommon there. Or a species like *Sturnus vulgaris*, not yet a granivorous pest (although damaging to fruits in some areas and noxious by its presence in others), could easily become one whether or not other commensals or pests are present. The starling would occupy a position in Fig. 8.5 somewhere between the *Corvus* group and the *Passer–Columba* group. The American Inca dove, *Scardafella inca* (Lesson) and a number of other species in the genera *Columba* and *Streptopelia* could also move into the *Passer–Columba* mode. We noted earlier that among the various finch groups, the carduelines are perhaps most readily predisposed toward pesthood, and the position of the carduelines included in our ordination analysis (Table 8.2) in the three-space of Fig. 8.5 lends support to this contention. The various pest and commensal species of the *Passer–Columba* and *Quelea* modes have minimum distance linkages to the 'non-pest' group through *Passer d. bactrianus* and *Petronia*, and thence to *Acanthis*, a cardueline which under modified environmental conditions

339

(New Zealand) has indeed become a pest. The distance network connection of the *Corvus* and *Agelaius–Spiza* modes to the 'non-pest' group is to *Carduelis tristis*, another cardueline that is not presently a pest or commensal. Readers with other experiences will no doubt have their own candidates for incipient pesthood or commensalism, in accordance with positionings in the ordination relative to the major modes of Fig. 8.5.

This synthesis emphasizes the multi-dimensional nature of adaptive responses to seeds as food resources, and the analysis is, at that, simplistic since we have considered only a sampling of adaptations and of species. The message, however, is clear. There is no single 'strategy' of granivory among birds, and the attainment of pesthood by species is a result of the interplay between features of its exploitation system and the agricultural or sociological practices man imposes upon its native habitats. As new areas are converted to agriculture, or as cropping or harvesting practices in existing agricultural biotopes are changed, attention must be given to the probable consequences of such actions upon the exploitation systems of birds which are active or incipient pests. Our future efforts to discern the patterns of pesthood must recognize that different species or species complexes have attainted such status in different ways, through different adaptive correlates of granivory. Resolution of these inherent adaptive responses is the key to effecting satisfactory management of pest or commensal bird populations.

The comments of the many colleagues who contributed to other chapters in this volume aided us in defining some of the many dimensions of granivorous adaptations. Calvin L. Cink prepared Fig. 8.3. This is contribution No. 52 of the Behavioral Ecology Laboratory, Oregon State University.

9. Epilogue

S. C. KENDEIGH, J. A. WIENS & J. PINOWSKI

The preceding chapters have summarized our current knowledge of granivorous birds, with special emphasis upon the house and European tree sparrows and a few other species. Although much work has been done, the information available is obviously sketchy and incomplete. If this is the case for the species considered here, selected in large part because of the apparent abundance of data concerning them, it must hold in greater degree for other species. Hopefully, this book will provide an impetus for filling existing gaps in our information and especially for the development of philosophies, procedures, equations and systems approaches for evaluating the role of birds in ecosystems.

How and when species originated and expanded their ranges in the geological past can only be speculative, but we can offer inferences, often with considerable confidence, if their present adaptations and adjustments are fully understood. Documentation of variations in the morphology, physiology, behavior and environmental relations of a species between localities or over a period of time may be especially rewarding and are critical to the development of sound management tactics. With continuous distribution through environmental gradients, one can expect to observe a continuum of characteristics from one extreme to another. Furthermore, there may be significant variation between individuals in their responses to the same environment. When individual variation in some trait is considerable, we should determine whether this represents differing individual sensitivities to proximate influences, substantial unchannelized genetic variation, or 'neutral' differences of little or no immediate importance. More effort should be made to breed native species under controlled conditions to determine the heritability of traits.

More detailed information on natality and mortality functions and on age and sex ratios is necessary before adequate life tables can be formulated. Our knowledge of mortality functions lags far behind that of natality, and mortality is the determinant of the survivorship component of life tables. Only if we know the nature, timing, magnitude and predictability of natural mortality can we develop intelligent management strategies based upon control or manipulation of population dynamics. Especially important here is the development of procedures for increasing the number of recaptured or recovered birds of known age, or of procedures other than mark–recapture techniques. In England, between 1954 and 1965 only 0.28% of banded house sparrows were retaken. In central

North America, retakes on over 45 000 icterids have amounted to only 1–4%.

Closely associated with the closer study of population dynamics is the need for information on life histories. Information is not available, for instance, on the role of the sexes in mating, nest-building, incubation and care of young even for the house sparrow, except to a very fragmentary extent. Likewise, we do not know the peak dates when the majority of individuals in a population initiates clutches or fledges young. We need more precise information on such matters as the number of broods attempted per female per year, their distribution in time, and the under-lying causal relationships. Before we can obtain better estimates of the energy cost of incubation, information is required on the length of attentive and inattentive periods, their number per day, and how they vary with time and weather. Along a different line, future studies should give more attention to the details of social organization and behavior, such as foraging tactics, the conditions underlying tendencies to aggregate and the interactions among individuals in feeding flocks.

Essential for evaluating populations of various species are measure-ments of their density in the different biotopes in which they occur. Since population densities in a given location vary with production of young, mortality, dispersal and migration, measurements are required for all times of the year. Special attention should be devoted to the problems of realistically measuring densities of populations which are highly aggregated in space or which vary in their dispersion on a daily basis, since the most exploitive granivorous species seem generally to be those predisposed to flocking or colonialism. For evaluating the ecosystem consequences of differences in population densities of different species, biomass needs to be considered, hence weight data are required.

The daily energy budget (DEB) of an individual varies with its weight, ambient temperature and season and may be calculated from the equations or coefficients that are provided in this volume. Further refinement is desirable, however, when the effects of exposure to incoming solar radi-ation, outgoing sky radiation, wind or rain, color of plumage, roosting site and roosting habits, nest exposure and hypothermy and hyperthermy on energy exchange and balance can be quantified. More precise information on the energy cost of free existence under natural conditions and of the various components of free existence, such as foraging, short flights, preening, singing and territorial behavior will probably have to await the further development of telemetry and the use of isotopes. Calculating the energy cost of incubation and care of young is especially complex, particularly as it is modified by the effectiveness of nest insulation, the number and weight of eggs and young, the ontogeny of endothermy and the role and behavior of the sexes. Attempts to gather information on these environmental and life-history parameters are just beginning. Once the energy costs of the various components of the daily activity budget are better known, conversion coefficients can be established for use with time–activity budgets. Such conversions at the present time are coarse and may be misleading.

To convert DEBs into food consumed, considerably more information and experimentation are required on the metabolizable calorific content of different kinds of food and how this varies with the condition of the food and weather and between species. This book has stressed the need of food as a source of energy, but birds may eat for specific nutrients as well as energy. Very little information is available on the nutritional requirements of birds for specific proteins, carbohydrates, fats, minerals and vitamins, and how these requirements vary with proximate resource conditions and stages of the annual cycle. We really know very little about the details of diet of most granivores and their 'preferences' for particular food items. In order to evaluate accurately the role of a granivorous bird population in a natural or agricultural ecosystem, we should also know a great deal about other members of the grain-eating guild, which may include mammals, insects and other forms.

Because of the multiplicity of factors involved, the possible variations in each factor and the interactions among the parameters, a synthesis of the whole process to show energy flow through a population and the consequent impact on the ecosystem can best be accomplished through use of computer simulation models. The potential usefulness of computer models is well shown in the analysis of changes in bird populations in Illinois over an interval of 50 years. The model presented here emphasizes generality and realism at the expense of precision. Precision will improve only with the development of greater accuracy and completeness of the input values and of the modelled functions. In general, however, the model appears fairly robust, permitting its use as a gaming device to estimate the relative importance of the various input parameters on the output estimates of impact in the ecosystem. Obviously, follow-up laboratory tests and field studies need to be undertaken to verify the model estimates and suggest refinements which may increase the accuracy of the predictions of the model functions. Especially important will be the incorporation into the model of the spatial dimensions of foraging, food consumption and impact about which we know very little. Such spatial considerations need to be studied in terms of underlying behavioral and environmental causes.

An accurate computer simulation model makes possible various applications, such as a comparison of the role of birds in natural ecosystems of different sorts or in the same ecosystem at different times or in different parts of the world. The use of computer models may open up

343

a fresh, new approach to the management of pest species in agricultural and other man-made ecosystems, especially if we can achieve a true documentation of 'importance' of all components in the cost–benefit ratio, including economic, political and sociological influences. Such models, once formed, should have special importance in less well-developed parts of the world, where human populations are large and where sophisticated studies have not, or cannot, be made quickly.

The use of computer models should make possible the determination of the most vulnerable stages or points of time in the yearly or life cycle of destructive species. These vulnerable periods, when population control may be exerted most effectively, may vary with the kind of crop involved, the prevailing weather conditions, the presence of other organisms and the degree of sociality expressed by the birds. Once these conditions are established, methods for population management can become more sophisticated and precise. All this goes beyond the treatment in this book, but the demonstrated possibilities of this procedure, if introduced into agriculture, may well be the major contribution of this volume to the objectives of the IBP i.e. to achieve that level of biological productivity which provides a rational basis for the sustenance, health and happiness of mankind.

The integration of population genetics, population dynamics, productivity, bioenergetics, social behavior and ecosystem interrelations leads to concepts of considerable theoretical interest. The adaptive strategies of granivory or the evolution of pest species are examples. Tools are provided here for analyzing the energetic adaptations of birds for living in different ecosystems or biomes or for playing different roles within these systems. A better understanding of the evolution of bird behavior can be developed with a realization that the DEB of a species is limited and that variations in the energy costs of one activity must be balanced by reciprocal variations in other activities. There can now be better and more complete explanation of variations in population densities from place to place and year to year, and the prediction of conditions under which outbreaks or catastrophes may occur should be of interest to the theoretical biologist as well as to the agriculturist and the sportsman.

If this volume stimulates a development of concepts along these lines and leads to a better understanding of how birds exist and function in their particular environments, it will have fulfilled the objectives established for it.

Appendix

Appendix 3.1. *Breeding density of* Passer domesticus

Breeding density individuals km^{-2}	Biotope type (see Table 3.2)	Size of biotope (ha)	Latitude	Longitude	Authority
3266	11	9.8	52° N	7° 45' E	Beckmann & Fröhlich (1967)
3810	11	10.5	52° N	7° 45' E	Beckmann & Fröhlich (1967)
120	11	89	52° 30' N	1° E	Benson & Williamson (1972)
875	14	—	50° 04' N	19° 57' E	Bocheński & Harmata (1962)
188	7	—	50° 04' N	19° 57' E	Bocheński & Harmata (1962)
1600	17	—	52° 05' N	5° 05' E	Bruijns (1961)
30	8	18	52° 05' N	5° 05' E	Bruijns (1961)
250	14	20	52° 05' N	5° 10' E	Bruijns (1961)
240	17	—			Bruijns (1961)
20	2	10.8	52° 05' N	16° 45' E	Czarnecki (1956)
533	11	—	40° 20' N	49° 30' E	Drozdov (1965)
296	14	—	59° N	18° E	Ehrström (1954)
1320	17	8.5	51° 30' N	7° 30' E	Erz (1959)
1240	17	8.5	51° 30' N	7° 30' E	Erz (1959)
1160	17	8.5	51° 30' N	7° 30' E	Erz (1959)
360	8	5.5	51° 30' N	7° 30' E	Erz (1959)
280	8	5.5	51° 30' N	7° 30' E	Erz (1959)
360	8	5.5	51° 30' N	7° 30' E	Erz (1959)
23	11	8.8	51° 30' N	7° 30' E	Erz (1959)
268	16	32.5	54° 20' N	10° 05' E	Erz (1964)
740	16	32.5	54° 20' N	10° 05' E	Erz (1964)
1020	17	224	54° 20' N	10° 05' E	Erz (1964)
980	17	224	54° 20' N	10° 05' E	Erz (1964)
500	14	44	54° 20' N	10° 05' E	Erz (1964)
580	14	44	54° 20' N	10° 05' E	Erz (1964)
120	8	8	54° 20' N	10° 05' E	Erz (1964)
120	8	8	54° 20' N	10° 05' E	Erz (1964)
120	7	12.7	54° 20' N	10° 05' E	Erz (1964)
120	7	12.7	54° 20' N	10° 05' E	Erz (1964)
40	9	23	54° 20' N	10° 05' E	Erz (1964)
40	9	23	54° 20' N	10° 05' E	Erz (1964)
680	16	72	51° 30' N	7° 30' E	Erz (1964)
680	16	72	51° 30' N	7° 30' E	Erz (1964)
1080	17	168	51° 30' N	7° 30' E	Erz (1964)
1080	17	168	51° 30' N	7° 30' E	Erz (1964)
1020	14	33	51° 30' N	7° 30' E	Erz (1964)
960	14	33	51° 30' N	7° 30' E	Erz (1964)
320	8	5.6	51° 30' N	7° 30' E	Erz (1964)
360	8	5.6	51° 30' N	7° 30' E	Erz (1964)
200	7	14.5	51° 30' N	7° 30' E	Erz (1964)
200	7	14.5	51° 30' N	7° 30' E	Erz (1964)
100	9	10	51° 30' N	7° 30' E	Erz (1964)
100	9	10	51° 30' N	7° 30' E	Erz (1964)

Appendix 3.1. (*cont.*)

Breeding density individuals km⁻²	Biotope type (see Table 3.2)	Size of biotope (ha)	Latitude	Longitude	Authority
38	13	—	48° 45′ N	18° 05′ E	Ferianc, Feriancová-Masárová & Brtek (1973)
45	13	—	48° 45′ N	18° 05′ E	Ferianc *et al.* (1973)
90	13	—	48° 45′ N	18° 05′ E	Ferianc *et al.* (1973)
42	7	1490	46° 30′ N	35° 20′ E	Filonov (1972)
223	11	15	47° 00′ N	28° 45′ E	Ganya (1965)
1167	11	12	50° 10′ N	20° 30′ E	Głowaciński (1975)
1000	14	—	54° 10′ N	12° 10′ E	Grempe (1971)
151	11	—	50° 10′ N	17° 10′ E	Havlín & Lelek (1957)
1062	16	—	51° 30′ N	0° 00′	Homes (1957)
988	16	—	51° 30′ N	0° 00′	Homes (1957)
675	17	—	51° 30′ N	7° E	Hyla (1967)
1600	16	—	51° 30′ N	7° E	Hyla (1967)
360	7	4	51° 30′ N	7° E	Hyla (1967)
60	7	33	51° 30′ N	7° E	Hyla (1967)
200	14	—	51° 30′ N	7° E	Hyla (1967)
1120	17	8	51° 30′ N	7° E	Hyla (1970)
160	3	150	47° 11′ N	25° 11′ E	Ion & Valenciuc (1967)
185	11	1553	—	—	Kashkarov *et al.* (1926)
3600	8	11	46° 45′ N	23° 45′ E	Korodi Gál (1960)
180	14	—	46° 40′ N	21° 20′ E	Korompai (1966)
870	14	40	51° 20′ N	20° 30′ E	Król (unpublished)
206	8	31	50° 05′ N	22° 15′ E	Kulczycki (1966)
200	8	31	50° 05′ N	22° 15′ E	Kulczycki (1966)
188	8	31	50° 05′ N	22° 15′ E	Kulczycki (1966)
188	8	31	50° 05′ N	22° 15′ E	Kulczycki (1966)
212	8	31	50° 05′ N	22° 15′ E	Kulczycki (1966)
26	8	15	52° 10′ N	22° 20′ E	Luniak (1974)
14	8	15	52° 10′ N	22° 20′ E	Luniak (1974)
82	8	9.8	52° 10′ N	21° 30′ E	Luniak (1974)
234	8	6	52° 05′ N	23° 05′ E	Luniak (1974)
114	8	5.3	51° 45′ N	22° 45′ E	Luniak (1974)
464	11	—	52° 21′ N	14° 51′ E	Mackowicz, Pinowski & Wieloch (1970)
215	11	—	52° 21′ N	14° 51′ E	Mackowicz *et al.* (1970)
168	8	38	60° 00′ N	30° 20′ E	Malchevskiï (1955)
152	8	38	60° 00′ N	30° 20′ E	Malchevskiï (1955)
157	8	38	60° 00′ N	30° 20′ E	Malchevskiï (1955)
122	8	115	52° 10′ N	8° 45′ E	Michels (1973)
190	2	4.49	51° 20′ N	18° 00′ E	Michocki (1974)
180	2	4.49	51° 20′ N	18° 00′ E	Michocki (1974)
80	2	4.49	51° 20′ N	18° 00′ E	Michocki (1974)
40	2	4.49	51° 20′ N	18° 00′ E	Michocki (1974)
140	2	4.49	51° 20′ N	18° 00′ E	Michocki (1974)
178	2	21	52° 30′ N	13° 00′ E	Michocki (1974)
228	2	37	52° 30′ N	13° 00′ E	Michocki (1974)
390	16	22	53° 45′ N	10° 00′ E	Mulsow (1967)
880	17	16	53° 45′ N	10° 00′ E	Mulsow (1967)
728	17	58	53° 45′ N	10° 00′ E	Mulsow (1967)

Appendix 3.1. (*cont.*)

Breeding density individuals km^{-2}	Biotope type (see Table 3.2)	Size of biotope (ha)	Latitude	Longitude	Authority
730	17	64	53° 45′ N	10° 00′ E	Mulsow (1967)
550	14	12	53° 45′ N	10° 00′ E	Mulsow (1967)
222	14	47	53° 45′ N	10° 00′ E	Mulsow (1967)
826	8	8	53° 45′ N	10° 00′ E	Mulsow (1967)
456	8	4.8	53° 45′ N	10° 00′ E	Mulsow (1967)
34	8	45	53° 45′ N	10° 00′ E	Mulsow (1967)
354	9	30	53° 45′ N	10° 00′ E	Mulsow (1967)
1235	8	—	51° 30′ N	0°	Nicholson (1951)
1714	18	2.8	50° 13′ N	17° 43′ E	Novotný (1970)
4714	18	2.8	50° 13′ N	17° 43′ E	Novotný (1970)
5428	18	2.8	50° 13′ N	17° 43′ E	Novotný (1970)
1150	11	7000	Hannover–Brunswick		Oelke (1963)
29	11	200	52° 20′ N	20° 50′ E	Pinowski & Wieloch (1973)
38	11	13.2	50° 04′ N	19° 57′ E	Pinowski & Wieloch (1973)
176	17	3.9	50° 04′ N	19° 57′ E	Pinowski & Wieloch (1973)
44	14	1200	61° 27′ N	24° 03′ E	Rassi (unpublished)
227	11	100	46° 10′ N	19° 20′ E	Rékási (unpublished)
14	11	88.5	54° 30′ N	2° 30′ W	Robson & Williamson (1972)
1062	16	16	51° 30′ N	0°	Summers-Smith (1963)
988	16	—	51° 30′ N	0°	Summers-Smith (1963)
494	14	8	55° 50′ N	4° 15′ W	Summers-Smith (1963)
247	14	94	54° 35′ N	1° 25′ W	Summers-Smith (1963)
49	11	250	54° 35′ N	1° 25′ W	Summers-Smith (1963)
123	11	518	51° 20′ N	1° 15′ W	Summers-Smith (1963)
7	Small island	1214	52° 30′ N	1° 35′ W	Summers-Smith (1963)
123	Small island	—	55° 15′ N	4° 30′ W	Summers-Smith (1963)
32	Small island	91	55° 15′ N	5° 10′ W	Summers-Smith (1963)
22	Small island	179	52° 45′ N	4° 40′ W	Summers-Smith (1963)
51	Small island	1553	59° 30′ N	2° 20′ W	Summers-Smith (1963)
24	Small island	50	56° 10′ N	2° 30′ W	Summers-Smith (1963)
20	Small island	423	51° 10′ N	4° 35′ W	Summers-Smith (1963)
37	Small island	242	55° 18′ N	5° 30′ W	Summers-Smith (1963)
208	2	15.3	49° 45′ N	21° 00′ E	Tomek (1973)
1057	17	2.27	49° 45′ N	21° 00′ E	Tomek (1973)
2000	11	3.5	49° 45′ N	21° 00′ E	Tomek (1973)
1452	17	31	51° 10′ N	16° 10′ E	Tomiałojć (1970)
1548	17	31	51° 10′ N	16° 10′ E	Tomiałojć (1970)
1454	17	33	51° 10′ N	16° 10′ E	Tomiałojć (1970)
1514	17	—	51° 10′ N	16° 10′ E	Tomiałojć (1970)
2266	17	30	51° 10′ N	16° 10′ E	Tomiałojć (1970)
1884	14	6.9	51° 10′ N	16° 10′ E	Tomiałojć (1970)
1192	14	13	51° 10′ N	16° 10′ E	Tomiałojć (1970)
126	8	35	51° 10′ N	16° 10′ E	Tomiałojć (1970)
24	7	34	51° 10′ N	16° 10′ E	Tomiałojć (1970)
24	7	34	51° 10′ N	16° 10′ E	Tomiałojć (1970)
24	7	—	51° 10′ N	16° 10′ E	Tomiałojć (1970)
10	9	18	51° 10′ N	16° 10′ E	Tomiałojć (1970)
128	2	15.6	51° 10′ N	16° 10′ E	Tomiałojć (1970)
46	2	8.4	51° 10′ N	16° 10′ E	Tomiałojć (1970)

Appendix 3.1. (*cont.*)

Breeding density individuals km⁻²	Biotope type (see Table 3.2)	Size of biotope (ha)	Latitude	Longitude	Authority
1292	11	64	51° 10′ N	16° 10′ E	Tomiałojć (1970)
24	2	8.4	51° 10′ N	16° 10′ E	Tomiałojć (1974)
133	2	15	52° 10′ N	20° 50′ E	Truszkowski (1963)
250	14	16	29° S	30° 10′ E	Vernon (1973)
120	11	—	43° 00′ N	44° 30′ E	Vtorov (1962)
888	2	25	53° 50′ N	12° 35′ E	Wendtland (unpublished)
680	2	25	53° 50′ N	12° 35′ E	Wendtland (unpublished)
872	2	25	53° 50′ N	12° 35′ E	Wendtland (unpublished)
976	2	25	53° 50′ N	12° 35′ E	Wendtland (unpublished)
1096	2	25	53° 50′ N	12° 35′ E	Wendtland (unpublished)
		P. domesticus bactrianus			
1670	16	—	40° 45′ N	72° 20′ E	Sharipov (1974)
1150	8	—	40° 45′ N	72° 20′ E	Sharipov (1974)
10	17	—	40° 45′ N	72° 20′ E	Sharipov (1974)
5	—	—	40° 45′ N	72° 20′ E	Sharipov (1974)
3020	17	—	40° 20′ N	71° 45′ E	Sharipov (1974)
780	16	—	40° 20′ N	71° 45′ E	Sharipov (1974)

Appendix 3.2. *Breeding density of* Passer montanus

Breeding density individuals km^{-2}	Biotope type (see Table 3.2)	Size of biotope (ha)	Latitude	Longitude	Authority
480	3	—	48° N	28° E	Averin & Ganya (1970)
760	3	—	47° N	29° E	Averin & Ganya (1970)
340	3	—	45° 34' N	28° 30' E	Averin & Ganya (1970)
895	1	11.4	51° 53' N	5° 50' E	Van Balen (unpublished)
1140	1	11.4	51° 53' N	5° 50' E	Van Balen (unpublished)
730	1	12.6	51° 53' N	5° 46' E	Van Balen (unpublished)
752	4	8.5	51° 57' N	5° 31' E	Van Balen (unpublished)
484	5	27.3	51° 50' N	12° E	Clausing (unpublished)
900	5	27.3	51° 50' N	12° E	Clausing (unpublished)
530	5	27.3	51° 50' N	12° E	Clausing (unpublished)
1000	1	1	48° 45' N	16° 50' E	Bauer (1974)
1400	1	1	48° 45' N	16° 50' E	Bauer (1974)
1200	1	1	48° 45' N	16° 50' E	Bauer (1974)
1400	1	1	48° 45' N	16° 50' E	Bauer (1974)
1200	1	1	48° 45' N	16° 50' E	Bauer (1974)
2000	1	1	48° 45' N	16° 50' E	Bauer (1974)
1000	1	1	49° N	16° 30' E	Bauer (1974)
1000	1	1	49° N	16° 30' E	Bauer (1974)
800	1	1	49° N	16° 30' E	Bauer (1974)
1000	1	1	49° N	16° 30' E	Bauer (1974)
1400	1	1	49° N	16° 30' E	Bauer (1974)
2000	1	1	49° N	16° 30' E	Bauer (1974)
524	2	4.5	52° 20' N	17° E	Bednorz (1970)
6	10	12.5	52° 20' N	17° E	Bednorz (1970)
8	7	10	53° 30' N	12° 30' E	Beitz (1972)
54	11	89	52° 30' N	1° E	Benson & Williamson (1972)
470	7	2.2	50° 04' N	19° 57' E	Bocheński & Harmata (1962)
70	14	—	50° 04' N	19° 57' E	Bocheński & Harmata (1962)
78	7	4.2	52° 05' N	16° 45' E	Czarnecki (1956)
72	2	10.8	52° 05' N	16° 45' E	Czarnecki (1956)
60	11	—	40° 20' N	49° 30' E	Drozdov (1965)
90	8	25	53° 05' N	18° 30' E	Dubicka (1957)
660	7	—	50° 10' N	51° E	Dubinin & Toropanova (1956a)
280	7	—	49° N	52° E	Dubinin & Toropanova (1956a)
620	7	—	47° N	52° E	Dubinin & Toropanova (1956a)
340	7	—	51° N	51° 30' E	Dubinin & Toropanova (1956b)
100	7	—	51° 30' N	53° E	Dubinin & Toropanova (1956b)
220	7	—	51° 30' N	53° E	Dubinin & Toropanova (1956b)
720	7	—	51° 45' N	56° E	Dubinin & Toropanova (1956b)
240	7	—	51° 58' N	58° E	Dubinin & Toropanova (1956b)
630	2	5.2	52° N	12° 30' E	Ebeling (unpublished)
960	2	5.2	52° N	12° 30' E	Ebeling (unpublished)
82	13	70	51° 50' N	36° 10' E	Eliseeva (1960)
122	1	70	51° 50' N	36° 10' E	Eliseeva (1960)
251	1	70	51° 50' N	36° 10' E	Eliseeva (1960)
217	1	70	51° 50' N	36° 10' E	Eliseeva (1960)
191	1	70	51° 50' N	36° 10' E	Eliseeva (1960)
185	1	70	51° 50' N	36° 10' E	Eliseeva (1960)
160	13	10	51° 50' N	36° 10' E	Eliseeva (1960)

Appendix 3.2. (*cont.*)

Breeding density individuals km^{-2}	Biotope type (see Table 3.2)	Size of biotope (ha)	Latitude	Longitude	Authority
2040	1	10	51° 50′ N	36° 10′ E	Eliseeva (1960)
2880	1	10	51° 50′ N	36° 10′ E	Eliseeva (1960)
2250	1	4	51° 50′ N	36° 10′ E	Eliseeva (1960)
3400	1	4	51° 50′ N	36° 10′ E	Eliseeva (1960)
3500	1	4	51° 50′ N	36° 10′ E	Eliseeva (1960)
1066	1	6	51° 50′ N	36° 10′ E	Eliseeva (1960)
2733	1	6	51° 50′ N	36° 10′ E	Eliseeva (1960)
2933	1	6	51° 50′ N	36° 10′ E	Eliseeva (1960)
490	1	20	51° 50′ N	36° 10′ E	Eliseeva (1960)
480	1	20	51° 50′ N	36° 10′ E	Eliseeva (1960)
380	1	20	51° 50′ N	36° 10′ E	Eliseeva (1960)
370	1	20	51° 50′ N	36° 10′ E	Eliseeva (1960)
1400	1	7	51° 50′ N	36° 10′ E	Eliseeva (1960)
2171	1	7	51° 50′ N	36° 10′ E	Eliseeva (1960)
2228	1	7	51° 50′ N	36° 10′ E	Eliseeva (1960)
1600	1	7	51° 50′ N	36° 10′ E	Eliseeva (1960)
1428	1	7	51° 50′ N	36° 10′ E	Eliseeva (1960)
1028	1	7	51° 50′ N	36° 10′ E	Eliseeva (1960)
833	1	6	51° 50′ N	36° 10′ E	Eliseeva (1960)
1700	1	6	51° 50′ N	36° 10′ E	Eliseeva (1960)
1833	1	6	51° 50′ N	36° 10′ E	Eliseeva (1960)
1600	1	6	51° 50′ N	36° 10′ E	Eliseeva (1960)
1266	1	6	51° 50′ N	36° 10′ E	Eliseeva (1960)
2	17	214	54° 20′ N	10° 05′ E	Erz (1964)
4	17	214	54° 20′ N	10° 05′ E	Erz (1964)
10	14	44	54° 20′ N	10° 05′ E	Erz (1964)
14	14	44	54° 20′ N	10° 05′ E	Erz (1964)
160	8	8	54° 20′ N	10° 05′ E	Erz (1964)
160	8	8	54° 20′ N	10° 05′ E	Erz (1964)
40	9	23	54° 20′ N	10° 05′ E	Erz (1964)
60	9	23	54° 20′ N	10° 05′ E	Erz (1964)
12	9	33	51° 30′ N	7° 30′ E	Erz (1964)
12	9	33	51° 30′ N	7° 30′ E	Erz (1964)
140	8	5.6	51° 30′ N	7° 30′ E	Erz (1964)
140	8	5.6	51° 30′ N	7° 30′ E	Erz (1964)
40	7	14.5	51° 30′ N	7° 30′ E	Erz (1964)
60	7	14.5	51° 30′ N	7° 30′ E	Erz (1964)
20	9	10	51° 30′ N	7° 30′ E	Erz (1964)
40	9	10	51° 30′ N	7° 30′ E	Erz (1964)
80	8	5.55	51° 30′ N	7° 30′ E	Erz (1956)
100	8	5.55	51° 30′ N	7° 30′ E	Erz (1956)
80	8	5.55	51° 30′ N	7° 30′ E	Erz (1956)
39	13	—	48° 45′ N	18° 05′ E	Ferianc *et al.* (1973)
81	13	—	48° 45′ N	18° 05′ E	Ferianc *et al.* (1973)
81	13	—	48° 45′ N	18° 05′ E	Ferianc *et al.* (1973)
17	13	34	49° N	17° E	Feriancová-Masárová & Brtek (1969)
26	13	30	49° N	17° E	Feriancová-Masárová & Brtek (1969)

Appendix 3.2. (*cont.*)

Breeding density individuals km^{-2}	Biotope type (see Table 3.2)	Size of biotope (ha)	Latitude	Longitude	Authority
7	13	30	49° N	17° E	Feriancová-Masárová & Brtek (1969)
20	13	20	49° N	17° E	Feriancová-Masárová & Brtek (1969)
50	13	20	49° N	17° E	Feriancová-Masárová & Brtek (1969)
33	13	24	49° N	17° E	Feriancová-Masárová & Brtek (1969)
5	13	40	49° N	17° E	Feriancová-Masárová & Brtek (1969)
75	13	8	49° N	17° E	Feriancová-Masárová & Brtek (1969)
20	13	10	49° N	17° E	Feriancová-Masárová & Brtek (1969)
18	13	22	49° N	17° E	Feriancová-Masárová & Brtek (1969)
20	7	1490	46° 30′ N	35° 20′ E	Filonov (1972)
340	3	10	47° N	28° 45′ E	Ganya (1965)
480	3	10	47° N	28° 45′ E	Ganya (1965)
760	3	10	47° N	28° 45′ E	Ganya (1965)
500	3	4	47° N	28° 45′ E	Ganya (1965)
32	13	24	47° N	28° 45′ E	Ganya (1965)
286	11	15	47° N	28° 45′ E	Ganya (1965)
266	3	3	47° N	28° 45′ E	Ganya (1965)
100	6	10	44° 30′ N	28° 30′ E	Ganya (1965)
150	6	—	44° 30′ N	28° 30′ E	Ganya (1965)
250	7	24	44° 30′ N	28° 30′ E	Ganya (1969)
250	7	20	44° 30′ N	28° 30′ E	Ganya (1969)
236	7	22	44° 30′ N	28° 30′ E	Ganya (1969)
84	11	12	50° 10′ N	20° 30′ E	Głowaciński (1975)
600	11	—	36° N	62° E	Gorelova (1974)
24	5	20	51° 30′ N	17° E	Graczyk, Chartanowicz, Fruziński & Małek (1966)
34	5	20	51° 30′ N	17° E	Graczyk et al. (1966)
12	5	17.6	51° 30′ N	17° E	Graczyk et al. (1966)
3	5	32	51° 30′ N	17° E	Graczyk et al. (1966)
100	4	45	51° 20′ N	17° E	Graczyk, Galiński & Klejnotowski (1967)
320	4	45	51° 20′ N	17° E	Graczyk et al. (1967)
45	4	45	51° 20′ N	17° E	Graczyk et al. (1967)
80	14	20	54° 10′ N	12° 10′ E	Grempe (1971)
1044	2	—	52° 10′ N	16° 45′ E	Gromadzki (1970)
912	2	—	52° 10′ N	16° 45′ E	Gromadzki (1970)
614	2	—	52° 10′ N	16° 45′ E	Gromadzki (1970)
630	2	—	52° 10′ N	16° 45′ E	Gromadzki (1970)
548	2	—	52° 10′ N	16° 45′ E	Gromadzki (1970)
178	2	—	52° 10′ N	16° 45′ E	Gromadzki (1970)
318	2	—	52° 10′ N	16° 45′ E	Gromadzki (1970)
296	2	—	52° 10′ N	16° 45′ E	Gromadzki (1970)
740	2	—	52° 10′ N	16° 45′ E	Gromadzki (1970)

Appendix 3.2

Appendix 3.2. (*cont.*)

Breeding density individuals km⁻²	Biotope type (see Table 3.2)	Size of biotope (ha)	Latitude	Longitude	Authority
32	2	—	52° 10′ N	16° 45′ E	Gromadzki (1970)
202	2	—	52° 10′ N	16° 45′ E	Gromadzki (1970)
40	7	—	51° 30′ N	7° E	Hyla (1967)
300	7	—	51° 30′ N	7° E	Hyla (1967)
60	7	—	51° 30′ N	7° E	Hyla (1967)
40	2	40	51° 30′ N	7° E	Hyla (1967)
140	3	150	47° 11′ N	25° 11′ E	Ion & Valenciuc (1967)
428	6	16.4	50° 30′ N	16° 50′ E	Jakubiec (1972)
272	6	33	56° 45′ N	10° 20′ E	Joensen (1965)
182	6	33	56° 45′ N	10° 20′ E	Joensen (1965)
12	6	17	56° 45′ N	10° 20′ E	Joensen (1965)
33	13	18	56° 45′ N	10° 20′ E	Joensen (1965)
520	2	5.2	52° N	12° 50′ E	Kaatz & Olberg (1975)
24	15	—	53° 30′ N	12° 50′ E	Kintzel (1972)
66	13	6	46° 45′ N	23° 40′ E	Korodi Gál (1957)
850	3	8.5	46° 45′ N	23° 45′ E	Korodi Gál (1958)
1100	3	8.5	46° 45′ N	23° 45′ E	Korodi Gál (1958)
7800	8	11	46° 46′ N	23° 34′ E	Korodi Gál (1960)
300	6	100	46° 45′ N	21° 45′ E	Korodi Gál (1964)
120	12	200	46° 45′ N	21° 45′ E	Korodi Gál (1964)
120	12	300	46° 45′ N	21° 45′ E	Korodi Gál (1964)
60	12	150	46° 45′ N	21° 45′ E	Korodi Gál (1964)
300	6	150	46° 45′ N	21° 45′ E	Korodi Gál (1964)
320	6	400	46° 45′ N	21° 45′ E	Korodi Gál (1964)
10	12	—	51° N	42° 06′ E	Korol'kova (1963)
26	14	—	46° 40′ N	21° 20′ E	Korompai (1966)
104	8	31	50° 05′ N	22° 15′ E	Kulczycki (1966)
116	8	31	50° 05′ N	22° 15′ E	Kulczycki (1966)
116	8	31	50° 05′ N	22° 15′ E	Kulczycki (1966)
132	8	31	50° 05′ N	22° 15′ E	Kulczycki (1966)
122	8	31	50° 05′ N	22° 15′ E	Kulczycki (1966)
21	2	—	48° N	23° E	Lugovoï & Talposh (1968)
254	8	15	52° 10′ N	22° 20′ E	Luniak (1974)
200	8	15	52° 10′ N	22° 20′ E	Luniak (1974)
82	8	9.8	52° 10′ N	21° 30′ E	Luniak (1974)
106	8	3.7	52° 10′ N	22° 20′ E	Luniak (1974)
134	8	6	52° 05′ N	23° 05′ E	Luniak (1974)
302	8	5.3	51° 45′ N	22° 45′ E	Luniak (1974)
63	11	—	52° 21′ N	14° 51′ E	Mackowicz *et al.* (1970)
75	11	—	52° 21′ N	14° 51′ E	Mackowicz *et al.* (1970)
105	8	—	60° N	30° 20′ E	Malchevskiï (1955)
105	8	—	60° N	30° 20′ E	Malchevskiï (1955)
157	8	—	60° N	30° 20′ E	Malchevskiï (1955)
2.8	—	—	54° N	70° E	Matyushkin (1967)
60	2	12	54° N	70° E	Matyushkin (1967)
40	2	12	54° N	70° E	Matyushkin (1967)
30	2	13	54° N	70° E	Matyushkin (1967)
22	8	115	52° 10′ N	8° 45′ E	Michels (1973)
140	2	4.49	51° 20′ N	18° E	Michocki (1974)

Appendix 3.2. (*cont.*)

Breeding density individuals km^{-2}	Biotope type (see Table 3.2)	Size of biotope (ha)	Latitude	Longitude	Authority
220	2	4.49	51° 20' N	18° E	Michocki (1974)
1160	2	4.49	51° 20' N	18° E	Michocki (1974)
1600	2	4.49	51° 20' N	18° E	Michocki (1974)
1360	2	4.49	51° 20' N	18° E	Michocki (1974)
1460	2	4.49	51° 20' N	18° E	Michocki (1974)
220	2	4.49	51° 20' N	18° E	Michocki (1974)
480	2	4.49	51° 20' N	18° E	Michocki (1974)
80	2	4.49	51° 20' N	18° E	Michocki (1974)
800	2	4.49	51° 20' N	18° E	Michocki (1974)
20	14	47	53° 45' N	10° E	Mulsow (1967)
84	8	4.8	53° 45' N	10° E	Mulsow (1967)
18	8	45	53° 45' N	10° E	Mulsow (1967)
26	9	30	53° 45' N	10° E	Mulsow (1967)
133	3	6	46° 20' N	26° 30' E	Munteanu (1963)
120	7	23.5	51° 30' N	16° 30' E	Ranoszek (1969)
280	2	23.5	51° 30' N	16° 30' E	Ranoszek (1969)
4	—	—	53° N	85° E	Ravkin (1973)
1	—	—	53° N	85° E	Ravkin (1973)
7	—	—	53° N	85° E	Ravkin (1973)
348	11	—	53° N	85° E	Ravkin (1973)
251	11	—	53° N	85° E	Ravkin (1973)
6.9	—	88.5	54° 30' N	2° 30' W	Robson & Williamson (1972)
132	2	13400	52° 25' N	10° 47' E	Scherner (1972)
80	2	8.5	51° 20' N	14° 05' E	Schlegel (1966)
100	2	—	51° 20' N	14° 05' E	Schlegel (1966)
160	2	—	51° 20' N	14° 05' E	Schlegel (1966)
227	2	2.2	49° N	14° 45' E	Šťastný (1973)
571	2	2.2	49° N	14° 45' E	Šťastný (1973)
1890	2	2.8	49° N	14° 45' E	Šťastný (1973)
2222	2	2.8	49° N	14° 45' E	Šťastný (1973)
120	7	2.8	49° N	14° 45' E	Šťastný (1973)
104	2	15.37	49° 45' N	21° E	Tomek (1973)
352	2	2.3	49° 45' N	21° E	Tomek (1973)
685	11	3.5	49° 45' N	21° E	Tomek (1973)
0	16	31	51° 10' N	16° 10' E	Tomiałojć (1970)
12	17	33	51° 10' N	16° 10' E	Tomiałojć (1970)
6	17	—	51° 10' N	16° 10' E	Tomiałojć (1970)
20	17	30	51° 10' N	16° 10' E	Tomiałojć (1970)
86	14	6.9	51° 10' N	16° 10' E	Tomiałojć (1970)
30	14	13	51° 10' N	16° 10' E	Tomiałojć (1970)
290	8	35	51° 10' N	16° 10' E	Tomiałojć (1970)
286	8	35	51° 10' N	16° 10' E	Tomiałojć (1970)
302	8	35	51° 10' N	16° 10' E	Tomiałojć (1970)
146	7	34	51° 10' N	16° 10' E	Tomiałojć (1970)
194	7	34	51° 10' N	16° 10' E	Tomiałojć (1970)
200	7	34	51° 10' N	16° 10' E	Tomiałojć (1970)
60	8	20	51° 10' N	16° 10' E	Tomiałojć (1970)
400	9	18	51° 10' N	16° 10' E	Tomiałojć (1970)
140	3	15.6	51° 10' N	16° 10' E	Tomiałojć (1970)

Appendix 3.2. (*cont.*)

Breeding density individuals km⁻²	Biotope type (see Table 3.2)	Size of biotope (ha)	Latitude	Longitude	Authority
46	3	8.4	51° 10′ N	16° 10′ E	Tomiałojć (1970)
80	11	64	51° 10′ N	16° 10′ E	Tomiałojć (1970)
286	2	8.4	51° 10′ N	16° 10′ E	Tomiałojć (1974)
170	6	25	51° 10′ N	16° 10′ E	Tomiałojć (1974)
94	6	14.9	51° 10′ N	16° 10′ E	Tomiałojć (1974)
36	12	11	51° 10′ N	16° 10′ E	Tomiałojć (1974)
20	13	9.9	51° 10′ N	16° 10′ E	Tomiałojć (1974)
60	12	10	51° 10′ N	16° 10′ E	Tomiałojć (1974)
40	12	14.6	51° 10′ N	16° 10′ E	Tomiałojć (1974)
98	10	26.6	51° 10′ N	16° 10′ E	Tomiałojć (1974)
20	15	20.3	51° 10′ N	16° 10′ E	Tomiałojć (1974)
218	2	—	51° 10′ N	16° 10′ E	Tomiałojć (1974)
21	12	—	51° 10′ N	16° 10′ E	Tomiałojć (1974)
146	8	15	52° 10′ N	20° 50′ E	Truszkowski (1963)
28	15	—	51° N	30° 30′ E	Vladÿshevskiĭ (1972)
65	10	—	51° N	30° 30′ E	Vladÿshevskiĭ (1972)
16	2	—	48° N	27° E	Volchanetskiĭ, Lisetskiĭ & Kholupyak (1970)
200	2	—	48° N	27° E	Volchanetskiĭ *et al.* (1970)
16	11	—	43° N	44° 30′ E	Vtorov (1962)
16	11	—	43° N	44° 30′ E	Vtorov (1962)
19	14	—	39° 54′ N	8° 34′ E	Walter & Mocci Demartis (1972)
16.5	11	2000	Wales, Great Britain		Williamson (1967)
16.4	13	—	Wales, Great Britain		Williamson (1975)
P. montanus pallidus					
4000	8	—	40° 45′ N	72° 20′ E	Sharipov (1974)
1140	16	—	40° 45′ N	72° 20′ E	Sharipov (1974)
300	3	—	40° 45′ N	72° 20′ E	Sharipov (1974)
40	17	—	40° 45′ N	72° 20′ E	Sharipov (1974)
3200	8	—	40° 20′ N	71° 45′ E	Sharipov (1974)
2800	16	—	40° 20′ N	71° 45′ E	Sharipov (1974)
2820	17	—	40° 20′ N	71° 45′ E	Sharipov (1974)
600	17	—	40° 20′ N	71° 45′ E	Sharipov (1974)
P. montanus saturatus					
53	8	—	—	—	Kuroda (1966 *b*)
66	8	—	—	—	Kuroda (1967)
60	8	—	—	—	Kuroda (1968)
98	8	—	—	—	Kuroda (1969)
37	8	—	—	—	Kuroda (1970)
63	8	—	—	—	Kuroda (1971)
55	8	—	—	—	Kuroda (1973 *b*)
242	8	—	—	—	Kuroda (1973 *a*)
625	8	—	—	—	Kuroda (1973 *a*)
404	8	—	—	—	Kuroda (1973 *a*)

Appendix 3.3. *Relationship between the onset of egg-laying and latitude in the house sparrow*

Coordinate[a]	1913	1960	1961	1962	1963	1964	1965	1966	1967	1968	1969	1970	1971	1972	1973	1974
(1) 22° 18'	—	—	—	—	—	—	—	—	—	10/02	10/02	13/02	24/01	8/02	17/02	—
(2) 31° 30'	—	—	—	—	—	—	—	—	—	14/03	14/03	—	—	—	—	—
(3) 34° 10'	—	—	—	—	—	—	—	—	—	16/03	—	—	—	—	—	—
(4) 38° 07'	—	—	—	—	—	—	—	2/04	2/04	27/03	—	—	—	—	—	—
(5) 38° 55'	—	—	—	—	—	—	—	—	—	—	10/04	8/04	3/04	14/04	—	—
(6) 42° 50'	—	—	—	—	—	—	—	—	—	—	—	9/04	—	—	—	—
(7) 45° 51'	—	—	—	—	—	—	—	—	2/04	25/03	—	—	—	—	—	—
(8) 47° 30'	—	—	—	—	—	—	—	—	2/04	24/03	—	—	—	—	—	—
(9) 48° 50'	—	—	—	20/04	10/04	13/04	9/04	—	—	—	—	—	—	—	—	—
(10) 50° 04'	—	—	—	—	—	—	—	—	13/04	4/04	28/04	11/04	8/04	30/03	31/03	—
(11) 50° 13'	—	—	—	—	13/04	13/04	7/05	—	—	—	—	—	—	—	—	—
(12) 51° 46'	—	—	2/04	10/04	12/04	—	—	—	—	—	—	—	—	—	—	—
(13) 52° 20'	—	20/04	2/05	23/04	15/04	16/04	3/05	8/04	10/04	6/04	13/04	22/04	9/04	9/04	4/04	—
(14) 52° 25'	—	—	—	—	—	—	—	—	—	—	—	—	27/04	—	—	—
(15) 52° 30'	—	—	—	—	—	12/04	4/04	—	—	—	—	—	—	—	—	—
(16) 54° 20'	—	—	—	—	—	—	—	—	—	—	—	—	20/04	12/04	11/04	—
(17) 54° 37'	—	—	—	—	—	—	—	—	—	—	—	1/05	16/04	11/04	22/03	17/04
(18) 61° 27'	—	—	—	28/04	30/04	2/05	28/04	6/05	7/05	28/04	27/04	30/04	4/05	—	—	—
(19) 65° 00'	—	—	—	—	—	—	—	—	—	7/05	7/05	9/05	8/05	9/05	9/05	4/05
(20) 67° 39'	5/05	—	—	—	—	—	—	—	—	—	—	—	—	—	—	—

[a] (1) Naik & Mistry (1973), Baroda, India; (2) Mirza (1973), Lahore, Pakistan; (3) Mitchell, Hayes, Holden & Hughes (1973), Plainview, Texas, USA; (4) Will (1969), McLeansboro, Illinois, USA; (5) Anderson (1973), Portage des Sioux, Missouri, USA; (6) North (1973a), Whitewater, USA; (7, 8) Ion (1973), Bujor Targ, Iaşi, Romania; (9) Barloy (1966), Paris, France; (10) Mackowicz et al. (1970), Kraków, Rzepin, Poland; (11) Novotný (1970), Slezké Rudoltice, Czechoslovakia; (12) Seel (1968a), Oxford, England; (13) Pinowska & Pinowski (1977), Dziekanów Leśny, Poland; (14) Wieloch & Fryska (1975); Turew, Poland; (15) Deckert (1969), Motzaner See, GDR; (16) Pinowska (1976), Wieniec, Poland; (17) Wieloch & Fryska (1975), Gdynia, Poland; (18) Rassi (in preparation), Kangasala, Finland; (19) Alatalo (1975), Oulu, Finland; (20) Montell (1917), Muonio, Finland.

Appendix 3.4

Appendix 3.4. *Relationship between the onset of egg-laying and latitude in the tree sparrow*

Coordinate[a]	1931	1932	1936	1938	1944	1952	1955	1956	1957	1958	1959	1960	1961
(1) 38° 55'	—	—	—	—	—	—	—	—	—	—	—	—	—
(2) 43° 20'	—	—	—	—	—	—	—	—	—	—	—	—	—
(3) 45° 51'	—	—	—	—	—	—	—	—	—	—	—	—	—
(4) 47° 11'	—	—	—	—	—	—	—	—	—	—	—	—	—
(5) 48° 45'	—	—	—	—	—	—	—	—	—	—	—	—	—
(6) 49° 00'	—	—	—	—	—	—	—	—	—	—	—	—	—
(7) 49° 00'	—	—	—	—	—	—	—	—	—	—	—	—	—
(8) 49° 20'	—	—	—	—	—	—	—	—	—	—	—	—	—
(9) 50° 04'	—	—	—	—	—	—	—	—	—	—	—	—	—
(10) 51° 20'	—	—	—	—	—	25/04	29/04	4/05	19/04	4/05	6/05	6/05	18/04
(11) 51° 30'	—	—	—	—	—	—	—	3/05	25/04	4/05	—	—	—
(12) 51° 46'	—	—	—	—	—	—	—	—	—	—	—	—	19/04
(13) 52° 20'	—	—	—	—	—	—	—	—	—	—	—	26/04	18/04
(14) 52° 21'	—	—	—	—	—	—	—	—	—	—	—	—	—
(15) 52° 25'	—	—	—	—	—	—	—	—	—	—	—	—	—
(16) 52° 25'	—	—	—	—	—	—	—	—	—	—	—	—	—
(17) 52° 30'	—	—	—	—	—	—	—	—	—	6/05	18/04	—	—
(18) 52° 30'	12/05	5/05	—	—	—	—	—	—	—	—	—	—	—
(19) 54° 37'	—	—	—	—	—	—	—	—	—	—	—	—	—
(20) 55° 31'	—	26/04	29/04	2/05	—	—	—	6/05	19/04	—	—	—	—

Appendix 3.4. (*cont.*)

Coordinate[a]	1962	1963	1964	1965	1966	1967	1968	1969	1970	1971	1972	1973	1974
(1) 38° 55'	—	—	—	—	—	—	—	26/04	21/04	18/04	16/04	22/04	15/04
(2) 43° 20'	20/04	27/04	—	—	—	—	—	—	—	—	—	—	—
(3) 45° 51'	—	—	—	—	—	11/04	3/04	—	—	—	—	—	—
(4) 47° 11'	—	—	—	—	—	17/04	12/04	—	—	—	—	—	—
(5) 48° 45'	—	—	—	—	—	20/04	24/04	1/05	24/04	16/04	14/04	—	—
(6) 49° 00'	—	—	—	—	—	—	—	25/04	24/04	17/04	12/04	—	—
(7) 49° 00'	—	—	—	—	17/04	21/04	22/04	30/04	25/04	24/04	14/04	21/04	10/04
(8) 49° 20'	—	—	—	—	—	—	19/04	29/04	25/04	18/04	21/04	14/04	10/04
(9) 50° 04'	—	—	—	—	—	14/04	6/04	21/04	20/04	15/04	10/04	10/04	—
(10) 51° 20'	—	—	—	—	—	—	—	—	—	—	—	—	—
(11) 51° 30'	—	—	—	—	—	—	—	—	—	—	—	—	—
(12) 51° 46'	—	27/04	28/08	—	—	—	—	—	—	—	—	—	—
(13) 52° 20'	19/04	22/04	19/04	25/04	23/04	—	—	30/04	—	—	—	—	—
(14) 52° 21'	—	—	—	—	25/04	16/04	—	—	—	—	—	—	—
(15) 52° 25'	—	—	—	—	—	—	—	—	30/04	12/04	19/04	—	—
(16) 52° 25'	—	—	—	—	—	—	23/04	1/05	—	—	—	—	—
(17) 52° 30'	—	—	—	—	—	—	—	—	—	—	—	—	—
(18) 52° 30'	—	—	—	—	—	—	—	—	—	—	—	—	—
(19) 54° 37'	—	—	—	—	—	—	—	—	—	28/04	1/05	—	—
(20) 55° 31'	—	—	—	—	—	—	—	—	—	—	—	—	—

[a] (1) Anderson (1973, 1977), Portage des Sioux, Missouri, USA; (2) Abé (1969), Sapporo, Japan; (3, 4) Ion (1973), Bujor Targ, Iaşi, Romania; (5, 6) Bauer (1974), Břeclav, Źidlochovice, Czechoslovakia; (7, 8) Balát (1971), Bzenec, Sokolnice, Czechoslovakia; (9) Mackowicz *et al.* (1970), Kraków, Poland; (10) de Bethune (1961), Marka, Belgium; (11) Eliseeva (1961), Kursk, USSR; (12) Seel (1968 a), Oxford, England; (13) Pinowska & Pinowski (1977), Dziekanów Leśny, Poland; (14) Mackowicz *et al.* (1970), Rzepin, Poland; (15) Wieloch & Fryska (1975), Turew, Poland; (16) Scherner (1972), Wolfsburg, GFR; (17) Deckert (1962), Motzaner See, GDR; (18) Boyd (1932), Great Budworth, England; (19) Wieloch & Fryska (1975), Gdańsk, Poland; (20) Ptushenko & Inozemtsev (1968), Moskwa, USSR.

Appendix 3.5

Appendix 3.5. *Egg loss and nestling mortality in the house sparrow for different regions of the world*

Locality	Years studied	Egg (no.)	Egg mortality (%)	Mortality of nestlings in relation to the number of nestlings hatched	Authority
New Zealand					
Christchurch (43° 42' S, 172° 38' E)	9	—	29	40	Dawson (1973)
Hawkes Bay (39° S, 177° E)	9	—	18	40	Dawson (1973)
Indian subcontinent					
Baroda, India (22° 18' N, 73° 13' E)	4	5460	38	59	Naik (1974)
Lahore, Pakistan (31° 30' N, 74° 10' E)	2	—	8	—	Mirza (1973)
North America					
Plainview, Texas (34° 10' N, 101° 43' W)	1	397	43	72	Mitchell et al. (1973)
Cloverlake Dairy, Plainview, Texas (34° 10' N, 101° 43' W)	1	692	39	33	Mitchell et al. (1973)
Stillwater, Oklahoma (35° 50' N, 94° 37' W)	2	434	50	35	North (1968)
McLeansboro, Illinois (38° 07' N, 88° 32' W)	3	1502	34	47	Will (1969)
Portage des Sioux, Missouri (38° 55' N, 90° 22' W)	3	2142	36	34	Anderson (1973)
Coldspring, Wisconsin (42° 50' N, 88° 32' W)	1	370	49	41.09	North (1973 a)
Continental Europe					
Bujor Targ, Romania (45° 51' N, 27° 57' E)	2	159	21	26	Ion (1973)
Cimpulung Moldovenesc, Romania (47° 30' N, 25° 30' E)	2	132	20	14	Ion (1973)
Iaşi, Romania (47° 10' N, 27° 35' E)	2	316	32	17	Ion (1973)
Veverska Bityska, Czechoslovakia (49° 05' N, 15° 20' E)	3	1009	23	21	Balát (1974)
Zakopane, Poland (49° 19' N, 19° 57' E)	2	207	27	21	Mackowicz et al. (1970)
Nowy Targ, Poland (49° 28' N, 20° 00' E)	4	162	5	11	Mackowicz et al. (1970)
Kraków, Poland (50° 04' N, 19° 57' E)	7	897	39	39	Pinowski & Wieloch (1973)
Slezké Rudoltice, Czechoslovakia (50° 13' N, 17° 43' E)	3	1903	19	15	Novotný (1970)
Dziekanów Leśny, Poland (52° 20' N, 20° 50' E)	14	6239	45	21	Pinowska & Pinowski (1977)
Berlin, GDR (52° 30' N, 13° 00' E)	1	407	12	46	Encke (1965)
Bulow, GDR (53° 30' N, 11° 30' E)	1	495	16	17	Encke (1965)

358

Appendix 3.5. (*cont.*)

Locality	Years studied	Egg (no.)	Egg mortality (%)	Mortality of nestlings in relation to the number of nestlings hatched	Authority
Wieniec, Poland (54° 20′ N, 18° 56′ E)	3	2561	41	30	Pinowska (1975)
Gdańsk, Poland (54° 20′ N, 18° 56′ E)	5	787	27	29	Strawiński & Wieloch (1972)
Sysert near Sverdlovsk, USSR (56° 30′ N, 61° 00′ E)	2	355	15	13	Nekrasov (1970)
Kangasala, Finland (61° 27′ N, 24° 03′ E)	5	637	15	46	Rassi(unpublished)
England Oxford	4	3008	15	55	Seel (1968*b*, 1970)
(51° 46′ N, 1° 15′ W)	1	915	20	48	Dawson (1972)

Appendix 3.6. *Egg loss and nestling mortality in the tree sparrow for different regions in the world*

Locality	Years studied	Egg (no.)	Egg mortality (%)	Mortality of nestlings in relation to the number of nestlings hatched	Authority
Asia					
Peking, China (39° 55′ N, 116° 25′ E)	4	—	15	—	Chia *et al.* (1963)
Sapporo, Japan (43° 20′ N, 141° 30′ E)	2	1165	7	13	Abé (1969)
North America					
Portage des Sioux, Missouri (38° 55′ N, 90° 22′ W)	4	776	24	27	Anderson (1973)
Continental Europe					
Bujor Targ, Romania (45° 51′ N, 27° 57′ E)	2	116	15	26	Ion (1973)
Iaşi, Romania (47° 10′ N, 27° 35′ E)	2	327	33	18	Ion (1973)
Cimpulung Moldovenesc, Romania (47° 30′ N, 25° 30′ E)	2	103	18	12	Ion (1973)
Sokolnice, Bzenec, Czechoslovakia (49° 01′ N, 17° 10′ E)	2	1202	12	36	Balát (1971)
Breclav, Czechoslovakia (48° 45′ N, 16° 50′ E)	6	659	8	11	Bauer (1974)
Nowy Targ, Poland (49° 28′ N, 20° 00′ E)	6	648	15	16	Mackowicz *et al.* (1970)
Kraków, Poland (50° 04′ N, 19° 57′ E)	7	883	27	22	Mackowicz *et al.* (1970)
Dresden, GDR (51° N, 13° 45′ E)	10	525	28	36	Creutz (1949)
Marka, Belgium (51° 20′ N, 2° 50′ E)	10	848	43	12	de Bethune (1961)
Kursk, USSR (51° 43′ N, 36° 15′ E)	3	—	15–19	12	Eliseeva (1961)
Laenen, the Netherlands (51° 52′ N, 5° 46′ E)	1	369	30	17	Van Balen (unpublished)
Oosterhout, the Netherlands (51° 52′ N, 5° 50′ E)	1	275	28	12	Van Balen (unpublished)
Turew, Poland (52° 04′ N, 16° 53′ E)	3	1443	44	26	Wieloch & Fryska (1975)
Dziekanów Leśny, Poland (52° 20′ N, 20° 50′ E)	14	11969	31	15	Pinowski (1968); Mackowicz *et al.* (1970); Pinowski & Wieloch (1973)
Wolfsburg, GFR (52° 25′ N, 10° 47′ E)	2	953	16	14	Scherner (1972)
Gdańsk, Poland (54° 20′ N, 48° 56′ E)	5	847	38	28	Wieloch & Fryska (1975)
Great Britain					
Oxford, England (51° 46′ N, 0° 15′ E)	4	1173	9	41	Seel (1968 *b*)
Entire country	1	2180	12	35	Seel (1964)

Appendix 3.7. *The average number of eggs laid and young birds raised per pair in one year by the house sparrow*

Locality	No. of eggs laid	No. of eggs lost	No. of nestlings lost	No. of total young leaving the nest	Authority
Baroda, India (22° 18′ N, 73° 13′ E)	18.49	7.03	6.76	4.7	Naik (1974)
Plainview, Texas, USA (34° 10′ N, 101° 43′ W)	16.38	7.04	6.72	2.62	Mitchell *et al.* (1973)
Cloverlake Dairy, Plainview, Texas, USA (34° 10′ N, 101° 43′ W)	12.6	4.91	2.54	5.15	Mitchell *et al.* (1973)
McLeansboro, Illinois, USA (38° 07′ N, 88° 32′ W)	10.12	3.44	3.14	3.54	Will (1969)
Portage des Sioux, Missouri, USA (38° 55′ N, 90° 22′ W)	10.8	3.88	2.35	4.57	Anderson (1973)
Coldspring, Wisconsin, USA (43° 50′ N, 88° 32′ W)	7.35	3.60	—	—	North (1973 a)
Kraków, Poland (50° 04′ N, 19° 57′ E)	9.6	3.74	2.28	3.58	Pinowski & Wieloch (1973)
Slezké Rudoltice, Czechoslovakia (50° 13′ N, 17° 43′ E)	11.18	2.12	1.36	7.7	Novotný (1970)
Dziekanów Leśny, Poland (52° 20′ N, 20° 50′ E)	9.87	4.44	1.14	4.29	Pinowska & Pinowski (1977)
Gdańsk, Poland (54° 20′ N, 18° 56′ E)	7.31	1.97	1.55	3.79	Strawiński & Wieloch (1972)
Kangasala, Finland (61° 20′ N, 24° 03′ E)	8.96	1.34	3.50	4.12	Rassi (unpublished)
Oxford, England (51° 46′ N, 1° 15′ W)	8.4	1.26	3.92	3.22	Dawson (1972)
Sysert near Sverdlovsk, USSR (56° 30′ N, 61° E) (1967)	9.07	0.93	0.5	4.4	Nekrasov (1970)
(1968)	9.52	1.74	1.39	6.4	
Average	10.69± 3.09	3.39± 1.92	2.85± 1.89	4.47± 1.31	

Appendix 3.8

Appendix 3.8. *The average number of eggs laid and young birds raised per pair in one year by the tree sparrow*

Locality	No. of eggs laid	No. of eggs lost	No. of nestlings lost	No. of total young leaving the nest	Authority
Portage des Sioux, Missouri, USA (38° 55′ N, 90° 22′ W)	13.0	3.12	2.67	7.21	Anderson (1973)
Sapporo, Japan (43° 20′ N, 141° 30′ E)	11.2	0.78	1.35	9.07	Abé (1969)
Bzenec, Czechoslovakia (49° 01′ N, 17° 10′ E)	9.26	1.11	2.93	5.22	Balát (1971)
Sokolnice, Czechoslovakia (49° 01′ N, 17° 10′ E)	9.12	1.09	2.89	5.14	Balát (1971)
Dresden, GDR (51° N, 13° 45′ E)	7.36	2.06	1.90	3.4	Creutz (1949)
Dziekanów Leśny, Poland (52° 20′ N, 20° 50′ E)	12.0	3.72	1.24	7.04	Pinowski (1968)
Wolfsburg, GFR (52° 25′ N, 10° 47′ E) (1968)	10.1	3.67	0.89	5.54	Scherner (1972)
(1969)	6.9	2.80	0.54	3.56	
Oxford, England (51° 46′ N, 0° 15′ E)	7.84	0.70	2.50	4.64	Seel (1968 *b*)
Average	9.64± 1.99	2.12± 1.17	1.88± 0.85	5.65± 1.73	

362

Appendix 5.1. *Standard and basal metabolic rates* (*SM and BM*).

Appendix 5.1. compiles all the known equations of SM ($M = $ kcal bird^{-1} day^{-1}) on temperature ($T = $ °C) classified taxonomically and arranged by increasing weight in each major taxon. Some weights are approximate. Most authors give metabolic equations in terms of rate of oxygen consumed. These are converted to kilocalories using 1 cc oxygen = 4.8 gcal unless otherwise indicated by the author. In some cases, equations had to be calculated from data or transcribed from figures given by the author. Some lower and upper critical temperatures likewise had to be determined from figures. Upper critical temperatures have frequently not been determined. The data for BM (SM in the zone of thermo-neutrality) are often slightly lower than can be calculated from the equation using the lower critical temperature. The reason for this is that many authors cite the lowest values obtained in the zone of the thermo-neutrality for BM rather than a mean for the zone. In three or four instances, it was not certain whether the birds were in a post-absorptive condition, but the values indicate that they were at least very close to it.

Special comment is desirable on the new data obtained by Dol'nik and Gavrilov, many of which have not previously been published or are not readily available to Western biologists. All European passerine species were captured as adults or sub-adults, except for *Corvus corax*, *C. corone* and *Pica pica* which were raised from nestlings. Most non-European and non-passerine species were obtained from the Moscow Zoo. All captured birds were kept in outdoor aviaries subject to natural temperatures and photoperiods for two to four weeks or more before measurements were taken. It is indicated whether the measurements were made during summer (S), winter (W), autumn (A), or spring (Sp). Tropical, sub-tropical and migrant species were housed at 15–22 °C from October to May. In the winter (December through February) only non-moulting birds were used; spring measurements on wild birds were made in late April and May near or at the end of northward migration; in the summer, measurements were made in late June and early July between the end of sexual activity and before moulting began, and in the autumn from September to November after moulting had ended and while migration southward was underway. Birds from the zoo were used during non-moulting and non-breeding periods.

Rates of oxygen consumption were measured only on fasting birds either during the day (D) or at night (N). For measurements at night, large birds were deprived of food from early morning; small birds for three to four hours before onset of darkness. Birds were placed singly into small cages and then into plexiglass boxes in the dark. The boxes varied in size from 3 to 25 l in volume to accommodate different sized birds. Air flow from the outside through the box was controlled and, after the temperature stabilized, shunted into the measuring equipment. Actual measurement of oxygen consumption never began earlier than one to three hours after onset of darkness and was always terminated one to two hours before onset of daylight. Each experiment lasted one to four hours and intervals of one to one and a half hours lapsed between consecutive measurements at two different temperatures during the same night. When several measurements were made on the same bird they were at progressive intervals of 5 °C. Some species were measured down to −20 °C, and several measurements were made in the neighborhood of the apparent lower critical temperature. Tests were made at the beginning and end of experiments to make certain the boxes were properly sealed.

Three types of thermostatically-controlled temperature chambers were available for different ranges of temperature. An air-conditioned room with controlled temperatures was also used, as was a room cooled by outside temperature and the temperature controlled by means of an electric fan. The number of individuals measured varied from two to four in some large or less available species, to eight to twelve in smaller species.

The regression of metabolic rate on temperature in some species became curvilinear at low temperatures. In these species the equations giving linear regressions were calculated only for temperatures above $-5\,°C$. Some adjustment of regression lines was commonly made so that M equalled zero when air temperatures equalled body temperatures (39–$40\,°C$).

The measurements of SM and BM obtained on 32 species of passerine species by Shilov (1968) did not become available until after the following list was compiled and analyzed. New species studied by Shilov included *Rhodospiza obsoleta, Alauda arvensis, Sitta europaea, Parus coeruleus, P. cristatus, Acrocephalus dumentorum, Sylvia cummunis* and *Luscinia luscinia*. Likewise, the daytime measurements of five species of thrushes (*Hylocichla, Catharus*) by Holmes & Sawyer (1975) came too late to be included in the following table.

Appendix 5.1 (cont.)

Species	Weight (g)	Season	Time of measurement	M (kcal bird^{-1} day^{-1})	Critical temperatures Lower (°C)	Upper (°C)	BM (kcal bird^{-1} day^{-1})	Authority
Sphenisciformes								
Peruvian penguin *Spheniscus humboldti*	3870		(N)	$317.0 - 13.999\ T$	6.5	25.0	196.0	Drent & Stonehouse (1971)
Anseriformes								
Wood duck *Aix sponsa*	448	S	N	$84.1 - 2.103\ T$	18.0	—	46.4	Gavrilov
Wood duck *Aix sponsa*	469	W	N	$85.3 - 2.133\ T$	17.0	—	49.1	Gavrilov
Wigeon *Anas penelope*	718	W	N	$88.6 - 2.215\ T$	12.0	—	62.2	Gavrilov
Wigeon *Anas penelope*	723	S	N	$96.7 - 2.417\ T$	16.0	—	58.3	Gavrilov
Black duck *Anas rubripes*	904	W	D	$124.8 - 2.69\ T$	19.0	—	73.7	Hartung (1967)
Mallard *Anas platyrhynchos*	1020	S	N	$128.0 - 3.200\ T$	14.0	—	84.0	Gavrilov
Mallard *Anas platyrhynchos*	1132	W	N	$130.0 - 3.250\ T$	8.0	—	104.0	Gavrilov
Mallard *Anas platyrhynchos*	1263	W	D	$183.1 - 3.57\ T$	22.0	—	104.6	Hartung (1967)
Mallard *Anas platyrhynchos* (♀)	1084	W	D	$222.3 - 6.494\ T$	20.0	—	92.4	Smith & Prince (1973)
Mallard *Anas platyrhynchos* (♂)	1248	W	D	$283.2 - 7.764\ T$	20.0	—	127.9	Smith & Prince (1973)
Greylag goose *Anser anser*	3250	S	N	$344.1 - 8.602\ T$	14.0	—	224.0	Gavrilov and Dol'nik
Falconiformes								
Sparrow hawk *Accipiter nisus*	135	S	N	$39.2 - 0.980\ T$	19.0	—	19.6	Gavrilov and Dol'nik
Hobby *Falco subbuteo*	208	A	N	$53.6 - 1.340\ T$	20.0	—	26.8	Gavrilov and Dol'nik
Honey buzzard *Pernis apivorus*	652	S	N	$95.6 - 2.390\ T$	20.0	—	48.3	Gavrilov and Dol'nik
Galliformes								
Quail *Coturnix coturnix*	97	S	N	$36.8 - 0.920\ T$	20.0	—	18.4	Gavrilov and Dol'nik
California quail *Lophortyx californicus*	138	SW	?	$44.4 - 1.075\ T$	27.3	37.5	15.7	Brush (1965)
White-tailed ptarmigan *Lagopus leucurus*	326	S	?	$60.1 - 1.765\ T$	6.5	38	48.8	Johnson (1968)
Rock ptarmigan *Lagopus mutus*	432	W	D	$68.7 - 1.784\ T$	-1.3	—	77.8	West (1972b)
Partridge *Perdix perdix*	483	S	N	$89.3 - 2.233\ T$	18.0	—	49.5	Gavrilov
Partridge *Perdix perdix*	501	W	N	$71.2 - 1.780\ T$	15.0	—	44.5	Gavrilov
Willow ptarmigan *Lagopus lagopus*	524	S	N	$98.1 - 2.452\ T$	14.0	—	64.2	Gavrilov
Willow ptarmigan *Lagopus lagopus*	567	W	N	$78.9 - 1.972\ T$	10.0	—	59.3	Gavrilov
Willow ptarmigan *Lagopus lagopus*	539	S	N	$97.2 - 2.532\ T$	7.7	—	78.8	West (1972b)
Willow ptarmigan *Lagopus lagopus*	590	W	D	$61.5 - 1.528\ T$	-6.3	—	70.5	West (1972b)
Rock partridge *Alectoris graeca*	620	S	N	$101.6 - 2.540\ T$	17	—	58.9	Gavrilov

Species	Weight (g)	Season	Time of measurement	M (kcal bird⁻¹ day⁻¹)	Critical temperatures		BM (kcal bird⁻¹ day⁻¹)	Authority
					Lower (°C)	Upper (°C)		
Rock partridge *Alectoris graeca*	633	W	N	80.2–2.005 *T*	14	—	52.3	Gavrilov
Ruffed grouse *Bonasa umbellus*	644	W	N	49.0–0.890 *T*	−0.3	—	49.2	Rasmussen & Brander (1973)
Domestic fowl *Gallus gallus*	2430	?	D	193.8–1.95 *T*	18.3	24	160.4	Barrott & Pringle (1946)
Capercaillie *Tetrao urogallus*	3900	S	N	328.0–8.200 *T*	10	—	246.0	Gavrilov
Capercaillie *Tetrao urogallus*	4010	W	N	270.0–6.750 *T*	4	—	244.0	Gavrilov
Gruiformes								
Corncrake *Crex crex*	96	S	N	36.1–0.902 *T*	22.0	—	16.3	Gavrilov and Dol'nik
European coot *Fulica atra*	412	S	N	69.4–1.760 *T*	16.0	—	42.1	Gavrilov
European coot *Fulica atra*	436	W	D	74.8–1.870 *T*	14.0	—	48.8	Gavrilov
European coot *Fulica atra*	780	W	N	100.8–3.4 *T*	13.0	—	57.6	J. A. L. Mertens (personal communication)
Charadriiformes								
Little ringed plover *Charadrius dubius*	36	S	N	19.1–0.478 *T*	22.0	—	8.6	Gavrilov and Dol'nik
Black-headed gull *Larus ridibundus*	285	S	N	69.3–1.758 *T*	16.0	—	41.4	Gavrilov
Black-headed gull *Larus ridibundus*	306	W	N	61.2–1.530 *T*	15.0	—	38.4	Gavrilov
Common gull *Larus canus*	428	S	N	82.4–2.060 *T*	15.0	—	48.0	Gavrilov
Common gull *Larus canus*	431	W	N	70.8–1.770 *T*	13.0	—	46.4	Gavrilov
Woodcock *Scolopax rusticola*	430	S	N	81.1–2.028 *T*	18.0	—	44.6	Gavrilov and Dol'nik
Columbiformes								
Western plumed pigeon *Laphophaps ferruginea*	81	S	N	29.9–0.672 *T*	33.0	39.0	8.2	Dawson & Bennett (1973)
Mourning dove *Zenaida macroura*	91	?	N	45.2–1.060 *T*	30.0	37.5	13.4	Hudson & Brush (1964)
Palm dove *Streptopelia senegalensis*	108	S	N	38.9–0.975 *T*	22.0	—	17.5	Gavrilov and Dol'nik
Turtle dove *Streptopelia turtur*	154	A	N	52.3–1.308 *T*	22.0	—	23.5	Gavrilov and Dol'nik
Collared dove *Streptopelia decaocto*	187	W	N	46.0–1.016 *T*	19.5	—	26.3	J. A. L. Mertens (personal communication)
Domestic pigeon *Columba livia*	314	W	D	64.1–1.52 *T*	19.0	>29.0	34.8	Kayser (1940)
Domestic pigeon *Columba livia*	315	?	N	58.3–0.942 *T*	23.0	36.5	35.9	Calder & Schmidt-Nielsen (1967)
Domestic pigeon *Columba livia*	353	W	N	66.1–1.653 *T*	17.0	—	38.3	Gavrilov and Dol'nik

Psittaciformes							
Parrot							
Melopsittacus undulatus	25.2	S	N	19.1−0.478 T	27.0	—	6.2	Gavrilov and Dol'nik
Agapornis roseicollis	48.1	S	N	27.4−0.685 T	26.0	—	9.6	Gavrilov and Dol'nik
Monk parakeet							
Myiopsitta monachus	81.5	?	?	38.5−0.892 T	25.0	40.0	16.2	Caccamise, Porreca & Weathers (1974)
Nymphicus hollandicus	85.6	S	N	35.8−0.895 T	24.0	—	14.2	Gavrilov and Dol'nik
Cuculiformes							
Cuckoo *Cuculus canorus*	111.6	S	N	38.5−0.963 T	22.0	—	17.3	Gavrilov and Dol'nik
Roadrunner *Geococcyx californianus*	284.7	?	N	57.1−0.981 T	27.0	36	30.2	Calder & Schmidt-Nielsen (1967)
Strigiformes							
Elf owl *Micrathene whitneyi*	45.9	?	?	15.3−0.317 T	27.0	38.0	5.3	Ligon (1968)
Burrowing owl *Speotyto cunicularia*	132.4	S	D	37.7−0.869 T	25.0	37	16.0	Coulombe (1970) (resting metabolism)
Burrowing owl *Speotyto cunicularia*	142.7	W	D	26.8−0.510 T	25	37	14.0	Coulombe (1970) (resting metabolism)
Long-eared owl *Asio otus*	236	S	N	49.1−1.228 T	18.0	—	27.0	Gavrilov and Dol'nik
Snowy owl *Nyctea scandiaca*	2024	W	D,N	98.4−2.740 T	4.2	> 18.5	87.0	Gessaman (1972, revised)
Caprimulgiformes							
Poor-will *Phalaenoptilus nuttalli*	40.0	?	?	18.7−0.429 T	35.0	44	3.7	Bartholomew, Hudson & Howell (1962)
Common nighthawk *Chordeiles minor*	72.0	?	D	31.5−0.829 T	27	—	9.1	Lasiewski & Dawson (1964)
Nightjar *Caprimulgus europaeus*	77.4	S	N	29.6−0.740 T	22.0	—	13.3	Gavrilov and Dol'nik
Papuan frogmouth *Podargus ocellatus*	145	S	D	33.4−0.735 T	30.0	43.0	11.7	Lasiewski, Dawson & Bartholomew (1970)
Apodiformes							
Costa's hummingbird *Calypte costae*	3.2	?	?	6.9−0.203 T	28.5	—	1.1	Lasiewski (1963)
Black-chinned hummingbird *Archilochus alexandri*	3.3	?	N,D	5.4−0.142 T	28.5	—	1.3	Lasiewski (1963)
Allen's hummingbird *Selasphorus sassin*	3.7	?	?	8.5−0.243 T	28.5	—	1.4	Lasiewski (1963)
Rufous hummingbird *Selasophorus rufus*	3.8	?	?	7.2−0.201 T	28.5	—	1.4	Lasiewski (1963)
Anna's hummingbird *Calypte anna*	4.8	?	?	9.1−0.243 T	28.5	—	2.4	Lasiewski (1963)
Rivoli's hummingbird *Eugenes fulgens*	6.6	?	?	9.1−0.228 T	31.0	—	2.1	Lasiewski & Lasiewski (1967)
Blue-throated hummingbird *Lampornis clemenciae*	7.9	?	?	8.5−0.209 T	31.0	—	2.1	Lasiewski & Lasiewski (1967)
West Indian hummingbird *Eulampis jugularis*	8.4	?	?	12.6−0.315 T	30.0	—	3.2	Hainsworth & Wolf (1970)
Giant hummingbird *Patagona gigas*	19.1	?	?	15.9−0.374 T	26.5	—	5.9	Lasiewski, Weathers & Berstein (1967)

Species	Weight (g)	Season	Time of measurement	M (kcal bird^{-1} day^{-1})	Critical temperatures		BM (kcal bird^{-1} day^{-1})	Authority
					Lower (°C)	Upper (°C)		
Swift *Apus apus*	44.9	S	N	$25.7 - 0.642\ T$	26.0	—	9.0	Gavrilov and Dol'nik
Coliiformes								
Speckled mousebird *Colius striatus*	52.5	?	D	$24.2 - 0.629\ T$	25.0	>40.0	7.3	Bartholomew & Trost (1970)
Coraciiformes								
Kingfisher *Alcedo atthis*	34.3	S	N	$20.9 - 0.522\ T$	25.0	—	7.8	Gavrilov and Dol'nik
Piciformes								
Wryneck *Jynx torquilla*	31.8	S	N	$18.6 - 0.465\ T$	24.0	—	7.4	Gavrilov and Dol'nik
Great spotted woodpecker *Dendrocopus major*	98.0	S	N	$33.5 - 0.837\ T$	18.0	—	18.5	Gavrilov
Great spotted woodpecker *Dendrocopus major*	117.0	W	N	$29.8 - 0.745\ T$	11.0	—	21.5	Gavrilov
Passeriformes								
Goldcrest *Regulus regulus*	5.5	A	N	$8.2 - 0.205\ T$	21.0	—	3.8	Gavrilov and Dol'nik
Black-rumped waxbill *Estrilda troglodytes*	6.4	?	D	$13.5 - 0.284\ T$	38.0	38.0	2.7	Cade, Tobin & Gold (1965)
Black-rumped waxbill *Estrilda troglodytes*	6.5	Sp	D	$11.1 - 0.300\ T$	28.0	35.0	2.6	Lasiewski, Hubbard & Moberly (1964)
Black-rumped waxbill *Estrilda troglodytes*	7.7	A	N	$13.0 - 0.325\ T$	30.0	—	3.2	Gavrilov and Dol'nik
Verdin *Auriparus flavicep*	6.8	S	D	$7.2 - 0.154\ T$	24.0	—	3.4	Goldstein (1974, modified)
Parula warbler *Parula americana*	7.0	?	N	$7.8 - 0.206\ T$	25.7	—	2.5	Yarbrough (1971)
Tiaris canora	7.0	A	N	$13.0 - 0.325\ T$	30.0	—	3.2	Gavrilov and Dol'nik
Anianiau *Viridonia parva*	8.0	?	D	$9.7 - 0.201\ T$	31.0	36.0	3.0	MacMillen (1974)
Chiffchaff *Phylloscopus collybita*	8.2	A	N	$9.8 - 0.245\ T$	25.5	—	3.4	Gavrilov and Dol'nik
Long-tailed tit *Aegithalos caudatus*	8.9	S	N	$10.2 - 0.255\ T$	24.0	—	4.1	Gavrilov V. M. (1974)
Long-tailed tit *Aegithalos caudatus*	8.8	W	N	$9.8 - 0.245\ T$	19.0	—	5.2	Gavrilov, V. M. (1974)
Wren *Troglodytes troglodytes*	9.0	S	N	$11.7 - 0.292\ T$	25.0	—	4.4	Gavrilov, V. M. (1974)
Troglodytes troglodytes	9.2	W	N	$11.0 - 0.275\ T$	21.5	—	5.0	Gavrilov, V. M. (1974)
Uraeginthus bengalis	9.2	A	N	$12.5 - 0.312\ T$	29.0	—	3.4	Gavrilov and Dol'nik

Species	(mass)							Reference
Wood warbler Phylloscopus sibilatrix	9.2	S	N	11.4−0.285 T	27.0	—	3.6	Gavrilov and Dol'nik
House wren Troglodytes aedon	9.7	S	D	14.6−0.227 T	37.8	37.8	6.0	Kendeigh (1939)
Yellow-throated warbler Dendroica dominica	9.8	?	N	9.1−0.228 T	25.2	—	3.3	Yarbrough (1971)
Palm warbler Dendroica palmarum	9.8	?	N	7.5−0.185 T	23.3	—	3.2	Yarbrough (1971)
Black-capped chickadee								
Parus atricapillus	(10.3)	S	N	13.3−0.316 T	26.0	—	5.2	Rising & Hudson (1974)
Parus atricapillus	10.3	W	N	10.5−0.202 T	(26.0)	—	5.2	Rising & Hudson (1974)
Lonchura striata	10.3	A	N	12.4−0.310 T	26.0	—	4.4	Gavrilov and Dol'nik
Vidua paradisaea	10.5	?	?	11.6−0.290 T	25.5	—	4.0	Terroine & Trautmann (1927)
Lesser whitethroat Sylvia curruca	10.6	S	N	11.8−0.295 T	26.5	—	4.1	Gavrilov and Dol'nik
Willow warbler Phylloscopus trochilus	10.7	A	N	12.4−0.310 T	26.0	—	4.3	Gavrilov and Dol'nik
Marsh warbler Acrocephalus palustris	10.8	S	N	13.8−0.345 T	28.0	—	4.2	Gavrilov and Dol'nik
Coal tit Parus ater	10.8	S	N	11.2−0.280 T	23.0	—	4.9	Gavrilov, V. M. (1974)
Coal tit Parus ater	11.0	W	N	10.6−0.265 T	19.0	—	5.6	Gavrilov, V. M. (1974)
Red-breasted nuthatch Sitta canadensis	11.2	SW	D	13.2−0.258 T	33	38	4.8	Mugas & Templeton (1970)
Zebra finch Poephila guttata	11.5	?	D	19.7−0.415 T	36	42	4.9	Cade et al. (1965)
Zebra finch Poephila guttata	11.7	?	?	16.3−0.391 T	29.5	40.0	4.4	Calder (1964)
Zebra finch Poephila guttata	11.8	A	N	15.8−0.395 T	28.0	—	4.8	Gavrilov and Dol'nik
Sedge warbler Acrocephalus schoenobaenus	11.5	S	N	13.5−0.337 T	27.0	—	4.5	Gavrilov and Dol'nik
Myrtle warbler Dendroica coronata	11.5	?	N	9.2−0.229 T	23.2	—	3.9	Yarbrough (1971)
Pied flycatcher Ficedula hypoleuca	11.7	A	N	12.0−0.300 T	24.0	—	4.8	Gavrilov and Dol'nik
Chipping sparrow Spizella passerina	11.9	?	N	9.8−0.240 T	24.4	—	4.0	Yarbrough (1971)
Pine warbler Dendroica pinus	12.0	?	N	9.0−0.224 T	23.4	—	3.7	Yarbrough (1971)
Hooded warbler Wilsonia citrina	12.0	?	N	11.0−0.264 T	24.9	—	4.4	Yarbrough (1971)
Acadian flycatcher Empidonax virescens	12.3	?	N	10.3−0.254 T	26.3	—	3.7	Yarbrough (1971)
Icterine warbler Hippolais icterina	12.5	S	N	12.9−0.322 T	24.0	—	5.2	Gavrilov and Dol'nik
Prothonotary warbler Protonotaria citrea	12.8	?	N	12.2−0.302 T	26.9	—	4.1	Yarbrough (1971)
Redstart								
Phoenicurus phoenicurus	13.0	S,A	N	12.0−0.300 T	24.0	—	4.8	Gavrilov and Dol'nik
Serinus canaria	13.3	A	N	11.8−0.295 T	24.0	—	4.7	Gavrilov and Dol'nik
Sand martin Riparia riparia	13.6	A	N	13.6−0.340 T	26.0	—	4.8	Gavrilov and Dol'nik
Grasshopper sparrow Ammodramus savannarum	13.8	?	N	10.7−0.277 T	25.3	—	3.7	Yarbrough (1971)
Black redstart Phoenicurus ochruros	13.9	S	N	12.5−0.312 T	24.0	—	5.0	Yarbrough (1971)
Pine siskin Carduelis pinus	14.0	S	N	12.6−0.315 T	21.0	—	6.0	Gavrilov, V. M. (1974)
Pine siskin Carduelis pinus	14.2	W	N	12.4−0.310 T	18.0	—	6.8	Gavrilov, V. M. (1974)
Common redpoll Acanthis flammea	12.8	S	N	9.7−0.196 T	21.0	—	5.4	Pohl & West (1973)
Common redpoll Acanthis flammea	13.2	S	D	12.0−0.249 T	23.5	—	5.9	West (1972a)

Species	Weight (g)	Season	Time of measurement	M (kcal bird^{-1} day^{-1})	Critical temperatures		BM (kcal bird^{-1} day^{-1})	Authority
					Lower (°C)	Upper (°C)		
Common redpoll *Acanthis flammea*	13.8	W	D	12.1−0.350 T	7.0	—	9.7	Pohl & West (1973)
Common redpoll *Acanthis flammea*	14.0	S	N	12.4−0.310 T	21.0	—	5.9	Gavrilov, V. M. (1974)
Common redpoll *Acanthis flammea*	14.3	W	N	12.3−0.307 T	17.0	—	7.0	Gavrilov, V. M. (1974)
Common redpoll *Acanthis flammea*	15.6	W	N	10.6−0.206 T	11.5	—	8.3	Pohl & West (1973)
Whinchat *Saxicola rubetra*	14.3	S	N	14.1−0.352 T	26.0	—	5.0	Gavrilov and Dol'nik
Spotted flycatcher *Muscicapa striata*	14.4	S	N	14.4−0.360 T	26.0	—	5.1	Gavrilov and Dol'nik
Yellow wagtail *Motacilla flava*	14.7	S	N	15.1−0.377 T	26.0	—	5.3	Gavrilov and Dol'nik
Swamp sparrow *Melospiza georgiana*	14.9	?	N	11.2−0.282 T	24.1	—	4.4	Yarbrough (1971)
Amakihi *Viridonia virens*	15.3	?	D	11.8−0.190 T	30.0	—	6.1	MacMillen (1974)
Savannah sparrow *Passerculus sandwichensis*	15.9	?	N	10.8−0.271 T	22.6	—	4.6	Yarbrough (1971)
Goldfinch *Carduelis carduelis*	16.5	W	N	12.7−0.317 T	16.5	—	7.2	Gavrilov and Dol'nik
Dunnock *Prunella modularis*	16.8	A	N	14.9−0.372 T	22.0	—	6.7	Gavrilov and Dol'nik
Linnet *Acanthis cannabina*	16.9	A	N	15.0−0.375 T	21.0	—	7.0	Gavrilov and Dol'nik
Great tit *Parus major*	16.4	S	N	14.9−0.372 T	22.0	—	6.8	Gavrilov, V. M. (1974)
Great tit *Parus major*	17.1	W	N	14.0−0.350 T	18.0	—	7.7	Gavrilov, V. M. (1974)
Great tit *Parus major*	18.0	W	N	10.3−0.204 T	15.0	33.0	6.7	J. A. L. Mertens (personal communication)
Great tit *Parus major*	18.4	W	?	12.1−0.127 T	28.0	31.0	7.8	Hissa & Palokangas (1970)
Great tit *Parus major*	19.0	S	?	20.4−0.527 T	29.0	33.0	7.6	Hissa & Palokangas (1970)
Reed bunting *Emberiza schoeniclus*	17.6	A	N	14.4−0.360 T	23.0	—	6.2	Gavrilov and Dol'nik
Robin *Erithacus rubecula*	17.6	S	N	16.7−0.417 T	25.0	—	6.2	Gavrilov
Robin *Erithacus rubecula*	17.6	W	N	15.6−0.390 T	25.0	—	5.8	Gavrilov
Swallow *Hirundo rustica*	18.0	S	N	22.6−0.565 T	29.5	—	6.5	Keskpaik (1968)
Pied wagtail *Motacilla alba*	18.0	S	N	15.4−0.385 T	24.0	—	6.2	Gavrilov, V. M. (1974)
Pied wagtail *Motacilla alba*	18.2	W	N	14.9−0.372 T	24.0	—	5.8	Gavrilov, V. M. (1974)
Northern waterthrush *Seiurus noveboracensis*	18.7	?	N	14.7−0.349 T	25.5	—	5.8	Yarbrough (1971)
Meadow pipit *Anthus pratensis*	18.9	S	N	15.8−0.395 T	24.0	—	6.2	Gavrilov and Dol'nik
Ovenbird *Seiurus aurocapillus*	19.0	?	N	11.1−0.267 T	22.9	—	5.0	Yarbrough (1971)
Song sparrow *Melospiza melodia*	19.1	?	N	13.0−0.317 T	24.5	—	5.2	Yarbrough (1971)
Tree pipit *Anthus trivialis*	19.7	A	N	17.6−0.400 T	24.0	—	7.0	Gavrilov and Dol'nik

Species								Reference
White-throated sparrow *Zonotrichia albicollis*	20.2	?	N	13.7−0.335 T	23.6	—	5.8	Yarbrough (1971)
White-throated sparrow *Zonotrichia albicollis*	22.5	?	N	14.8−0.435 T	20.0	35.0	7.1	Hudson & Kimzey (1964)
House martin *Delichon urbica*	20.5	S	N	22.1−0.552 T	29.5	—	7.3	Keskpaik (1968)
Bluethroat *Luscinia svecica*	20.8	S	N	17.5−0.438 T	23.0	—	7.4	Gavrilov and Dol'nik
Chaffinch *Fringilla coelebs*	20.8	W	N	18.4−0.460 T	20.0	—	9.1	Gavrilov, V. M. (1974)
Chaffinch *Fringilla coelebs*	21.0	S	N	17.5−0.438 T	22.5	—	7.7	Gavrilov, V. M. (1974)
Brambling *Fringilla montifringilla*	21.0	A	N	16.0−0.400 T	20.0	—	7.9	Gavrilov and Dol'nik
Barred warbler *Sylvia nisoria*	21.3	S	N	18.6−0.462 T	23.0	—	7.9	Gavrilov and Dol'nik
Scarlet rosefinch *Carpodacus erythrinus*	21.2	S	N	20.3−0.507 T	25.0	—	7.6	Gavrilov, V. M. (1974)
Scarlet rosefinch *Carpodacus erythrinus*	21.6	W	N	18.7−0.467 T	24.0	—	7.4	Gavrilov, V. M. (1974)
Vesper sparrow *Pooecetes gramineus*	21.5	?	N	13.2−0.334 T	22.5	—	5.6	Yarbrough (1971)
Eastern phoebe *Sayornis phoebe*	21.6	?	N	16.5−0.406 T	23.3	—	7.1	Yarbrough (1971)
Tawny pipit *Anthus campestris*	21.8	S	N	17.5−0.438 T	22.0	—	7.9	Gavrilov and Dol'nik
Blackcap *Sylvia atricapilla*	21.9	A	N	17.0−0.425 T	20.0	—	8.6	Gavrilov and Dol'nik
Ortolan bunting *Emberiza hortulana*	22.0	W	?	18.7−0.379 T	25.0	—	8.7	Wallgren (1954)
Ortolan bunting *Emberiza hortulana*	24.3	S	N	20.7−0.502 T	23.0	—	8.6	Gavrilov, V. M. (1974)
Ortolan bunting *Emberiza hortulana*	27.0	W	N	18.8−0.470 T	22.0	—	8.4	Gavrilov, V. M. (1974)
European tree sparrow *Passer montanus*	22.3	A	N	16.7−0.417 T	20.0	—	8.4	Gavrilov and Dol'nik
Garden warbler *Sylvia borin*	24.8	A	N	19.0−0.475 T	22.0	—	8.6	Gavrilov and Dol'nik
House sparrow								
Passer d. bactrianus	23.0	S	N	18.0−0.450 T	23.0	—	7.6	Gavrilov, V. M. (1974)
Passer d. bactrianus	23.2	W	N	18.0−0.450 T	23.0	—	7.6	Gavrilov, V. M. (1974)
Passer domesticus	25.1	W	D	22.5−0.356 T	37	37	9.3	Kendeigh (1944)
Passer domesticus	25.3	?	N	20.7−0.568 T	22	37	8.4	Hudson & Kimzey (1966)
Passer domesticus	25.5	?	N	15.5−0.424 T	21	37	6.9	Hudson & Kimzey (1966)
Passer domesticus	26.5	S	N	22.0−0.550 T	22.5	—	9.8	Gavrilov, V. M. (1974)
Passer domesticus	26.4	W	N	21.6−0.540 T	21.0	—	10.1	Gavrilov, V. M. (1974)
Horned lark *Eremophila alpestris*	26	S	D	15.9−0.270 T	35	35	6.4	Trost (1972)
Horned lark *Eremophila alpestris*	26	S	N	13.4−0.300 T	<25.0	—	6.8	Trost (1972)
White-crowned sparrow *Zonotrichia leucophrys*	26.1	?	N	14.4−0.358 T	20.6	—	7.0	Yarbrough (1971)
White-crowned sparrow *Zonotrichia leucophrys*	28.6	W	N	17.4−0.412 T	23	37	8.0	King (1964)
Yellowhammer *Emberiza citrinella*	26.4	W	?	16.4−0.293 T	25	—	9.1	Wallgren (1954)
Yellowhammer *Emberiza citrinella*	26.8	S	N	18.5−0.463 T	20.5	—	9.0	Gavrilov, V. M. (1974)
Yellowhammer *Emberiza citrinella*	27.4	W	N	17.1−0.428 T	16.0	—	10.3	Gavrilov, V. M. (1974)
Red-backed shrike *Lanius collurio*	27.0	S	N	19.8−0.495 T	24.0	—	7.9	Gavrilov and Dol'nik
Greenfinch *Carduelis chloris*	28.2	S	N	19.4−0.485 T	20.0	—	9.8	Gavrilov
Greenfinch *Carduelis chloris*	29.0	W	N	19.2−0.480 T	16.0	—	11.5	Gavrilov

371

Species	Weight (g)	Season	Time of measure-ment	M (kcal bird⁻¹ day⁻¹)	Critical temperatures Lower (°C)	Upper (°C)	BM (kcal bird⁻¹ day⁻¹)	Authority
Pyrrhuloxia *Cardinalis sinuata*	32.0	S	N	$23.8-0.527\ T$	29.7	42.8	8.1	Hinds & Calder (1973)
Red crossbill *Loxia curvirostra*	29.4	W	N	$16.6-0.406\ T$	15.0	> 28.5	10.5	Dawson & Tordoff (1964)
Red crossbill *Loxia curvirostra*	39.4	S	N	$21.5-0.538\ T$	17.0	—	12.4	Gavrilov, V. M. (1974)
Red crossbill *Loxia curvirostra*	42.7	W	N	$20.6-0.515\ T$	13.0	—	13.9	Gavrilov, V. M. (1974)
White-winged crossbill *Loxia leucoptera*	29.8	W	N	$14.9-0.378\ T$	14.0	—	9.6	Dawson & Tordoff (1964)
Bullfinch *Pyrrhula pyrrhula*	30.4	W	N	$19.1-0.478\ T$	15.0	—	11.4	Gavrilov and Dol'nik
Harris sparrow *Zonotrichia querula*	33.0	A,Sp	?	$35.1-0.616\ T$	> 33	37	13.3	Rising (1968)
Harris sparrow *Zonotrichia querula*	33.3	?	N	$18.7-0.457\ T$	20.7	—	9.3	Yarbrough (1971)
Wood lark *Lullula arborea*	33.2	A	N	$20.6-0.515\ T$	20.5	—	10.1	Gavrilov and Dol'nik (1971)
Great crested flycatcher *Myiarchus crinitus*	33.9	?	N	$19.3-0.469\ T$	24.3	—	7.9	Yarbrough (1971)
Cardinal *Cardinalis cardinalis*	41.0	W	?	$20.7-0.463\ T$	18.0	33	12.2	Dawson (1958)
Cardinal *Cardinalis cardinalis*	41.0	S	N	$30.5-0.704\ T$	28.7	42.6	10.4	Hinds & Calder (1973)
Brown-headed cowbird *Molothrus ater*	42.5	?	?	$33.5-0.588\ T$	35	—	12.9	Lustick (1969)
Brown towhee *Pipilo fuscus*	43.7	?	?	$30.0-0.750\ T$	23	34	13.7	Dawson (1954)
Abert's towhee *Pipilo aberti*	46.6	?	?	$37.2-0.930\ T$	25	35	15.0	Dawson (1954)
Red-winged blackbird (♀) *Agelaius phoeniceus*	46.5	W	D	$42.9-0.591\ T$	—	45+	—	Lewies & Dyer (1969)
Red-winged blackbird (♀) *Agelaius phoeniceus*	46.5	W	N	$17.5-0.101\ T$	—	—	—	Lewies & Dyer (1969)
Red-winged blackbird (♂) *Agelaius phoeniceus*	70.3	W	D	$45.7-0.769\ T$	27.5	40	24.4	Lewies & Dyer (1969)
Red-winged blackbird (♂) *Agelaius phoeniceus*	70.3	W	N	$29.4-0.312\ T$	—	—	—	Lewies & Dyer (1969)
Hawfinch *Coccothraustes coccothraustes*	48.3	A	N	$25.1-0.628\ T$	17.0	—	14.4	Gavrilov and Dol'nik
Dipper *Cinclus mexicanus*	50.2	?	?	$13.9-0.376\ T$	11.5	34	9.5	Murrish (1970)
Parrot crossbill *Loxia pytiopsittacus*	53.7	W	N	$24.0-0.600\ T$	12.5	—	16.5	Gavrilov and Dol'nik
Evening grossbeak *Coccothraustes vespertinus*	54.5	SWASp	N	$28.3-0.535\ T$	25	35	14.9	West & Hart (1966)
Evening grossbeak *Coccothraustes vespertinus*	54.5	—	D	$46.0-0.602\ T$	—	—	—	West & Hart (1966)

Evening grossbeak *Coccothraustes vespertina*	55.3	W	N	25.6-0.624 T	16	>34	15.6	Dawson & Tordoff (1959)
Redwing *Turdus iliacus*	58.0	A	Z	30.0-0.750 T	20.0	—	14.9	Gavrilov and Dol'nik
Song thrush *Turdus philomelos*	62.8	S	Z	31.6-0.790 T	21.0	—	15.0	Gavrilov, V. M. (1974)
Song thrush *Turdus philomelos*	64.0	W	Z	30.0-0.750 T	19.5	—	15.6	Gavrilov, V. M. (1974)
Golden oriole *Oriolus oriolus*	64.9	S	Z	29.6-0.740 T	22.0	—	13.4	Gavrilov and Dol'nik
Gray jay *Perisoreus canadensis*	71.2	S	?	25.8-0.233 T	36	—	14.0	Veghte (1964)
Gray jay *Perisoreus canadensis*	71.2	A	?	23.5-0.466 T	9	—	14.0	Veghte (1964)
Gray jay *Perisoreus canadensis*	71.2	W	?	20.0-0.458 T	7	—	14.0	Veghte (1964)
Gray jay *Perisoreus canadensis*	71.2	Sp	?	21.2-0.554 T	7	—	14.0	Veghte (1964)
Great grey shrike *Lanius excubitor*	72.4	A	Z	30.5-0.763 T	18.0	—	16.8	Gavrilov and Dol'nik
Bohemian waxwing *Bombycilla garrulus*	72.5	A	Z	29.8-0.745 T	14.0	—	19.7	Gavrilov and Dol'nik
Starling *Sturnus vulgaris*	75.0	A	Z	32.6-0.815 T	17.0	—	18.5	Gavrilov and Dol'nik
Blue jay *Cyanocitta cristata*	80.8	?	?	24.6-0.410 T	18.0	—	17.2	Misch (1960)
Blackbird *Turdus merula*	82.6	S	Z	33.2-0.830 T	17.0	—	19.2	Gavrilov, V. M. (1974)
Blackbird *Turdus merula*	83.0	W	Z	32.9-0.823 T	14.0	—	21.4	Gavrilov, V. M. (1974)
Mistle thrush *Turdus viscivorus*	108.2	A	Z	37.2-0.930 T	16.0	—	22.8	Gavrilov and Dol'nik
Nutcracker *Nucifraga caryocatactes*	147.0	W	Z	45.2-1.130 T	15.0	—	27.8	Gavrilov and Dol'nik
Jay *Garrulus glandarius*	153.0	W	Z	45.4-1.135 T	15.0	—	28.6	Gavrilov and Dol'nik
Magpie *Pica pica*	202.0	A	Z	52.8-1.320 T	13.0	—	35.5	Gavrilov and Dol'nik
Alpine chough *Pyrrhocorax graculus*	206.4	W	Z	50.4-1.260 T	11.0	—	35.4	Gavrilov and Dol'nik
Jackdaw *Coleus monedula*	215.0	W	Z	53.2-1.330 T	11.0	38.4	—	Gavrilov and Dol'nik
Rook *Corvus frugilegus*	390.0	W	Z	72.3-1.808 T	10.0	—	54.0	Gavrilov and Dol'nik
Carrion crow *Corvus corone*	518.0	S	Z	94.2-2.355 T	11.0	—	68.5	Gavrilov, V. M. (1974)
Carrion crow *Corvus corone*	540.0	W	Z	92.7-2.318 T	6.0	—	79.0	Gavrilov, V. M. (1974)
Corvus ruficollis	660.0	W	Z	93.5-2.338 T	10.0	—	70.1	Gavrilov and Dol'nik
Raven *Corvus corax*	1203.0	S	Z	150.4-3.760 T	10.0	—	113.7	Gavrilov, V. M. (1974)
Raven *Corvus corax*	1208.0	W	Z	137.2-3.430 T	4.0	—	123.8	Gavrilov, V. M. (1974)

Appendix 5.2. *Existence metabolic rates* (*EM*)

This appendix includes all known equations of EM (M = kcal bird^{-1} day^{-1}) on temperature (T = °C) from the literature, unpublished theses, and a large number of new data obtained by Dol'nik and Gavrilov on wild caught and 'zoo' birds. Measurements on these birds were obtained at three to six different temperatures between −20 and +30 °C. Small birds were confined in cages 40×30×25 cm in size and medium-sized birds in cages 1.5×0.75×0.75 m. Experiments lasted 3–4 days in small birds and 10 days or longer in large birds. Weights were taken early in the morning at the beginning and end of the test. If the weight change was small, corrections were made using a calorific equivalent of 6 kcal g^{-1}. If the weight change was large, the data were not used. Insectivorous species were maintained on chopped boiled chicken's eggs, granivorous species on seeds, piscivorous species on fish flesh, and predators on horsemeat. The weighing of food, determining its water and calorific content, and the collecting, drying and bombing of the excreta followed standardized procedures.

The EM of large birds was measured under greater difficulties. The birds were tested in zoo cages, protected from disturbance by people, other caged birds and wild birds. The routine, mixed diets of these birds were analyzed separately for each component. In some cases, where calorific determinations of the excrement could not be made, a value of 3.5 kcal^{-1} g^{-1} dry weight was assumed. For birds with fluid excreta, EM was simply calculated at 0.8 times the gross energy intake. Not all birds could be weighed a second time. Obviously, the values obtained on some of the larger species are less trustworthy than on the smaller species, so caution is required in evaluating the results.

Measurements were commonly made at two different photoperiods, but there was considerable variation between species in the actual photoperiods used. These were often chosen to represent the extreme photoperiods to which the species was exposed during the winter and during the summer. The longer photoperiods were run in the summer, the shorter ones in the winter, so that the birds were acclimatized to the prevailing temperatures. The records are grouped for easier analysis into short photoperiods ranging from 7 to 12 hours, averaging 10± hours, and long photoperiods ranging from 15 to 24, averaging 15+ hours.

| | | Photoperiod | | |
| | | 7–12 hours | 15–24 hours | |
Species	Weight (g)	M (kcal bird^{-1} day^{-1})	M (kcal bird^{-1} day^{-1})	Authority
Sphenisciformes				
Penguin				
Eudyptes chrysolophus	3860	—	364.3−2.500 T	Gavrilov and Dol'nik
Aptenodytes patagonica	11080	—	602.8−2.070 T	Gavrilov and Dol'nik
Struthioniformes				
Ostrich *Struthio camelus*	122000	—	2787.7−5.167 T	Gavrilov and Dol'nik
Rheiformes				
Rhea *Rhea americana*	32000	—	1156.4−1.923 T	Gavrilov and Dol'nik
Casuariiformes				
Emu *Dromiceus novae-hollandiae*	45000	—	1463.2−2.413 T	Gavrilov and Dol'nik
Cassowary *Casuarius casuarius*	52000	—	1604.3−3.900 T	Gavrilov and Dol'nik
Ciconiiformes				
Bittern *Botaurus stellaris*	1010	—	184.5−2.873 T	Gavrilov and Dol'nik

Species	Weight (g)	Photoperiod 7–12 hours M (kcal bird^{-1} day^{-1})	15–24 hours M (kcal bird^{-1} day^{-1})	Authority
Flamingo *Phoenicopterus roseus*	3 980	—	387.3−4.000 T	Gavrilov and Dol'nik
Flamingo *Leptoptilus crumeniferus*	5 760	—	448.4−3.953 T	Gavrilov and Dol'nik
Jabiru *Jabiru mycteria*	6 020	—	452.4−3.460 T	Gavrilov and Dol'nik
Anseriformes				
Blue-winged teal *Anas discors* (♀)	309	106.9−1.952 T	—	Owen (1970)
Blue-winged teal *Anas discors* (♂)	363	116.8−2.212 T	—	Owen (1970)
Wood duck *Aix sponsa*	448	112.4−1.760 T	111.7−1.763 T	Gavrilov
Black-bellied tree duck				
Dendrocygna autumnalis	707	191.3−3.29 T	228.3−3.93 T	Cain (1973)
Wigeon *Anas penelope*	723	139.7−1.553 T	158.6−1.724 T	Gavrilov
Red-crested pochard *Netta rufina*	1 020	—	134.3−0.717 T	Gavrilov and Dol'nik
Mallard *Anas platyrhynchos*	1 070	164.6−2.017 T	184.7−2.287 T	Gavrilov
Red-breasted goose *Branta ruficollis*	1 120	179.3−2.000 T	—	Gavrilov and Dol'nik
Magellan goose *Chloëphaga leucoptera*	2 725	—	318.8−2.683 T	Gavrilov and Dol'nik
Greylag goose *Anser anser*	3 250	—	352.1−2.773 T	Gavrilov and Dol'nik
Canada goose				
Branta canadensis	4 300	506.6−5.392 T	515.5−4.574 T	Williams (1965)
Cereopsis novae-hollandiae	4 500	384.5−3.267 T	—	Gavrilov and Dol'nik
Falconiformes				
Sparrow hawk *Accipiter nisus*	149	59.3−0.970 T	—	Gavrilov and Dol'nik
Honey buzzard *Pernis apivorus*	649	—	139.4−2.023 T	Gavrilov and Dol'nik
Buzzard *Buteo buteo*	983	—	160.2−1.693 T	Gavrilov and Dol'nik
Egyptian vulture *Neophron perenopterus*	2 063	—	280.4−3.670 T	Gavrilov and Dol'nik
King vulture				
Sarcoramphus papa	3 650	—	351.7−3.443 T	Gavrilov and Dol'nik
Harpia harpyja	4 300	419.3−3.687 T	—	Gavrilov and Dol'nik
Galliformes				
Quail *Excalfactoria chinensis*	39	—	33.8−0.723 T	Gavrilov and Dol'nik
Quail *Coturnix coturnix*	97	—	42.3−0.640 T	Gavrilov and Dol'nik
Bobwhite *Colinus virginianus*	189	49.0−0.840 T	65.5−1.195 T	Case & Robel (1974)
Partridge *Perdix perdix*	483	108.7−1.073 T	112.4−1.252 T	Gavrilov
Willow ptarmigan *Lagopus lagopus*	524	111.4−0.811 T	105.3−0.797 T	Gavrilov
Rock partridge *Alectoris graeca*	620	111.3−0.837 T	116.4−0.932 T	Gavrilov
Japanese green pheasant				
Phasianus versicolor (♀)	800	116.8−1.831 T	160.1−2.646 T	Moore (1961)
Phasianus versicolor (♂)	1 000	172.5−2.738 T	181.5−2.457 T	Moore (1961)
Ring-necked pheasant				
Phasianus colchicus (♀)	800	105.4−0.632 T	137.3−1.971 T	H. C. Seibert (personal communication)
Phasianus colchicus (♂)	1 400	143.0−0.567 T	162.9−2.187 T	H. C. Seibert (personal communication)
Reeve's pheasant				
Syrmaticus reevesii (♀)	1 000	131.8−1.134 T	152.0−2.565 T	Seibert (1963)
Syrmaticus reevesii (♂)	1 300	205.8−2.576 T	206.3−2.484 T	Seibert (1963)
Crossoptilon crossoptilon	1 560	—	237.3−2.950 T	Gavrilov and Dol'nik
Crossoptilon auritum	1 580	—	251.2−3.293 T	Gavrilov and Dol'nik
Guinea-fowl *Numida meleagris*	1 610	—	231.3−3.000 T	Gavrilov and Dol'nik
Domestic fowl *Gallus gallus*	2 330	—	288.7−3.910 T	Gavrilov and Dol'nik
Capercaillie *Tetrao urogallus*	3 900	328.7−0.981 T	340.1−1.067 T	Gavrilov
Pavo cristatus	4 250	—	407.3−4.390 T	Gavrilov and Dol'nik
Gruiformes				
Corncrake *Crex crex*	162	68.1−1.190 T	—	Gavrilov and Dol'nik
Coot *Fulica atra*	407	112.4−1.703 T	118.8−1.697 T	Gavrilov

Appendix 5.2

Species	Weight (g)	Photoperiod 7–12 hours M (kcal bird^{-1} day^{-1})	Photoperiod 15–24 hours M (kcal bird^{-1} day^{-1})	Authority
Porphyrio poliocephalus	1 820	—	263.1−3.756 T	Gavrilov and Dol'nik
Anthropoides virgo	2 500	—	324.3−4.253 T	Gavrilov and Dol'nik
Balearica pavonina	3 450	—	362.4−3.037 T	Gavrilov and Dol'nik
Great bustard *Otis tarda*	4 420	406.2−3.127 T	—	Gavrilov and Dol'nik
Crane				
Grus grus	4 480	370.3−2.553 T	—	Gavrilov and Dol'nik
Grus japonensis	8 130	526.3−3.220 T	—	Gavrilov and Dol'nik
Sarus crane *Grus antigona*	9 460	601.2−3.763 T	—	Gavrilov and Dol'nik
Charadriiformes				
Little ringed plover *Charadrius dubius*	36.4	—	26.4−0.540 T	Gavrilov and Dol'nik
Ringed plover *Charadrius hiaticula*	52.4	—	33.2−0.537 T	Gavrilov and Dol'nik
Common sandpiper *Tringa hypoleucos*	53.8	—	34.1−0.577 T	Gavrilov and Dol'nik
Wood sandpiper *Tringa glareola*	61.2	—	36.3−0.633 T	Gavrilov and Dol'nik
Ruff *Phylomachus pugnax*	167	—	62.3−0.940 T	Gavrilov and Dol'nik
Lapwing *Vanellus vanellus*	220	—	72.3−1.063 T	Gavrilov and Dol'nik
Black-headed gull *Larus ridibundus*	285	92.4−1.510 T	98.1−1.521 T	Gavrilov
Oystercatcher *Haematopus ostralegus*	385	89.7−0.970 T	—	Gavrilov and Dol'nik
Common gull *Larus canus*	424	114.3−1.750 T	119.3−1.743 T	Gavrilov
Woodcock *Scolopax rusticola*	429	106.5−1.407 T	—	Gavrilov and Dol'nik
Black guillemot *Cepphus grylle*	482	126.3−1.940 T	—	Gavrilov and Dol'nik
Curlew *Numenius arquata*	520	102.5−0.947 T	—	Gavrilov and Dol'nik
Brunnick's guillemot *Uria lomvia*	1 320	159.6−1.007 T	—	Gavrilov and Dol'nik
Lesser black-backed gull				
Larus fuscus	1 650	—	207.3−2.167 T	Gavrilov and Dol'nik
Columbiformes				
Geopelia cuneata	29.6	—	28.2−0.627 T	Gavrilov and Dol'nik
Palm dove *Streptopelia senegalensis*	108	—	52.8−0.940 T	Gavrilov and Dol'nik
Turtle dove *Streptopelia turtur*	162	—	61.7−1.027 T	Gavrilov and Dol'nik
Domestic pigeon *Columba livia*	353	—	97.6−1.380 T	Gavrilov and Dol'nik
Wood pigeon *Columba palumbus*	593	119.8−1.433 T	—	Gavrilov and Dol'nik
Psittaciformes				
Melopsittacus undulatus	25.2	—	28.8−0.680 T	Gavrilov and Dol'nik
Agapornis roseicollis	48.1	—	44.1−1.030 T	Gavrilov and Dol'nik
Nymphicus hollandicus	85.6	—	48.8−0.967 T	Gavrilov and Dol'nik
Calyptorhynchus funereus	670	—	154.2−2.563 T	Gavrilov and Dol'nik
Probosciger atterimus	680	—	152.1−2.417 T	Gavrilov and Dol'nik
Cuculiformes				
Cuckoo *Cuculus canorus*	106	—	58.9−1.227 T	Gavrilov and Dol'nik
Tauraco corythaix	302	—	98.4−1.887 T	Gavrilov and Dol'nik
Strigiformes				
Tengmalm's owl *Aegolius funereus*	138	58.1−0.943 T	—	Gavrilov and Dol'nik
Barn owl *Tyto alba*	246	—	82.4−1.400 T	Gavrilov and Dol'nik
Long-eared owl *Asio otus*	283	—	90.6−1.487 T	Gavrilov and Dol'nik
Snowy owl *Nyctea scandiaca*	1 970	—	191.7−5.69 T	Gessaman (1972)
Coraciiformes				
Hoopoe *Upupa epops*	49.9	—	36.0−0.730 T	Gavrilov and Dol'nik
Dacelo novaeguinea	350	—	101.7−1.950 T	Gavrilov and Dol'nik
Piciformes				
Downy woodpecker *Dendrocopus pubescens*	27.3	30.1−0.403 T	—	Koplin (1972)
Wryneck *Jynx torquilla*	31.8	—	27.2−0.587 T	Gavrilov and Dol'nik
Hairy woodpecker *Dencrocopus villosus*	63.1	48.3−0.613 T	—	Koplin (1972)

376

Species	Weight (g)	Photoperiod 7–12 hours M (kcal bird^{-1} day^{-1})	Photoperiod 15–24 hours M (kcal bird^{-1} day^{-1})	Authority
Great spotted woodpecker *Dendrocopus major*	98.0	$42.1-0.510\ T$	$46.8-0.541\ T$	Gavrilov
Passeriformes				
Goldcrest *Regulus regulus*	5.5	$10.8-0.187\ T$	—	Gavrilov and Dol'nik
Chiffchaff *Phylloscopus collybita*	6.5	$12.4-0.227\ T$	—	Gavrilov and Dol'nik
Wood warbler *Phylloscopus sibilatrix*	7.6	—	$13.8-0.273\ T$	Gavrilov and Dol'nik
Black-rumped waxbill *Estrilda troglodytes*	7.7	—	$16.0-0.353\ T$	Gavrilov and Dol'nik
Tiaris canora	7.8	—	$16.1-0.343\ T$	Gavrilov and Dol'nik
Long-tailed tit *Aegithalos caudatus*	8.8	$14.5-0.260\ T$	$14.8-0.291\ T$	Gavrilov
Wren *Troglodytes troglodytes*	9.0	$14.0-0.230\ T$	$14.9-0.283\ T$	Gavrilov
Uraeginthus bengalis	9.2	—	$16.1-0.323\ T$	Gavrilov and Dol'nik
Yellow-bellied seedeater *Sporophila nigricollis*	9.3	$14.2-0.262\ T$	$18.0-0.401\ T$	Cox (1961)
Blue-black grassquit *Volatinia jacarina*	9.4	$13.9-0.259\ T$	$16.3-0.324\ T$	Cox (1961)
Coal tit *Parus ater*	9.7	$14.8-0.250\ T$	$15.6-0.294\ T$	Gavrilov
Willow warbler *Phylloscopus trochilus*	9.8	$15.8-0.293\ T$	—	Gavrilov and Dol'nik
Lonchura striata	10.3	—	$17.1-0.327\ T$	Gavrilov and Dol'nik
Lesser whitethroat *Sylvia curruca*	10.3	—	$16.3-0.297\ T$	Gavrilov and Dol'nik
Variable seedeater *Sporophila aurita*	10.7	$16.0-0.282\ T$	$18.0-0.340\ T$	Cox (1961)
Marsh tit *Parus palustris*	11.1	—	$16.2-0.280\ T$	Gavrilov and Dol'nik
Pied flycatcher *Ficedula hypoleuca*	11.2	$16.8-0.310\ T$	—	Gavrilov and Dol'nik
Sedge warbler *Acrocephalus schoenobenus*	11.5	—	$17.4-0.320\ T$	Gavrilov and Dol'nik
Redstart *Phoenicurus phoenicurus*	11.6	$16.5-0.283\ T$	—	Gavrilov and Dol'nik
Pine siskin *Carduelis pinus*	12.0	$15.9-0.237\ T$	$16.4\ \ 0.262\ T$	Gavrilov
Zebra finch *Peophila guttata*	12.1	$18.0-0.380\ T$	—	El-Wailly (1966)
Serinus canaria	12.2	$16.8-0.250\ T$	—	Gavrilov and Dol'nik
Whitethroat *Sylvia communis*	13.1	$17.5-0.280\ T$	—	Gavrilov and Dol'nik
Field sparrow *Spizella pusilla* (♀)	13.2	$14.7-0.261\ T$	$16.2-0.288\ T$	Olson (1965)
Field sparrow *Spizella pusilla* (♂)	13.9	$15.6-0.285\ T$	$16.6-0.295\ T$	Olson (1965)
Common redpoll *Acanthis flammea*	13.5	$16.1-0.240\ T$	$17.0-0.252\ T$	Gavrilov
Common redpoll *Acanthis flammea*	14.6	$15.6-0.215\ T$	$20.2-0.341\ T$	Brooks (1968)
Black redstart *Phoenicurus ochruros*	14.0	—	$18.6-0.327\ T$	Gavrilov and Dol'nik
Robin *Erithacus rubecula*	14.0	$18.2-0.319\ T$	$18.8-0.347\ T$	Gavrilov
Icterine warbler *Hippolais icterina*	14.1	—	$19.2-0.347\ T$	Gavrilov and Dol'nik
Linnet *Acanthis cannabina*	14.5	$19.0-0.320\ T$	—	Gavrilov and Dol'nik
Hoary redpoll *Acanthis hornemanni*	14.5	$15.6-0.255\ T$	$19.6-0.324\ T$	Brooks (1968)
Yellow wagtail *Montacilla flava*	15.4	—	$19.5-0.330\ T$	Gavrilov and Dol'nik
Whinchat *Saxicola rubetra*	15.7	—	$19.8-0.333\ T$	Gavrilov and Dol'nik
Goldfinch *Carduelis carduelis*	15.9	$18.8-0.283\ T$	—	Gavrilov and Dol'nik
Dunnock *Prunella modularis*	16.5	$19.8-0.323\ T$	—	Gavrilov and Dol'nik
Great tit *Parus major*	17.0	$19.2-0.307\ T$	$20.4-0.321\ T$	Gavrilov
Meadow pipit *Anthus pratensis*	17.5	—	$21.4-0.380\ T$	Gavrilov and Dol'nik
Pied wagtail *Motacilla alba*	18.0	$19.6-0.324\ T$	$20.8-0.342\ T$	Gavrilov
Reed bunting *Emberiza schoeniclus*	18.2	$20.8-0.333\ T$	—	Gavrilov and Dol'nik
Bluethroat *Luscinia svecica*	18.2	$21.6-0.380\ T$	—	Gavrilov and Dol'nik
American tree sparrow *Spizella arborea*	19.0	$18.6-0.247\ T$	$22.6-0.366\ T$	West (1960)
Dark-eyed junco *Junco hyemalis*	19.4	$17.2-0.152\ T$	$21.4-0.259\ T$	Seibert (1949)
Blackcap *Sylvia atricapilla*	20.0	—	$23.2-0.417\ T$	Gavrilov and Dol'nik
Chaffinch *Fringilla coelebs*	20.0	$21.3-0.363\ T$	$24.3-0.413\ T$	Gavrilov
Chaffinch *Fringilla coelebs*	20.0	$20.5-0.535\ T$	$21.0-0.333\ T$	Dol'nik (1974 *b*)

Species	Weight (g)	Photoperiod 7–12 hours M (kcal bird^{-1} day^{-1})	Photoperiod 15–24 hours M (kcal bird^{-1} day^{-1})	Authority
Barred warbler *Sylvia nisoria*	20.6	—	$24.2-0.433\ T$	Gavrilov and Dol'nık
Brambling *Fringilla montifringilla*	21.0	$22.6-0.327\ T$	—	Gavrilov and Dol'nik
Scarlet rosefinch *Carpodacus erythrinus*	21.0	$23.5-0.383\ T$	$26.4-0.417\ T$	Gavrilov
Garden warbler *Sylvia borin*	22.0	$22.9-0.347\ T$	—	Gavrilov and Dol'nik
House sparrow *Passer d. bactrianus*	22.2	$23.1-0.373\ T$	$26.3-0.423\ T$	Gavrilov
House sparrow *Passer domesticus*	25.0	$23.6-0.273\ T$	$24.7-0.312\ T$	Gavrilov
House sparrow *Passer domesticus*	25.2	$24.7-0.272\ T$	$25.7-0.320\ T$	Kendeigh (1949); Davis (1955)
European tree sparrow *Passer montanus*	22.5	$23.2-0.373\ T$	—	Gavrilov and Dol'nik
Pine bunting *Emberiza leucocephala*	22.5	$22.8-0.373\ T$	—	Gavrilov and Dol'nik
Tawny pipit *Anthus campestris*	23.1	$24.2-0.380\ T$	—	Gavrilov and Dol'nik
Tree pipit *Anthus trivialis*	23.2	$24.9-0.427\ T$	—	Gavrilov and Dol'nik
Wood lark *Lullula arborea*	25.0	$24.2-0.313\ T$	—	Gavrilov and Dol'nik
Red-backed shrike *Lanius collurio*	25.3	—	$24.2-0.333\ T$	Gavrilov and Dol'nik
Greenfinch *Carduelis chloris*	26.5	$23.5-0.308\ T$	$25.8-0.327\ T$	Gavrilov
Ortolan bunting *Emberiza hortulana*	26.5	$25.6-0.380\ T$	$27.4-0.415\ T$	Gavrilov
Yellowhammer *Emberiza citrinella*	27.1	$24.3-0.310\ T$	$26.9-0.413\ T$	Gavrilov
White-throated sparrow *Zonotrichia albicollis*	27.4	$23.0-0.38\ \ T$	—	Kontogiannis (1968)
Bullfinch *Pyrrhula pyrrhula*	30.4	$26.3-0.373\ T$	—	Gavrilov and Dol'nik
Dickcissel *Spiza americana* (♀)	29.6	$24.4-0.518\ T$	$29.2-0.545\ T$	Zimmerman (1965 a)
Dickcissel *Spiza americana* (♂)	31.6	$25.9-0.537\ T$	$28.5-0.517\ T$	Zimmerman (1965 a)
Snow finch *Montifringilla nivalis*	30.7	—	$26.1-0.353\ T$	Gavrilov and Dol'nik
Rock sparrow *Petronia petronia*	32.8	—	$27.6-0.403\ T$	Gavrilov and Dol'nik
Horned lark *Eremophila alpestris*	36.6	—	$29.5-0.427\ T$	Gavrilov and Dol'nik
Green-backed sparrow *Arremonops conirostris*	37.0	$29.4-0.554\ T$	$30.4-0.555\ T$	Cox (1961)
Sky lark *Alauda arvensis*	39.4	—	$31.2-0.440\ T$	Gavrilov and Dol'nik
Red crossbill *Loxia curvirostra*	44.7	$30.6-0.297\ T$	$32.4-0.333\ T$	Gavrilov
Hawfinch *Coccothraustes coccothraustes*	48.2	$33.9-0.393\ T$	—	Gavrilov and Dol'nik
Corn bunting *Emberiza calandra*	48.8	—	$34.5-0.487\ T$	Gavrilov and Dol'nik
Parrot crossbill *Loxia pytiopsittacus*	53.4	$33.0-0.277\ T$	—	Gavrilov and Dol'nik
Evening grosbeak *Coccothraustes vespertinus*	54.5	$37.2-0.612\ T$	—	West & Hart (1966)
Evening grosbeak *Coccothraustes vespertinus*	57.0	—	$40.5-0.540\ T$	J. E. Williams (personal communication)
Redwing *Turdus iliacus*	57.0	$37.5-0.500\ T$	—	Gavrilov and Dol'nik
Golden oriole *Oriolus oriolus*	65.1	—	$40.4-0.530\ T$	Gavrilov and Dol'nik
Bohemian waxwing *Bombycilla garrulus*	66.5	$40.5-0.517\ T$	—	Gavrilov and Dol'nik
Great gray shrike *Lanius excubitor*	71.3	$42.3-0.543\ T$	—	Gavrilov and Dol'nik
Starling *Sturnus vulgaris*	78.0	$42.8-0.490\ T$	—	Gavrilov and Dol'nik
Song thrush *Turdus philomelos*	80.0	$44.7-0.610\ T$	$52.1-0.697\ T$	Gavrilov
Blackbird *Turdus merula*	83.0	$43.2-0.457\ T$	$46.3-0.462\ T$	Gavrilov
Blue jay *Cyanocitta cristata*	84.5	$57.4-0.829\ T$	$58.3-0.485\ T$	Clemans (1974)
Mistle thrush *Turdus viscivorus*	112	$53.4-0.603\ T$	—	Gavrilov and Dol'nik
Nutcracker *Nucifraga caryocatactes*	147	$59.0-0.540\ T$	—	Gavrilov and Dol'nik
Jay *Garrulus glandarius*	153	$61.0-0.656\ T$	—	Gavrilov and Dol'nik
Magpie *Pica pica*	205	$71.1-0.620\ T$	—	Gavrilov and Dol'nik
Alpine chough *Pyrrhocorax graculus*	208	$71.2-0.653\ T$	—	Gavrilov and Dol'nik
Jackdaw *Corvus monedula*	215	$73.0-0.727\ T$	—	Gavrilov and Dol'nik
Rook *Corvus frugilegus*	390	$98.0-0.520\ T$	—	Gavrilov and Dol'nik
Carrion crow *Corvus corone*	540	$115.5-0.703\ T$	$118.1-0.712\ T$	Gavrilov
Corvus ruficollis	660	$127.1-0.337\ T$		Gavrilov and Dol'nik
Raven *Corvus corax*	1 203	$187.2-0.633\ T$	$194.6-0.731\ T$	Gavrilov

References[1]

Abé, M. T. (1969). Ecological studies on *Passer montanus kaibatoi* Munsterhjelm. *Bulletin of the Government Forest Experiment Station*, No. **220**, 11–57. (In Japanese with English summary.)

Abé, M. T. (1970). Some factors influencing growth of nestlings of *Passer montanus kaibatoi* Munsterhjelm. *International Studies on Sparrows*, **4**, 7–10.

Akhmedov, K. P. (1949). Biologiya ispanskogo vorob'ya v Tadzhikistane i prinosimoï im vred. [Biology of the Spanish sparrow in Tadzhikistan and its harmfulness.] *Soobshcheniya Tadzhikskogo Filiala Akademii Nauk SSSR*, **12**, 29–31.

Alatalo, R. (1975). Varpusen *Passer domesticus* pesimäbiologiasta Oulun Sanginsuussa. [On the breeding biology of the house sparrow at Oulu.] *Lintumies*, **10**, 1–7.

Alcock, J. (1973). Cues used in searching for food by red-winged blackbirds (*Agelaius phoeniceus*). *Behaviour*, **46**, 174–87.

Allen, A. A. (1914). The red-winged blackbird: a study in the ecology of a cat-tail marsh. *Abstracts of the Proceedings of the Linnaean Society of New York*, **24–25**, 43–128.

Anderson, D. R., Kimball, C. F. & Fiehrer, F. R. (1974). A computer program for estimating survival and recovery rates. *Journal of Wildlife Management*, **38**, 369–70.

Anderson, T. R. (1973). A comparative ecological study of the house sparrow and the tree sparrow near Portage des Sioux, Missouri. Ph.D. Thesis, Saint Louis University, Missouri, 148 pp.

Anderson, T. R. (1975). Fecundity of the house sparrow and the tree sparrow near Portage des Sioux, Missouri, USA. *International Studies on Sparrows*, **8**, 6–23.

Anderson, T. R. (1977). Population studies of European sparrows in North America. *University of Kansas Publications, Museum of Natural History*, **70**. (In Press.)

Anonymous (1967). Blackbird depredation in agriculture. A report on the 1967 North American Conference. Agricultural Science Review, Cooperative State Research Service, US Department of Agriculture, **5**, 15–22.

AOU Check-list Committee (1957). *Check-list of North American birds*, xiii+691 pp. Baltimore: Lord Baltimore Press.

Ar, A., Paganelli, C. V., Reeves, R. B., Greene, D. G. & Rahn, H. (1974). The avian egg: water vapor conductance, shell thickness, and functional pore area. *Condor*, **76**, 153–8.

Arbib, R. & Heilbrun, L. H. (1974). The 74th Christmas bird count. *American Birds*, **28**, 135–540.

Aschoff, J. & Pohl, H. (1970a). Rhythmic variations in energy metabolism. *Federation Proceedings*, **29**, 1541–52.

Aschoff, J. & Pohl, H. (1970b). Der Ruheumsatz von Vögeln als Funktion der Tageszeit und der Körpergrösse. *Journal für Ornithologie*, **111**, 38–47.

Ašmera, J. (1962). Studie über die Nahrung des Haussperlings (*Passer domesticus* (L.)) und Feldsperlings (*Passer montanus* (L.)). *Acta Rerum Naturalium Districtus Silesiae*, **23**, 207–24.

Aulie, A. (1971). Body temperatures in pigeons and budgerigars during sustained flight. *Comparative Biochemistry and Physiology*, **39** (2A), 173–6.

[1] The preparation of these references (by H. Dominas, M.Sc.) was partly financed by the Committee of Ecology, Polish Academy of Sciences.

379

References

Austin, G. T. (1974). Nesting success of the cactus wren in relation to nest orientation. *Condor*, **76**, 216–17.

Averin, Yu. V. & Ganya, I. M. (1970). *Ptitsỹ Moldavii*. [*Birds of Moldavia*.] Vol. 1, 240 pp. Kishinev: Akademiya Nauk Moldavskoï SSR, Institut Zoologii.

Balát, F. (1971). Clutch size and breeding success of the tree sparrow, *Passer montanus* (L.), in central and southern Moravia. *Zoologické Listy*, **20**, 265–80.

Balát, F. (1972 a). Ptačí složka biocenózy dvou drubežáren a její úloha v udržování a rozšiřování aviární tuberkulózy. [The avian component of the biocenosis of two poultry farms and its role in maintaining and spreading avian tuberculosis.] *Veterinární Medicína*, **17**, 713–24.

Balát, F. (1972 b). Zur Frage des Legebeginns bei dem Feldsperling, *Passer montanus* (L.). *Zoologické Listy*, **21**, 235–44.

Balát, F. (1973). Die zwischenartlichen Brutbeziehungen zwischen dem Haussperling, *Passer domesticus* (L.) und der Mehlschwalbe, *Delichon urbica* (L.). *Zoologické Listy*, **22**, 213–22.

Balát, F. (1974). Zur Frage der Nistkonkurrenz des Feldsperlings, *Passer montanus* (L.). *Zoologické Listy*, **23**, 123–35.

Balát, F. (1975). Die Alterstruktur der Brutpopulation des Feldsperlings, *Passer montanus* (L.). *Zoologické Listy*, **24**, 137–47.

Balát, F. & Toušková, I. (1972). Zur Erkenntnis der Biomasse-Produktion der Nachkommenschaft des Feldsperlings, *Passer montanus* (L.). *Zoologické Listy*, **21**, 444–555.

Baldwin, P. H. (1973). The feeding regime of granivorous birds in shortgrass prairie in Colorado, USA. In *Productivity, population dynamics and systematics of granivorous birds*, ed. S. C. Kendeigh & J. Pinowski, pp. 237–47. Warszawa: PWN-Polish Scientific Publishers.

Baldwin, S. P. & Kendeigh, S. C. (1938). Variations in the weight of birds. *Auk*, **55**, 416–67.

Barloy, J.-J. (1966). Recherches sur le moineau domestique *Passer d. domesticus* (Linné) 1758: cycle biologique, ecologie, dynamique de population; sa place dans l'avifaune Parisienne. Thèses présentées à la Faculté des Sciences de l'Université de Paris pour obtenir le grade de Docteur des Sciences Naturelles, Paris, 211 pp.

Barnett, L. B. (1970). Seasonal changes in temperature acclimatization of the house sparrow, *Passer domesticus*. *Comparative Biochemistry and Physiology*, **33**, 559–78.

Barott, H. G. & Pringle, E. M. (1946). Energy and gaseous metabolism of the chicken from hatching to maturity as affected by temperature. *Journal of Nutrition*, **31**, 35–50.

Barrows, W. B. (1889). The English sparrow (*Passer domesticus*) in North America. *US Department of Agriculture, Division of Economic Ornithology and Mammalogy Bulletin*, No. **1**, 405 pp.

Bartholomew, G. A., Hudson, J. W. & Howell, T. R. (1962). Body temperature, oxygen consumption, evaporative water loss, and heart rate in the poor-will. *Condor*, **64**, 117–25.

Bartholomew, G. A. & Trost, C. H. (1970). Temperature regulation in the speckled mousebird, *Colius striatus*. *Condor*, **72**, 141–6.

Bar-Yosef, O. & Tchernov, E. (1966). Archaeological finds and the fossil faunas of the Natufian and Microlithic industries at Hayonim Cave (western Galillee, Israel). *Israel Journal of Zoology*, **15**, 104–40.

Bauer, Z. (1974). *Ptaci složka a jej životni projevy ve skupine geobiocenoz Ulmi-*

References

Fraxineta Carpini. [*Avicenoza and its living processes in Ulmi-Fraxineta Carpini community.*] 63 pp. Brno: Vysoká Škola Zemedelska v Brne, Fakulta Lesnicka.

Bauer, Z. (1975). The biomass production of the tree sparrow, *Passer m. montanus* (L.), population in the conditions of the flood forest. *International Studies on Sparrows*, **8**, 124–39.

Beal, F. E. L. (1900). *Food of the bobolink, blackbirds, and grackles.* US Department of Agriculture, Division of Biological Survey Bulletin, No. **13**, 77 pp.

Beckmann, B. & Fröhlich, H. (1967). Quantitative Untersuchungen der Avifauna von zwei unterschiedlichen Dörfern in Münsterland. *Natur und Heimat*, **27**, 82–8.

Bednorz, J. (1970). Awifauna Maltańskiego klina zieleni w Poznaniu. [Avifauna of the Maltański forest in Poznań.] *Poznańskie Towarzystwo Przyjaciół Nauk, Wydział Matematyczno-Przyrodniczy, Prace Komisji Biologicznej*, **33**, 267–312.

Beecher, W. J. (1951). Adaptations for food-getting in the American blackbirds. *Auk*, **68**, 411–40.

Beer, J. R. (1965). Variation in red-winged blackbird eggs. *Passenger Pigeon*, **27**, 60–4.

Beer, J. R. & Tibbitts, S. D. (1950). Nesting behavior of the red-wing blackbird. *Flicker*, **22**, 61–77.

Beimborn, D. A. (1967). Population ecology of the English sparrow in North America. M.Sc. Thesis, University of Wisconsin, Milwaukee, 138 pp.

Beitz, W. (1972). Siedlungsdichteaufnahmen in einem Ufergehölz. *Ornithologischer Rundbrief Mecklenburgs, Neue Folge*, **13**, 38–42.

Benson, G. B. G. & Williamson, K. (1972). Breeding birds of a mixed farm in Suffolk. *Bird Study*, **19**, 34–50.

Berger, M. & Hart, J. S. (1974). Physiology and energetics of flight. In *Avian biology*, ed. D. S. Farner & J. R. King, vol. 4, pp. 415–77. New York: Academic Press.

Berger, R. (1959). Untersuchungen über die Ernährungsweise der Nestlinge des Feldsperlings (*Passer m. montanus*). M.A. Thesis, Halle University, Halle, 57 pp.

Bernstein, M. H., Thomas, S. P. & Schmidt-Nielsen, K. (1973). Power input during flight of the fish crow, *Corvus ossifragus*. *Journal of Experimental Biology*, **58**, 401–10.

Besser, J. F. (1973). Protecting seeded rice from blackbirds with methiocarb. *International Rice Commission Newsletter*, **22**, 9–14.

Besser, J. F., DeGrazio, J. W. & Guarino, J. L. (1968). Costs of wintering starlings and red-winged blackbirds at feedlots. *Journal of Wildlife Management*, **32**, 179–80.

Bethune, G. de (1961). Notes sur le moineau friquet, *Passer montanus* (L.). *Le Gerfaut*, **51**, 387–98.

Birkebak, R. C. (1966). Heat transfer in biological systems. *International Review of General and Experimental Zoology*, **2**, 269–344.

Blackith, R. E. & Reyment, R. A. (1971). *Multivariate morphometrics*, ix+412 pp. London: Academic Press.

Blackmore, F. H. (1969). The effect of temperature, photoperiod and molt on the energy requirements of the house sparrow, *Passer domesticus. Comparative Biochemistry and Physiology*, **30**, 433–44.

Blagosklonov, K. N. (1950). Biologiya i sel'skokhozyaĭstvennoe znachenie pol-

381

References

evogo vorob'ya v polezashchitnӯkh lesonasazhdeniyakh Yugo-Vostoka. [The biology and agricultural significance of tree sparrows in forest shelterbelts of southeast.] *Zoologicheskiĭ Zhurnal*, **29**, 244–54.

Blem, C. R. (1973 *a*). Geographic variation in the bioenergetics of the house sparrow. *Ornithological Monographs*, **14**, 96–121.

Blem, C. R. (1973 *b*). Laboratory measurements of metabolized energy in some passerine nestlings. *Auk*, **90**, 895–7.

Blem, C. R. (1974). Geographic variation of thermal conductance in the house sparrow. *Comparative Biochemistry and Physiology*, **47A**, 101–8.

Blem, C. R. (1975). Energetics of nestling house sparrows *Passer domesticus*. *Comparative Biochemistry and Physiology*, **52A**, 305–12.

Blyumental', T. I. & Dol'nik, V. R. (1970). Ves tela, dlina krӯla, zhirovӯe otlozheniya i polet ptits. [Body weight, wing length, fat deposits, and flight in birds.] *Zoologicheskiĭ Zhurnal*, **49**, 1069–72.

Blyumental', T. I., Gavrilov, V. M. & Dol'nik, V. R. (1967). O prichinakh volnoobraznosti migratsii zyablika. [On the causes of the wave-like migration of chaffinches.] In *Soobshcheniya Pribaltiĭskoĭ Komissii po Izucheniyu Migratsiĭ Ptits*, ed. E. Kumari, vol. 4, pp. 69–80. Tallin: 'Valgus'.

Bocheński, Z. & Harmata, W. (1962). Ptaki południowego krańca Jury Krakowsko-Wieluńskiej. [The birds of the southern border of the Kraków-Wieluń Jurassic Ridge.] *Acta Zoologica Cracoviensia*, **7**, 483–574.

Bortoli, L. (1969). Contribution à l'étude du problème des oiseaux granivores en Tunisie. I. Les moineaux (Aves, Ploceidae). *Bulletin de la Faculté d'Agronomie*, **22–23**, 33–153.

Bortoli, L. (1973). Sparrows in Tunisia. In *Productivity, population dynamics and systematics of granivorous birds*, ed. S. C. Kendeigh & J. Pinowski, pp. 249–52. Warszawa: PWN-Polish Scientific Publishers.

Bortoli, L. (1974). Les oiseaux granivores en Afrique tropicale avec référence speciale à *Quelea quelea* – le milieu et les dégâts. *International Studies on Sparrows*, **7**, 37–75.

Botkin, D. B. & Miller, R. S. (1974). Mortality rates and survival of birds. *American Naturalist*, **108**, 181–92.

Boudreau, G. W. (1975). How to win the war with pest birds. *Wildlife Technology*, 174 pp. Hollister, California.

Bowers, D. E. (1956). A study of methods of color determination. *Systematic Zoology*, **5**, 147–60; concluded on p. 182.

Boyd, A. W. (1932). Notes on the tree sparrow. *British Birds*, **25**, 278–85.

Brenner, F. J. (1964). Growth, fat deposition and development of endothermy in nestling red-winged blackbirds. *Journal of the Scientific Laboratories of Denison University*, **46**, 81–9.

Brenner, F. J. (1965). Metabolism and survival time of grouped starlings at various temperatures. *Wilson Bulletin*, **77**, 388–95.

Brenner, F. J. (1966). The influence of drought on reproduction in a breeding population of redwinged blackbirds. *American Midland Naturalist*, **76**, 201–10.

Brenner, F. J. (1968). Energy flow in two breeding populations of redwinged blackbirds. *American Midland Naturalist*, **79**, 289–310.

Brisbin, I. L. Jr (1969). Bioenergetics of the breeding cycle of the ring dove. *Auk*, **86**, 54–74.

Brisbin, I. L. Jr & Tally, L. J. (1973). Age-specific changes in the major body components and caloric value of growing Japanese quail. *Auk*, **90**, 624–35.

Brody, S. (1945). *Bioenergetics and growth*. xii+1023 pp. New York: Reinhold Publishing Company.

References

Brooks, W. S. (1968). Comparative adaptations of the Alaskan redpolls to the arctic environment. *Wilson Bulletin*, **80**, 253–80.

Brown, J. L. (1974). Alternate routes to sociality in jays – with a theory for the evolution of altruism and communal breeding. *American Zoologist*, **14**, 63–80.

Bruijns, M. F. M. (1961). De dichtheid van broedvogelbevolkingen in bebouwde kommen. [The density of breeding bird populations in built-up areas.] *Levende Natuur*, **64**, 193–9.

Brush, A. H. (1965). Energetics, temperature regulation and circulation in resting, active and defeathered California quail, *Lophortyx californicus*. *Comparative Biochemistry and Physiology*, **15**, 399–421.

Bull, P. C. (1973). The starling: friend or foe? *New Zealand Journal of Agriculture*, **127**, 55–9.

Bumpus, H. C. (1899). The elimination of the unfit as illustrated by the introduced sparrow, *Passer domesticus*. *Biological Bulletin. Marine Biological Laboratory, Woods Hole*, 209–26.

Burckard, E. L., Dontcheff, L. & Kayser, C. (1933). Le rythme nycthéméral chez le pigeon. *Annales de Physiologie et de Physicochimie Biologique*, **9**, 303–68.

Busse, P. (1962). Zmienność wielkości, kształtu i ubarwienia jaj w podwarszawskiej populacji mazurka *Passer montanus* (L.). [Variability of largeness, form and colour of eggs in a population of tree sparrows, *Passer montanus* (L.) near Warsaw.] *Notatki Ornitologiczne*, **3**, 23–33.

Caccamise, D. F., Porreca, F. J. & Weathers, W. W. (1974). Temperature regulation and the distribution of monk parrakeets. *Bulletin of the Ecological Society of America*, **55**, 16.

Cade, T. J., Tobin, C. A. & Gold, A. (1965). Water economy and metabolism of two estrildine finches. *Physiological Zoology*, **38**, 9–33.

Cain, B. W. (1970). Growth and plumage development of the black-bellied tree duck, *Dendrocygna autumnalis* (Linnaeus). *Taius* (Texas A. & I. University Studies), **3**, 25–48.

Cain, B. W. (1973). Effect of temperature on energy requirements and northward distribution of the black-bellied tree duck. *Wilson Bulletin*, **85**, 308–17.

Cain, B. W. (1976). Energetics of growth for black-bellied tree ducks. *Condor*, **78**, 124–8.

Calder, W. A. (1964). Gaseous metabolism and water relations of the zebra finch, *Taeniopygia castanotis*. *Physiological Zoology*, **37**, 400–13.

Calder, W. A. (1973 a). An estimate of the heat balance of a nesting hummingbird in a chilling climate. *Comparative Biochemistry and Physiology*, **46A**, 291–300.

Calder, W. A. (1973 b). Microhabitat selection during nesting of hummingbirds in the Rocky Mountains. *Ecology*, **54**, 127–34.

Calder, W. A. (1974). The thermal and radiant environment of a winter hummingbird nest. *Condor*, **76**, 268–73.

Calder, W. A. & Booser, J. (1973). Hypothermia of broad-tailed hummingbirds during incubation in nature with ecological correlations. *Science*, **180**, 751–3.

Calder, W. A. & King, J. R. (1974). Thermal and caloric relations of birds. In *Avian biology*, ed. D. S. Farner & J. R. King, vol. 4, pp. 259–413. New York: Academic Press.

Calder, W. A. & Schmidt-Nielsen, K. (1967). Temperature regulation and evaporation in the pigeon and the roadrunner. *American Journal of Physiology*, **213**, 883–9.

Caldwell, P. J. & Cornwell, G. W. (1975). Incubation behavior and temperatures of the mallard duck. *Auk*, **92**, 706–31.

Calhoun, J. B. (1947). The role of temperature and natural selection in relation

References

to the variations in the size of the English sparrow in the United States. *American Naturalist*, **81**, 203–28.

Carleton, A. R. & Owre, O. T. (1975). The red-whiskered bulbul in Florida: 1960–71. *Auk*, **92**, 40–57.

Case, N. A. & Hewitt, O. H. (1963). Nesting and productivity of the red-winged blackbird in relation to habitat. *Living Bird*, **2**, 7–20.

Case, R. M. (1973). Bioenergetics of a covey of bobwhites. *Wilson Bulletin*, **85**, 52–9.

Case, R. M. & Robel, R. J. (1974). Bioenergetics of the bobwhite. *Journal of Wildlife Management*, **38**, 638–52.

Cătuneanu, I. & Theis, F. (1965). Cercetări asupra reproducerii la vrăbii (*Passer domesticus* (L.) şi *Passer montanus* (L.)) în Romania. [Investigation on sparrow (*Passer domesticus* (L.) and *Passer montanus* (L.)) reproduction in Romania.] *Analele Secţiei de Protecţia Plantelor*, **3**, 329–36.

Chaplin, S. B. (1974). Daily energetics of the black-capped chickadee, *Parus atricapillus*, in winter. *Journal of Cellular and Comparative Physiology*, **89**, 321–30.

Chew, R. M. (1974). Consumers as regulators of ecosystems: an alternative to energetics. *Ohio Journal of Science*, **74**, 359–70.

Chia, H. K., Bei, T. H., Chen, T. Y. & Cheng, T. (1963). Preliminary studies on the breeding behaviour of the tree sparrow (*Passer montanus saturatus*). *Acta Zoologica Sinica*, **15**, 527–36.

Clausing, P. (1975). Vergleichende Analyse der Gelegegrösse von Populationen des Feldsperlings (*Passer montanus* (L.)) in der DDR. *Zoologische Jahrbücher, Abteilung für Systematik, Ökologie und Geographie der Tiere*, **102**, 89–100.

Clemans, R. J. (1974). The bioenergetics of the blue jay in central Illinois. *Condor*, **76**, 358–60.

Cody, M. L. (1971). Finch flocks in the Mohave Desert. *Theoretical Population Biology*, **2**, 142–58.

Cody, M. L. (1974). *Competition and the structure of bird communities. Monographs in Population Biology*, No. 7, 318 pp. Princeton, New Jersey: Princeton University Press.

Coleman, J. D. (1974). Breakdown rates of foods ingested by starlings. *Journal of Wildlife Management*, **38**, 910–12.

Collias, N. E. & Collias, E. C. (1969). Size of breeding colony related to attraction of mates in a tropical passerine bird. *Ecology*, **50**, 481–8.

Collier, G. (1968). Annual cycle and behavioral relationships in the red-winged and tri-colored blackbirds of southern California. Ph.D. Thesis, University of California, Los Angeles, 390 pp.

Collinge, W. E. (1924–1927). *The food of some British wild birds: a study in economic ornithology*, 2nd edn, revised. 427 pp. York (published by author).

Collins, J. M. (1968). The effects of environmental temperature on the rate of development of embryonic red-winged blackbirds. M.Sc. Thesis, University of Guelph, Guelph, 36 pp.

Cooley, W. W. & Lohnes, P. R. (1971). *Multivariate data analysis*, x+364 pp. New York: John Wiley & Sons.

Coulombe, H. N. (1970). Physiological and physical aspects of temperature regulation in the burrowing owl *Speotyto cunicularia*. *Comparative Biochemistry and Physiology*, **35**, 307–37.

Coulson, J. C. (1963). Egg size and shape in the kittiwake (*Rissa tridactyla*) and their use in estimating age composition of populations. *Proceedings of the Zoological Society of London*, **140**, 211–27.

Cox, G. W. (1961). The relation of energy requirements of tropical finches to distribution and migration. *Ecology*, **42**, 253–66.

Creutz, G. (1949). Untersuchungen zur Brutbiologie des Feldsperlings (*Passer m. montanus* (L.)). *Zoologische Jahrbücher, Abteilung für Systematik, Ökologie und Geographie der Tiere*, **78**, 133–72.

Crook, J. H. (1962). The adaptive significance of pair formation types in weaver birds. *Symposia of the Zoological Society of London*, **8**, 57–70.

Crook, J. H. (1963). Comparative studies on the reproductive behaviour of two closely related weaver species (*Ploceus cucullatus* and *Ploceus nigerrimus*) and their races. *Behaviour*, **21**, 177–232.

Crook, J. H. (1964). The evolution of social organisation and visual communication in the weaver birds (Ploceinae). *Behaviour*, Supplement **10**, 1–178.

Crook, J. H. (1965). The adaptive significance of avian social organizations. *Symposia of the Zoological Society of London*, **14**, 181–218.

Crook, J. H. & Butterfield, P. A. (1970). Gender role in the social system of *Quelea*. In *Social behaviour in birds and mammals*, ed. J. H. Crook, pp. 211–48. New York: Academic Press.

Croze, H. (1970). Searching image in carrion crows. *Zeitschrift für Tierpsychologie, Beiheft*, **5**, 1–85.

Cummins, K. W. (1967). *Calorific equivalents for studies in ecological energetics*, 52 pp. University of Pittsburg, Pittsburgh, Pennsylvania. Mimeographed.

Cummins, K. W. & Wuycheck, J. C. (1971). Caloric equivalents for investigations in ecological energetics. *Mitteilungen, Internationale Vereinigung für Theoretische und Angewandte Limnologie*, No. **18**, 1–158.

Custer, T. W. & Pitelka, F. A. (1975). Correction factors for digestion rates for prey taken by snow buntings (*Plectrophenax nivalis*). *Condor*, **77**, 210–12.

Czarnecki, Z. (1956). Materiały do ekologii ptaków gnieżdżących się w śródpolnych kępach drzew. [Material illustrating the ecology of birds nesting in clumps of trees surrounded by open fields.] *Ekologia Polska*, Ser. A, **4**, 374–417.

Danilov, N., Nekrasov, E., Dobrinskiĭ, L. & Kopein, K. (1969). K voprosu ob izmenchivosti populyatsiĭ *Passer domesticus* (L.) i *P. montanus* (L.). [The variability of populations of *Passer domesticus* (L.) and *P. montanus* (L.).] *Ekologia Polska*, Ser. A, **17**, 489–501.

Dargol'ts, V. G. (1973). Kaloricheskiĭ ékvivalent izmeneniya vesa tela u gomoĭotermnȳkh zhivotnȳkh: zavisimost' ot okislyaemȳkh veshchestv i ot isparitel'noĭ teplootdachi. [The caloric value equivalent of body weight changes of homoiothermal animals: relationship between oxidized substances and heat loss through evaporation.] *Zhurnal Obshcheĭ Biologii*, **34**, 887–99.

Dargol'ts, V. G. (1974). Vliyanie mȳshechnoĭ nagruzki i temperaturȳ sredȳ na kaloricheskiĭ ékvivalent poter' vesa u v'yurka i yulȳ. [Effect of exercises and ambient temperature on the calorific equivalent of body weight in two passerine birds.] In *Materialȳ VI Vsesoyuznoĭ Ornitologicheskoĭ Konferentsii*, ed. R. L. Beme & V. E. Flint, vol. 1, pp. 139–40. Moskva: Izdatel'stvo Moskovskogo Universiteta.

Darling, F. F. (1952). Social behavior and survival. *Auk*, **69**, 183–91.

Darlington, P. J. Jr (1957). *Zoogeography: the geographic distribution of animals*, xi+675 pp. New York: John Wiley & Sons.

Davis, D. E. (1962). Gross effects of triethylenemelamine on gonads of starlings. *Anatomical Record*, **142**, 353–7.

References

Davis, E. A. (1955). Seasonal variation in the energy resources of the English sparrow. *Auk*, **72**, 385–411.

Davis, J. (1973). Field notes concerning dispersal of house sparrows at Hastings Natural History Reservation, California. Manuscript (Star Rt., Box 80, Carmel Valley, California 93924, USA).

Dawson, D. G. (1964). The eggs of the house sparrow. *Notornis*, **11**, 187–9.

Dawson, D. G. (1967). Roosting sparrows *Passer domesticus* killed by rainstorm, Hawkes Bay, New Zealand. *Notornis*, **14**, 208–10.

Dawson, D. G. (1970). Estimation of grain loss due to sparrows (*Passer domesticus*) in New Zealand. *New Zealand Journal of Agricultural Research*, **13**, 681–8.

Dawson, D. G. (1972). The breeding ecology of house sparrows. Ph.D. Thesis, University of Oxford, Oxford, 112 pp.

Dawson, D. G. (1973). House sparrow, *Passer domesticus* (L.), breeding in New Zealand. In *Productivity, population dynamics and systematics of granivorous birds*, ed. S. C. Kendeigh & J. Pinowski, pp. 129–32. Warszawa: PWN-Polish Scientific Publishers.

Dawson, D. G. & Bull, P. C. (1970). A questionnaire survey of bird damage to fruit. *New Zealand Journal of Agricultural Research*, **13**, 681–8.

Dawson, W. R. (1954). Temperature regulation and water requirements of the brown and Abert towhees, *Pipilo fuscus* and *Pipilo aberti*. *University of California Publications in Zoology*, **59**, 81–124.

Dawson, W. R. (1958). Relation of oxygen consumption and evaporative water loss in the cardinal. *Physiological Zoology*, **31**, 37–48.

Dawson, W. R. & Bennett, A. F. (1973). Roles of metabolic level and temperature regulation in the adjustment of western plumed pigeons (*Lophophaps ferruginea*) to desert conditions. *Comparative Biochemistry and Physiology*, **44A**, 249–66.

Dawson, W. R. & Hudson, J. W. (1970). Birds. In *Comparative physiology of thermoregulation*, ed. G. C. Whittow, vol. 1, pp. 223–310. New York: Pergamon Press.

Dawson, W. R. & Tordoff, H. B. (1959). Relation of oxygen consumption to temperature in the evening grosbeak. *Condor*, **61**, 388–96.

Dawson, W. R. & Tordoff, H. B. (1964). Relation of oxygen consumption to temperature in the red- and white-winged crossbills. *Auk*, **81**, 26–35.

Deckert, G. (1962). Zur Ethologie des Feldsperlings (*Passer m. montanus* (L.)). *Journal für Ornithologie*, **103**, 428–86.

Deckert, G. (1969). Zur Ethologie und Okologie des Haussperlings (*Passer d. domesticus* (L.)). *Beiträge zur Vogelkunde*, **15**, 1–84.

Deevey, E. S. Jr (1947). Life tables for natural populations of animals. *Quarterly Review of Biology*, **22**, 283–314.

DeGrazio, J. W. & Besser, J. F. (1970). Bird damage problems in Latin America. In *Proceedings of fourth vertebrate pest conference*, ed. R. H. Dana, pp. 162–7. West Sacramento, California.

DeGrazio, J. W., Besser, J. F., DeCino, T. J., Guarino, J. L. & Starr, R. I. (1971). Use of 4-aminopyridine to protect ripening corn from blackbirds. *Journal of Wildlife Management*, **35**, 565–9.

DeHaven, R. W., Crase, F. T. & Woronecki, P. P. (1975). Movements of tricolored blackbirds banded in the Central Valley of California, 1965–1972. *Bird-Banding*, **46**, 220–9.

Diehl, B. (1971). Energy requirement in nestling and fledgling red-backed shrike (*Lanius collurio* L.). *Ekologia Polska*, **19**, 235–48.

References

Diehl, B. & Myrcha, A. (1973). Bioenergetics of nestling red-backed shrikes (*Lanius collurio*). *Condor*, **75**, 259–64.

Dolbeer, R. A. (1976). Reproductive rate and temporal spacing of nesting of red-winged blackbirds in upland habitat. *Auk*, **93**, 343–55.

Dolbeer, R. A., Ingram, C. R., Seubert, J. L., Stickley, A. R. Jr & Mitchell, R. T. (1976). 4-aminopyridine effectiveness in sweet corn related to blackbird population density. *Journal of Wildlife Management*, **40**, 564–70.

Dolbeer, R. A., Ingram, C. R. & Stickley, A. R. Jr (1973). A field test of methiocarb efficacy in reducing bird damage to Michigan blueberries. In *Proceedings of sixth bird control seminar*, ed. H. N. Cones, Jr & W. B. Jackson, pp. 28–40. Bowling Green, Ohio: Bowling Green State University.

Dol'nik, T. V. (1972). Énergeticheskiĭ balans pri vȳkarmlivanii ptentsov zyablika. [Energy balance during the rearing of chaffinch nestings.] *Ékologiya*, **4**, 43–7.

Dol'nik, V. R. (1965). Bioénergetika lin'ki vorob'inȳkh ptits kak adaptatsiya k migratsii. [Bioenergetics of moult in finches as adaptation for migration.] In *Novosti ornitologii*, pp. 124–6. Alma-Ata: Izdatel'stvo 'Nauka' Kazakhskoĭ SSR.

Dol'nik, V. R. (1967 *a*). Godovȳe tsiklȳ bioénergeticheskikh prisposobleniĭ k usloviyam sushchestvovaniya u 16 vidov Passeriformes. [Annual cycles of bioenergetic adaptations to environmental conditions in 16 passerine species.] In *Migratsii ptits Pribaltiki*, *Trudȳ Zoologicheskogo Instituta*, **40**, 115–63. Leningrad: Izdatel'stvo Nauka.

Dol'nik, V. R. (1967 *b*). Bioenergetische Anpassungen der Vögel und die Überwinterung in verschiedenen Breiten. *Der Falke*, **14**, 305–6, 347–9.

Dol'nik, V. R. (1968). Caloric value of the daily variation of the body weight in birds. *International Studies on Sparrows*, **2**, 89–95.

Dol'nik, V. R. (1969). Bioénergetika letyashcheĭ ptitsȳ. [Bioenergetics of flying birds.] *Zhurnal Obshcheĭ Biologii*, **30**, 273–91.

Dol'nik, V. R. (1971 *a*). Énergetika pereletov ptits. [Energetics of bird migration.] In *Itogi Nauki, Seria Biologii, Voprosȳ Ornitologii*, pp. 59–81. Moskva: Izdatel'stvo VINITI.

Dol'nik, V. R. (1971 *b*). Sravnenie énergeticheskikh raskhodov na migratsiyu i zimovku u ptits. [A comparison of values of energy expenditures for migration and wintering.] *Ékologiya*, **3**, 88–9.

Dol'nik, V. R. (1974 *a*). Énergeticheskiĭ metabolizm vorona *Corvus corax corax*. [Energy metabolism in *Corvus corax corax*.] *Ékologiya*, **5**, 56–62.

Dol'nik, V. R. (1974 *b*). The energy requirements for existence and for migration, moult and breeding in chaffinches, *Fringilla coelebs* L. *International Studies on Sparrows*, **7**, 11–20.

Dol'nik, V. R. & Blyumental', T. I. (1964). Bioénergetika migratsiĭ. [Bioenergetics of bird migration.] *Uspekhi Sovremennoĭ Biologii*, **58**, 280–301.

Dol'nik, V. R. & Blyumental', T. I. (1967). Autumnal premigratory and migratory periods in the chaffinch (*Fringilla coelebs coelebs*) and some other temperate-zone passerine birds. *Condor*, **69**, 435–68.

Dol'nik, V. R. & Gavrilov, V. M. (1971 *a*). Raskhod énergii na polet u nekotorȳkh vorob'inȳkh ptits. [The energy metabolism of flight in some passerine birds.] In *Ékologicheskie i fiziologicheskie aspektȳ pereletov ptits*, *Trudȳ Zoologicheskogo Instituta*, **50**, 236–41. Leningrad: Izdatel'stvo Nauka.

Dol'nik, V. R. & Gavrilov, V. M. (1971 *b*). Kaloricheskiĭ ékvivalent izmeneniya vesa tela u zyablika (*Fringilla coelebs*). [Calorific density of body weight variations in the chaffinch, *Fringilla coelebs*.] *In Ékologicheskie i fiziologicheskie aspektȳ pereletov ptits*, *Trudȳ Zoologicheskogo Instituta*, **50**, 226–35. Leningrad: Izdatel'stvo Nauka.

387

References

Dol'nik, V. R. & Gavrilov, V. M. (1975). A comparison of the seasonal and daily variations of bioenergetics, locomotor activities and major body composition in the sedentary house sparrow (*Passer d. domesticus* (L.)) and the migratory 'Hindian' sparrow (*P. d. bactrianus* Zar. et Kudasch). *Ekologia Polska*, **23**, 211–26.

Dol'nik, V. R., Keskpaïk, Yu. É. & Gavrilov, V. M. (1969). Bioénergetika osennego predmigratsionnogo perioda u zyablika. [Bioenergetics of autumnal premigratory period in the chaffinch.] In *Soobshcheniya Pribaltiïskoï Komissii po Izucheniyu Migratsiï Ptits, Tartu,* **6**, 125–47.

Domaniewski, J. (1933). Sprawozdanie z działalności Stacji Badania Wędrówek Ptaków za rok 1932. [Compte rendu de l'activité de la station pour l'étude des migrations des oiseaux pour l'année 1932.] *Acta Ornithologica*, **1**, 123–46.

Domaniewski, J. (1934). Sprawozdanie z działalności Stacji Badania Wędrówek Ptaków za rok 1933. [Compte rendu de l'activité de la station pour l'étude des migrations des oiseaux pour l'année 1933.] *Acta Ornithologica*, **1**, 321–64.

Domaniewski, J. & Kreczmer, B. (1936). Sprawozdanie z działalności Stacji Badania Wędrówek Ptaków za rok 1934. [Compte rendu de l'activité de la station pour l'étude des migrations des oiseaux pour l'année 1934.] *Acta Ornithologica*, **1**, 467–510.

Domaniewski, J. & Kreczmer, B. (1937). Sprawozdanie z działalności Stacji Badania Wędrówek Ptaków za rok 1935. [Compte rendu de l'activité de la station pour l'étude des migrations des oiseaux pour l'année 1935.] *Acta Ornithologica*, **2**, 87–131.

Dombrowski, R. (1912). *Ornis Romaniae. Die Vogelwelt Rumäniens. Systematisch und biologisch-geographisch beschrieben*, 924 pp. Bucureşti: Staatsdruckerei.

Dontcheff, L. & Kayser, C. (1934). Le rythme saisonnier du métabolisme de base chez le pigeon en fonction de la témperature moyenne du milieu. *Annales de Physiologie et de Physicochimie Biologique*, **10**, 285–300.

Dontcheff, L., Kayser, C. & Reiss, P. (1935). Le rythme nycthérméral de la production de chaleur chez le pigeon et des rapports avec l'excitabilité des centres thermorégulateurs. *Annales de Physiologie et de Physicochimie Biologique*, **11**, 1185–208.

Drent, R. H. (1967). *Functional aspects of incubation in the herring gull (Larus argentatus* Pont.), pp. 1–132. Leiden, Netherlands: University of Groningen.

Drent, R. H. (1970). Functional aspects of incubation in the herring gull (*Larus argentatus* Pont.). *Behaviour*, Supplement **17**, 1–132.

Drent, R. H. & Stonehouse, B. (1971). Thermoregulatory responses of the Peruvian penguin, *Spheniscus humboldti*. *Comparative Biochemistry and Physiology*, **40A**, 689–710.

Drozdov, N. N. (1965). Geografiya letnego naseleniya ptits v izbrannȳkh landshaftakh Azerbaïdzhana. [The geography of summer populations of bird species in selected regions of Azerbaijan.] *Ornitologiya*, **7**, 166–99.

Dubicka, H. (1957). Ptaki parku miejskiego w Toruniu. [Birds of the municipal park in Toruń.] *Ochrona Przyrody*, **24**, 382–95.

Dubinin, N. P. & Toropanova, T. A. (1956a). Ptitsȳ lesov dolinȳ r. Ural. Chast' II. [Forest birds of the river Ural Valley. Part II.] *Trudȳ Instituta Lesa*, **32**, 7–108.

Dubinin, N. P. & Toropanova, T. A. (1956b). Ptitsȳ lesov dolinȳ r. Ural. Chast' III. [Forest birds of the river Ural Valley. Part III.] *Trudȳ Instituta Lesa*, **32**, 109–306.

Dunn, E. H. (1975a). Caloric intake of nestling double-crested cormorants. *Auk*, **92**, 553–65.

Dunn, E. H. (1975 b). Growth, body components and energy content of nestling double-crested cormorants. *Condor*, **77**, 431–8.

Dunn, E. H. (1975 c). The timing of endothermy in the development of altricial birds. *Condor*, **77**, 288–93.

Dyer, M. I. (1964). Radar and morphometric studies on transient red-winged blackbird populations. Ph.D. Thesis, University of Minnesota, Minneapolis, 156 pp.

Dyer, M. I. (1967). An analysis of blackbird flock feeding behavior. *Canadian Journal of Zoology*, **45**, 765–72.

Dyer, M. I. (1968). *Blackbird and starling research program, 1964–1968*, 29 pp. Toronto, Canada: Ontario Department of Agriculture and Food.

Dyer, M. I. (1970). Territorial male red-winged blackbird distribution in Wood County, Ohio. In *Proceedings of fifth bird control seminar*, ed. D. L. Rintamaa & W. B. Jackson, pp. 185–94. Bowling Green, Ohio: Bowling Green State University.

Dyer, M. I. (1973). Plant–animal interactions: The effects of redwinged blackbirds on corn growth. In *Proceedings of sixth bird control seminar*, ed. H. N. Cones, Jr & W. B. Jackson, pp. 229–41. Bowling Green, Ohio: Bowling Green State University.

Dyer, M. I. (1975). The effects of red-winged blackbirds (*Agelaius phoeniceus* L.) on biomass production of corn grains (*Zea mays* L.). *Journal of Applied Ecology*, **12**, 719–26.

Dyer, M. I., Siniff, D. B., Curtis, S. G. & Webb, J. S. (1973). Distribution of red-winged blackbird (*Agelaius phoeniceus* L.) breeding populations in the Lake Erie region of the United States and Canada. In *Productivity, population dynamics and systematics of granivorous birds*, ed. S. C. Kendeigh & J. Pinowski, pp. 213–34. Warszawa: PWN-Polish Scientific Publishers.

Eberhardt, L. L. (1971). Population analysis. In *Wildlife management techniques*, ed. R. H. Giles, Jr, 3rd edn, pp. 457–95. Washington D.C.: The Wildlife Society.

Eberhardt, L. L. (1972). Some problems in estimating survival from banding data. In *Population ecology of migratory birds*. Papers from a symposium held at the Migratory Bird Populations Station, Laurel, Maryland, 9–10 October 1969. United States Department of the Interior, Fish and Wildlife Service, Bureau of Sport Fisheries and Wildlife. *Wildlife Research Report*, **2**, 153–71.

Ehrström, C. (1954). Vinterfågelbeståndet i Sigtuna. [Winter bird censuses in a garden city.] *Vår Fågelvärld*, **13**, 76–83.

Elder, W. H. (1964). Chemical inhibitors of ovulation in the pigeon. *Journal of Wildlife Management*, **28**, 556–75.

Eliseeva, V. I. (1960). Vzaimootnosheniya mezhdu polevým vorob'em i melkimi duplognezdnikami pri zaselenii iskusstvennýkh gnezdoviĭ. [Interrelations between tree sparrows and other small tree-hole nesting birds when occupying artificial nesting places.] *Trudý Tsentral'no-Chernozemnogo Gosudarstvennogo Zapovednika im. Prof. V. V. Alekhina*, **6**, 321–31.

Eliseeva, V. I. (1961). O razmnozhenii polevogo vorob'ya v iskusstvennýkh gnezdov'yakh. [Nesting of tree sparrows in nest-boxes.] *Zoologicheskiĭ Zhurnal*, **40**, 583–91.

Ellis, J. E., Wiens, J. A., Rodell, C. F. & Anway, J. C. (1976). A conceptual model of diet selection as an ecosystem process. *Journal of Theoretical Biology*, **60**, 93–108.

El-Wailly, A. J. (1966). Energy requirements for egg-laying and incubation in the zebra finch, *Taeniopygia castanotis*. *Condor*, **68**, 582–94.

References

Emlen, J. M. (1973). *Ecology: An evolutionary approach*, 493 pp. Reading, Massachusetts: Addison-Wesley Publishing Company.

Emlen, J. T. Jr (1952). Flocking behavior in birds. *Auk*, **69**, 160–70.

Emlen, J. T. & Wiens, J. A. (1965). The dickcissel invasion of 1964 in southern Wisconsin. *Passenger Pigeon*, **27**, 51–9.

Encke, F. W. (1965). Über Gelege-, Schlupf- und Ausflugsstärken des Haussperlings (*Passer d. domesticus*) in Abhängigkeit von Biotop und Brutperiode. *Beiträge zur Vogelkunde*, **10**, 268–87.

Erskine, A. J. (1971 a). Some new perspectives on the breeding ecology of common grackles. *Wilson Bulletin*, **83**, 352–70.

Erskine, A. J. (1971 b). A preliminary catalogue of bird census studies in Canada. *Canadian Wildlife Service, Progress Notes*, No. 20, 78 pp.

Erz, W. (1956). Der Vogelbestand eines Großstadtparkes im westfälischen Industriegebiet. *Ornithologische Mitteilungen*, **8**, 221–5.

Erz, W. (1959). Der Vogelbestand in Wohnviertel einer Großstadt im westfälischen Industriegebiet. *Ornithologische Mitteilungen*, **11**, 221–7.

Erz, W. (1964). Populationsökologische Untersuchungen an der Avifauna zweier nordwestdeutscher Großstädte. *Zeitschrift für Wissenschaftliche Zoologie*, **170**, 1–111.

Eyster, M. B. (1954). Quantitative measurement of the influence of photoperiod, temperature, and season on the activity of captive songbirds. *Ecological Monographs*, **24**, 1–28.

Fallet, M. (1958 a). Der Jahresrhythmus eines großstädtischen Bestandes des Haussperlings (*Passer domesticus* (L.)). *Schriften des Naturwissenschaftlichen Vereins für Schleswig-Holstein*, **29**, 39–46.

Fallet, M. (1958 b). Zum sozialverhalten des Haussperlings. *Zoologischer Anzeiger*, **161**, 178–87.

Fankhauser, D. P. (1967). Survival rates in red-winged blackbirds. *Bird-Banding*, **38**, 139–42.

Fankhauser, D. P. (1971 a). Annual adult survival rates of blackbirds and starlings. *Bird-Banding*, **42**, 36–42.

Fankhauser, D. P. (1971 b). Percentages of grackles taken in subsequent breeding seasons in a different breeding area from the area where banded. *Bird-Banding*, **42**, 43–5.

Farner, D. D. (1955). Bird banding in the study of population dynamics. In *Recent studies in avian biology*, ed. A. Wolfson, pp. 397–449. Urbana: University of Illinois Press.

Farner, D. S. (1975). Photoperiodic controls in the secretion of gonadotropins in birds. *American Zoologist*, **15** (Supplement 1), 117–35.

Feare, C. J., Dunnett, G. M. & Patterson, I. J. (1974). Ecological studies of the rook (*Corvus frugilegus* L.) in north-east Scotland: food intake and feeding behaviour. *Journal of Applied Ecology*, **11**, 867–96.

Ferianc, O., Feriancová-Masárová, Z. & Brtek, V. (1973). Vogelsynusien des Eichen-Hainbuchenwalds in Báb bei Nitra. *Acta Facultatis Rerum Naturalium Universitatis Comenianae, Zoologia*, **18**, 177–211.

Feriancová-Masárová, Z. & Brtek, V. (1969). Vtactvo južnej časti Malych Karpat. [Birds of the southern part of the Malé Karpaty.] *Biologické Prace*, **15**, 7–76.

ffrench, R. P. (1967). The dickcissel on its wintering grounds in Trinidad. *Living Bird*, **6**, 123–40.

Filonov, K. P. (1972). Chislennost' ptits v razlichnykh landshaftakh severnogo Priazov'ya. [Numbers of birds in various regions of northern Priazovie.] *Vestnik Zoologii*, **6** (**4**), 20–7.

References

Flannery, K. V. (1965). The ecology of early food production in Mesopotamia. *Science*, **147**, 1247–56.

Folk, C. & Novotný, I. (1970). Variation in body weight and wing length in the house sparrow, *Passer domesticus* (L.), in the course of a year. *Zoologické Listy*, **19**, 333–42.

Follett, B. K. (1973). The neuroendocrine regulation of gonadotropin secretion in avian reproduction. In *Breeding biology of birds*, ed. D. S. Farner, pp. 209–43. Washington D.C.: National Academy of Sciences.

Forbush, E. H. (1929). *Birds of Massachusetts and other New England States*, part III, 466 pp. Norwood, Massachusetts: Massachusetts Department of Agriculture.

Formozov, A. N. (1944). Zametki ob ékologii i sel'skokhozyaïstvennom znachenii vorob'ev. (*Passer domesticus bactrianus* Zar. et Kudasch i *Passer montanus pallidus* Zar.) v Yuzhnoï Turkmenii. [Note on the ecology of the sparrows, *Passer domesticus bactrianus* Zar. et Kudasch and *Passer montanus pallidus* Zar., and on their role in the agriculture of southern Turkmenistan.] *Zoologicheskiï Zhurnal*, **23**, 342–50.

Foster, M. S. (1975). Temporal patterns of resource allocation and life history phenomena. *Florida Scientist*, **38**, 129–39.

Fretwell, S. D. (1972 a). *Populations in a seasonal environment. Monographs in Population Biology*, No. 5, xxiii+217 pp. Princeton, New Jersey: Princeton University Press.

Fretwell, S. D. (1972 b). The regulation of bird populations on Konza Prairie. The effects of events of the prairie. In *Third midwest prairie conference proceedings*, pp. 71–6. Manhattan: Kansas State University.

Fretwell, S. D., Francis, M. & Shane, T. (1974). Dickcissels nesting in Mesquite, Texas. *Bulletin of the Texas Ornithological Society*, **7**, 5–6.

Fry, C. H., Ash, J. S. & Ferguson-Less, I. J. (1970). Spring weights of some Palaearctic migrants at Lake Chad. *Ibis*, **112**, 58–82.

Gadgil, M. & Solbrig, O. T. (1972). The concept of r and K selection: evidence from wild flowers and some theoretical considerations. *American Naturalist*, **106**, 461–71.

Gage, S. H., Miller, C. A. & Mook, L. J. (1970). The feeding response of some forest birds to the black-headed budworm. *Canadian Journal of Zoology*, **48**, 359–66.

Gaginskaya, E. R. (1967). O migratsiyakh ptits na yuzhnom poberezh'e Finskogo Zaliva. [Migration of birds along the southern coasts of the Bay of Finland.] In *Itogi ornitologicheskikh issledovanii v Pribaltike* (*Trudȳ V Pribaltiïskoï Ornitologicheskoï Konferentsii, Tartu, 5–10 iyulya 1963 goda*), ed. É. V. Kumari, pp. 191–8. Tallin: Izdatel'stvo 'VALGUS'.

Ganya, I. M. (1965). Kolichestvennaya kharakteristika ornitofaunȳ sadov v Pridnestrov'e Moldavii. [Bird populations in the orchards of Moldavia along the banks of the Dniester River.] *Ornitologiya*, **7**, 290–308.

Ganya, I. M. (1969). Ptitsȳ skalistȳkh beregov Dnestra v predelakh Moldavii. [Birds of the rocky banks of the Dniester River in Moldavia.] *Voprosȳ ékologii i prakticheskogo znacheniya ptits i mlekopitayushchikh Moldavii*, **3**, 3–17.

Gashwiler, J. S. (1970). Further study of conifer seed survival in a western Oregon clearcut. *Ecology*, **51**, 849–54.

Gatehouse, S. N. & Markham, B. J. (1970). Respiratory metabolism of three species of raptors. *Auk*, **87**, 738–41.

Gavrilov, É. I. (1962). Biologiya ispanskogo vorob'ya (*Passer hispaniolensis*

References

(Temm.)) i merȳ bor'bȳ s nim v Kazakhstane. [The biology of the Spanish sparrow, *Passer hispaniolensis* (Temm.), and how its harmful effect is counteracted in Kazakhstan.] *Trudȳ Nauchno-Issledovatel'skogo Instituta Zashchitȳ Rastenii̇, 7*, 459–528.

Gavrilov, É. I. (1963). The biology of the eastern Spanish sparrow in Kazakhstan. *Journal of the Bombay Natural History Society*, **60**, 301–17.

Gavrilov, É. I. (1965). La lutte contre les moineaux au Kazakhstan. In *Congrès de la Protection des Cultures Tropicales*, Compte rendu des travaux, Chambre de Commerce et d'Industrie de Marseille 23–27 mars 1965, pp. 581–5.

Gavrilov, É. I. (1974). Semeǐstvo Tkachikovȳe – Ploceidae (Family Ploceidae). In *Ptitsȳ Kazakhstana*, ed. A. F. Kovshar', vol. 5, pp. 363–406. Alma-Ata: Izdatel'stvo 'Nauka' Kazakhskoǐ SSR.

Gavrilov, É. I. & Korelov, M. N. (1968). O vidovoǐ samostoyatel'nosti indiǐskogo vorob'ya. [The Indian sparrow as a distinct good species.] *Byulleten' Moskovskogo Obshchestva Ispȳtateleǐ Prirody, Otdel Biologicheskii̇*, **73**, 115–22.

Gavrilov, V. M. (1972). Zavisimost' metabolizma ot povedeniya u melkikh vorob'inȳkh ptits. [Effects of behavior on the energy metabolism in passerine birds.] In *Pervoe Vsesoyuznoe Coveshchanie po Ékologicheskim i Évolyutsionnȳm Aspektam Povedeniya Zhivotnȳkh*, ed. B. P. Manteǐfel', pp. 241–3. Moskva: Izdatel'stvo 'Nauka'.

Gavrilov, V. M. (1974). Sezonnȳe i sutochnȳe izmeneniya urovnya standartnogo metabolizma u vorob'inȳkh ptits. [Seasonal and daily variations of standard metabolic rate in passerine birds.] In *Materialȳ VI Vsesoyuznoǐ Ornitologicheskoǐ Konferentsii, Moskva – 1–5 fevralya 1974 goda*, ed. R. L. Beme & V. E. Flint, chast' I, pp. 134–6. Moskva: Izdatel'stvo Moskovskogo Universiteta.

Gavrilov, V. M. & Dol'nik, V. R. (1974). Bioénergetika i regulyatsiya poslebrachnoǐ i postyuvenal'noǐ linek u zyablikov (*Fringilla coelebs coelebs* L.). [Bioenergetics and regulations of the postnuptial and postjuvenal moult in chaffinches (*Fringilla coelebs coelebs* L.).] *Issledovaniya po biologii ptits, Trudȳ Zoologicheskogo Instituta, Leningrad*, **55**, 14–61.

Gavrilov, V. M., Dol'nik, V. R. & Keskpaǐk, Yu. É. (1970). Énergeticheskii̇ metabolizm zyablika v zimnii̇ period. [The winter energy metabolism in the chaffinch.] *Eesti NSV Teaduste Akadeemia Toimetised, Biologia*, **19**, 211–18.

Geiler, H. (1959). Geschlechterverhältnis, Körpergewicht und Flügellänge der Individuen einer mitteldeutschen Sperlingspopulation. *Beiträge zur Vogelkunde*, **6**, 359–66.

Gersdorf, E. (1951). Zur Sperlingsbekämpfung. *Zeitschrift für Pflanzenkrankheiten, Pflanzenpathologie und Pflanzenschutz*, **58**, 188–90.

Gersdorf, E. (1955). Die Verbreitung des Haussperlings im Gebiet der Landwirtschaftskammer Hannover. *Beiträge zur Naturkunde Niedersachsens*, **8**, 12–18.

Gerstell, R. (1942). The place of winter feeding in practical wildlife management. *Pennsylvania Game Commission Research Bulletin*, **3**, 1–121.

Gessaman, J. A. (1972). Bioenergetics of the snowy owl (*Nyctea scandiaca*). *Arctic and Alpine Research*, **4**, 223–38.

Gessaman, J. A. (1973). Ecological energetics of homeotherms. *Utah State University Press, Logan, Monograph Series*, **20**, 1–155.

Gibb, J. (1957). Food requirements and other observations of captive tits. *Bird Study*, **4**, 207–15.

Gibb, J. A. (1962). L. Tinbergen's hypothesis of the role of specific search images. *Ibis*, **104**, 106–11.

Głowaciński, Z. (1975). Ptaki Puszczy Niepołomickiej. [Birds of the Niepołomice Forest.] *Acta Zoologica Cracoviensia*, **20**, 1–87.

Goddard, S. V. & Board, V. V. (1967). Reproductive success of red-winged blackbirds in north central Oklahoma. *Wilson Bulletin*, **79**, 283–9.

Godfrey, W. E. (1966). *The birds of Canada. Bulletin of the National Museum of Canada*, No. 203, *Biological Series*, No. 73, 428 pp.

Goldstein, R. B. (1974). Relation of metabolism to ambient temperature in the verdin. *Condor*, **76**, 116–19.

Golley, F. B. (1961). Energy values of ecological materials. *Ecology*, **42**, 581–4.

Golovanova, É. N. (1966). O vrede i pol'ze vorob'ev. [Economic significance of sparrows.] *Zashchita Rastenii*, **7**, 42–3.

Golovanova, É. N. & Zusmanovich, T. G. (1961). Otravlennye zernovye primanki protiv vorob'ev. [Poisoned grain baits against sparrows.] *Zashchita Rastenii ot Vrediteleï i Bolezneï*, **6**, 34.

Goodhue, L. D. & Baumgartner, F. M. (1965). Applications of new bird control chemicals. *Journal of Wildlife Management*, **29**, 830–7.

Gorelova, R. I. (1974). K kolichestvennoï kharakteristike avifauny doliny r. Kushki. [Bird populations in the Kushka River Valley.] In *Materialy VI Vsesoyuznoï Ornitologicheskoï Konferentsii, Moskva – 1–5 fevralya 1974 goda*, ed. R. L. Beme & V. E. Flint, chast' I, pp. 269–70. Moskva: Izdatel'stvo Moskovskogo Universiteta.

Gould, S. J. & Johnston, R. F. (1972). Geographic variation. *Annual Review of Ecology and Systematics*, **3**, 457–98.

Graber, R. R. & Graber, J. W. (1963). A comparative study of bird populations in Illinois, 1906–1909 and 1956–1958. *Illinois Natural History Survey Bulletin*, **28**, 383–528.

Graczyk, R. (1961). Obserwacje nad zbiorowymi miejscami noclegowymi wróbla, *Passer domesticus* (L.). [Observations on collective night's lodgings of sparrow, *Passer domesticus* (L.).] *Przegląd Zoologiczny*, **5**, 241–5.

Graczyk, R., Chartanowicz, W., Fruziński, B. & Małek, J. (1966). Wpływ praktycznej ochrony na zasiedlenie ptaków w drzewostanach leśnych. [The influence of practical protection on the inhabitation of birds in the forest-stands.] *Poznańskie Towarzystwo Przyjaciół Nauk, Wydział Nauk Rolniczych i Leśnych, Prace Komisji Nauk Rolniczych i Komisji Nauk Leśnych*, **20**, 45–78.

Graczyk, R., Galiński, T. & Klejnotowski, Z. (1967). Ptaki gnieżdżące się w skrzynkach lęgowych na terenie sadu doświadczalnego w Przybrodzie (woj. poznańskie) w latach 1965 i 1966. [Birds nesting in nesting-boxes on the terrain of Przybrodzie experimental orchard (Poznań Voivodship) in the years 1965 and 1966.] *Roczniki Wyższej Szkoły Rolniczej w Poznaniu, Wydział Zootechniczny, Ornitologia Stosowana*, **38**, 49–59.

Granett, P., Trout, J. R., Messersmith, D. H. & Stockdale, T. M. (1974). Sampling corn for bird damage. *Journal of Wildlife Management*, **38**, 903–9.

Grant, P. (1972). Centripetal selection and the house sparrow. *Systematic Zoology*, **21**, 23–30.

Grempe, G. (1971). Der Brutvogelbestand eines Villenviertels in Rostock. *Ornithologischer Rundbrief Mecklenburgs, Neue Folge*, **12**, 49–54.

Grimm, H. (1954). Biometrische Bemerkungen über mitteldeutsche und westdeutsche Sperlingspopulationen. *Journal für Ornithologie*, **95**, 306–18.

Grinnell, J. (1919). The English sparrow has arrived in Death Valley: an experiment in nature. *American Naturalist*, **53**, 468–73.

Grodziński, W. & Sawicka-Kapusta, K. (1970). Energy values of tree-seeds eaten by small mammals. *Oikos*, **21**, 52–8.

References

Groebbels, F. & Möbert, F. (1927). Oologische Studien. II. Mitteilungen über die künstliche Bestimmung der Brutdauer einiger Vogelarten mit besonderer Berücksichtigung des Eigewichts. *Verhandlungen der Ornithologischen Gesellschaft in Bayern*, **17**, 198–204.

Gromadzki, M. (1966). Variability of egg-size of some species of the forest birds. *Ekologia Polska*, Ser. A, **14**, 99–109.

Gromadzki, M. (1970). Breeding communities of birds in mid-field afforested areas. *Ekologia Polska*, **18**, 307–50.

Grün, G. (1964 *a*). Untersuchungen zur Ökologie und wirtschaftlichen Bedeutung des Feldsperlings, *Passer montanus* (L.), unter besonderer Berücksichtigung seiner Ernährungsweise. Ph.D. Thesis, University of Greifswald, Greifswald.

Grün, G. (1964 *b*). Schäden durch Feldsperlinge (*Passer montanus* (L.)) an Kulturursaaten. *Aufsätze zu Vogelschutz und Vogelkunde*, **1**, 42–7.

Grün, G. (1968). Comments on the programme and methods of international studies on sparrows. *International Studies on Sparrows*, **2**, 11–15.

Grün, G. (1975). Die Ernährung der Sperlinge *Passer domesticus* (L.) und *Passer montanus* (L.) unter verschiedenen Umweltbedingungen. *International Studies on Sparrows*, **8**, 24–103.

Guarino, J. L. & Forbes, J. E. (1970). Preventing bird damage to sprouting corn with a carbamate repellent. *New York Fish Game Journal*, **17**, 117–20.

Haartman, L. von, Hildén, O., Linkola, P., Suomalainen, P. & Tenovuo, R. (eds.) (1963–1972). *Pohjolan linnut värikuvin. [Birds of the North in colours.]* Helsinki: Otava.

Haftorn, S. (1972). Hypothermia of tits in the arctic winter. *Ornis Scandinavica*, **3**, 153–66.

Hagar, D. C. (1960). The interrelationships of logging, birds, and timber regeneration in the douglas-fir region of northwestern California. *Ecology*, **41**, 116–25.

Hainsworth, F. R. & Wolf, L. L. (1970). Regulation of oxygen consumption and body temperature during torpor in a hummingbird, *Eulampis jugularis*. *Science*, **168**, 368–9.

Hainsworth, F. R. & Wolf, L. L. (1972). Power for hovering flight in relation to body size in hummingbirds. *American Naturalist*, **106**, 589–96.

Hairston, N. G., Tinkle, D. W. & Wilber, H. M. (1970). Natural selection and the parameters of population growth. *Journal of Wildlife Management*, **34**, 681–90.

Haldane, J. B. S. (1955). The calculation of mortality rates from ringing data. In *Acta XI Congressus Internationalis Ornithologici, Basel 29.v.–5.vi.1954, Experientia, Supplementum III*, ed. A. Portmann & E. Sutter, pp. 454–8. Basel, Stuttgart: Birkhäuser Verlag.

Hall, B. P. & Moreau, R. E. (eds.) (1970). *An atlas of speciation in African passerine birds*, xv pp. London: Trustees of the British Museum (Natural History).

Hamilton, S. (1974). Variation in body and bill sizes in North American and European house sparrows. M.A. Thesis, University of Kansas, Lawrence.

Hamilton, S. & Johnston, R. F. (1977). Evolution in the house sparrow. VI. Variability and niche width. *Auk*, **95**. (In Press.)

Hamilton, W. D. (1971). Geometry for the selfish herd. *Journal of Theoretical Biology*, **31**, 295–311.

Hamilton, W. J. III & Gilbert, W. M. (1969). Starling dispersal from a winter roost. *Ecology*, **50**, 886–98.

References

Hamilton, W. J. III, Gilbert, W. M., Heppner, F. H. & Planck, R. J. (1967). Starling roost dispersal and a hypothetical mechanism regulating rhythmical animal movement to and from dispersal centers. *Ecology*, **48**, 825–33.

Hamilton, W. J. III & Watt, K. E. F. (1970). Refuging. *Annual Review of Ecology and Systematics*, **1**, 263–86.

Hammer, M. (1948). Investigations on the feeding habits of the house sparrow (*Passer domesticus*) and the tree sparrow (*Passer montanus*). *Danish Review of Game Biology*, **1**, 1–59.

Hardin, G. (1968). The tragedy of the commons. *Science*, **162**, 1243–8.

Harlan, J. R. (1971). Agricultural origins: centers and noncenters. *Science*, **174**, 468–74.

Harlan, J. R. & Zohary, D. (1966). Distribution of wild wheats and barley. *Science*, **153**, 1074–80.

Harper, J. L. & White, J. (1974). The demography of plants. *Annual Review of Ecology and Systematics*, **5**, 419–63.

Harris, L. E. (1960). Glossary of energy terms. *National Academy of Sciences, National Research Council, Publication*, **1040**, pp. 1–10, Washington D.C.

Hart, J. S. & Roy, O. Z. (1967). Temperature regulation during flight in pigeons. *American Journal of Physiology*, **213**, 1311–16.

Hartert, E. [J. O.] (1903–1910). *Die Vögel der paläarktischen Fauna. Systematische Übersicht der in Europa, Nord Asien und der Mittelmeerregion vorkommenden Vögel*, vol. 1, xlix+832 pp. Berlin: R. Friedländer.

Hartung, R. (1967). Energy metabolism in oil-covered ducks. *Journal of Wildlife Management*, **31**, 798–804.

Hasse, H. (1962). Ein frisches Ei vom Feldsperling (*Passer montanus*) im November. *Ornithologische Mitteilungen*, **14**, 214.

Havlín, J. (1974). Vom Haussperling (*Passer domesticus*) und Feldsperling (*P. montanus*) an reifenden Getreidepflanzen verursachte Schäden. *Zoologické Listy*, **23**, 241–59.

Havlín, J. & Lelck, A. (1957). Příspěvek k poznáni příčin kvalitativního i kvantitativniho rozšiřeni ptactva v Jeseníkách. [*The contribution to recognition of the causes of qualitative and quantitative distribution of birds in Jeseniki.*] 54 pp. Opava: Slezský Studijni Ústav.

Hayne, D. W. (1946). The relationship between number of ears opened and the amount of grain taken by redwings in cornfields. *Journal of Agricultural Research*, **72**, 289–95.

Headley, J. C. (1972). Defining the economic threshold. In *Pest control strategies for the future*, ed. R. Metcalf, pp. 100–8. Washington D.C.: National Academy of Sciences.

Heinroth, O. (1922). Beziehungen zwischen Vogelgewicht, Eigewicht, Gelegegewicht und Brutdauer. *Journal für Ornithologie*, **70**, 172–285.

Hémery, G. & Le Toquin, A. (1975 a). Déterminisme énergétique des concentrations de pinsons du nord (*Fringilla montifringilla*) en relation avec l'évolution de la culture du maïs (*Zea mais*) en France de 1955 à 1973. *Comptes Rendus Hebdomadaires des Séances de l'Académie des Sciences*, Ser. D, **281**, 835–8.

Hémery, G. & Le Toquin, A. (1975 b). Utilisation de la méthode des modèles pour l'étude des dépenses énergétiques des populations de pinsons du nord (*Fringilla montifringilla*) en période internuptiale. *Comptes Rendus Hebdomadaires des Séances de l'Académie des Sciences*, Ser. D, **280**, 1153–6.

Henny, C. J., Overton, W. S. & Wight, H. M. (1970). Determining parameters for populations by using structural models. *Journal of Wildlife Management*, **34**, 690–703.

References

Henny, C. J. & Wight, H. M. (1969). An endangered osprey population: estimates of mortality and production. *Auk*, **86**, 188–98.

Henny, C. J. & Wight, H. M. (1972). Population ecology and environmental pollution: red-tailed and Cooper's hawks. In *Population ecology of migratory birds*. Papers from a symposium held at the Migratory Bird Populations Station, Laurel, Maryland, 9–10 October 1969. United Stated Department of the Interior, Fish and Wildlife Service, Bureau of Sport Fisheries and Wildlife. *Wildlife Research Report*, **2**, 229–49.

Heppner, F. H. (1970). The metabolic significance of differential absorption of radiant energy by black and white birds. *Condor*, **72**, 50–9.

Herreid, C. F. H. & Kessel, B. (1967). Thermal conductance in birds and mammals. *Comparative Biochemistry and Physiology*, **21**, 405–14.

Hespenheide, H. A. (1973). Ecological inferences from morphological data. *Annual Review of Ecology and Systematics*, **4**, 213–29.

Hesse, W. & Lustick, S. (1972). A comparison of the water requirements of marsh and upland redwing blackbirds (*Agelaius phoeniceus*). *Physiological Zoology*, **45**, 196–203.

Hickey, J. J. (1952). *Survival studies of banded birds. Special Scientific Report: Wildlife*, No. **15**, 177 pp. United States Department of the Interior, Fish and Wildlife Service, Bureau of Sport Fisheries and Wildlife, Washington D.C.

Hill, E. F. (1972). Avoidance of lethal dietary concentrations of insecticide by house sparrows. *Journal of Wildlife Management*, **36**, 635–9.

Hinde, R. A. (1961). Behaviour. In *Biology and comparative physiology of birds*, ed. A. J. Marshall, vol. 2, pp. 373–411. New York: Academic Press.

Hinds, D. S. & Calder, W. A. (1973). Temperature regulation of the pyrrhuloxia and the Arizona cardinal. *Physiological Zoology*, **46**, 55–71.

Hintz, J. V. & Dyer, M. I. (1970). Daily rhythm and seasonal change in the summer diet of adult red-winged blackbirds. *Journal of Wildlife Management*, **34**, 789–99.

Hirshfield, M. F. & Tinkle, D. W. (1975). Natural selection and the evolution of reproductive effort. *Proceedings of the National Academy of Sciences of the United States of America*, **72**, 2227–31.

Hissa, R. & Palokangas, R. (1970). Thermoregulation in the tit-mouse (*Parus major* L.). *Comparative Biochemistry and Physiology*, **33**, 941–53.

Holcomb, L. & Twiest, G. (1968). Red-winged blackbird nestling growth compared to adult size and differential development of structures. *Ohio Journal of Science*, **68**, 277–84.

Hole, F. (1966). Investigating the origins of Mesopotamian civilization. *Science*, **153**, 605–11.

Holling, C. S. (1965). The functional response of predators to prey density and its role in mimicry and population regulation. *Memoirs of the Entomological Society of Canada*, **45**, 1–60.

Holm, C. H. (1973). Breeding sex ratios, territoriality, and reproductive success in the red-winged blackbird (*Agelaius phoeniceus*). *Ecology*, **54**, 356–65.

Holmes, R. T. & Sawyer, R. H. (1975). Oxygen consumption in relation to ambient temperature in five species of forest-dwelling thrushes (*Hylocichla* and *Catharus*). *Comparative Biochemistry and Physiology*, **50A**, 527–31.

Holmes, R. T. & Sturges, F. W. (1973). Annual energy expenditure by the avifauna of a northern hardwoods ecosystem. *Oikos*, **24**, 24–9.

Holmes, R. T. & Sturges, F. W. (1975). Bird community dynamics and energetics in a northern hardwoods ecosystem. *Journal of Animal Ecology*, **44**, 175–200.

396

References

Homes, R. C. (ed.) (1957). *The birds of the London area since 1900*, 305 pp. London: Collins.

Hoogerwerf, A. (1949). Bijdrage tot de oölogie van Java. [Contribution to the oology of Java.] *Limosa*, **22**, 1–279.

Horn, H. S. (1968). The adaptive significance of colonial nesting in the Brewer's blackbird (*Euphagus cyanocephalus*). *Ecology*, **49**, 682–94.

Hudson, J. W. & Brush, A. H. (1964). A comparative study of the cardiac and metabolic performance of the dove, *Zenaidura macroura*, and the quail, *Laphortxy californicus*. *Comparative Biochemistry and Physiology*, **12**, 157–70.

Hudson, J. W. & Kimzey, S. L. (1964). Body temperature and metabolism cycles in the house sparrow, *Passer domesticus*, compared with the white-throated sparrow, *Zonotrichia albicollis*. *American Zoologist*, **4**, 294–5.

Hudson, J. W. & Kimzey, S. L. (1966). Temperature regulation and metabolic rhythms in populations of the house sparrow, *Passer domesticus*. *Comparative Biochemistry and Physiology*, **17**, 203–17.

Huggins, R. A. (1941). Egg temperatures of wild birds under natural conditions. *Ecology*, **22**, 148–57.

Hughes, B. O. (1972). A circadian rhythm of calcium intake in the domestic fowl. *Poultry Science*, **51**, 485–93.

Hurley, R. J. & Franks, E. C. (1976). Changes in the breeding ranges of two grassland birds. *Auk*, **93**, 108–15.

Hyla, W. (1967). Quantitative Bestandsaufnahme und Vorkommen der Vögel in der Stadtlandschaft Oberhausen. *Charadrius*, **3**, 84–119.

Hyla, W. (1970). Der Vogelbestand einer um die Jahrhundertwende errichteten Werksiedlung. *Ornithologische Mitteilungen*, **22**, 80–1.

Il'enko, A. I. (1958). Faktorȳ, opredelyayushchie nachalo razmnozheniya v populyatsii domovȳkh vorob'ev (*Passer domesticus* (L.)) g. Moskvȳ. [Factors determining the reproduction onset in the population of house sparrows (*Passer domesticus* (L.)) in Moscow.] *Zoologicheskii Zhurnal*, **37**, 1867–73.

Il'enko, A. I. (1962). K izucheniyu sezonnȳkh izmenenii vesa melkikh ptits. [Seasonal variation of body weight in small birds.] *Ornitologiya*, **4**, 427–30.

Immelmann, K. (1973). Role of the environment in reproduction as source of 'predictive' information. In *Breeding biology of birds*, ed. D. S. Farner, pp. 121–47. Washington D.C.: National Academy of Sciences.

Innis, G. S., Wiens, J. A., Chuculate, C. A. & Miskimins, R. (1974). BIRD model description and documentation. *US IBP Grassland Biome Technical Report*, No. **246**, 133 pp.

Ion, I. (1971). Studiu asupra compoziţiei hranei consumată de puii vrăbiei de casă – *Passer domesticus* (L.) şi vrăbiei de cîmp – *Passer montanus* (L.). [Étude sur la composition des aliments consommés par les petits du moineau domestique (*Passer domesticus* (L.)) et du friquet (*Passer montanus* (L.)).] *Muzeul de Ştiinţele Naturii Bacău, Studii şi Comunicări*, **1971**, 263–76.

Ion, I. (1973). Studiu biostatistic şi ecologic asupra populatiilor celor doua specii de vrabii (*Passer d. domesticus* (L.) şi *Passer m. montanus* (L.)) din Moldova. [Biostatistic and ecological researches on the populations of two sparrow species (*Passer d. domesticus* (L.) and *Passer m. montanus* (L.)) in Moldavia.] Ph.D. Thesis, Iaşi University, Iaşi, 45 pp.

Ion, I. & Saracu, S. (1971). Dezboltarea postembrionara a puilor de *Passer montanus* (L.) şi *Passer domesticus* (L.). [Le développment postembryonnaire des poussins du *Passer montanus* (L.) et *Passer domesticus* (L.).] *Muzeul Judetean Suceava, Studii şi Communicări, Ştiinţele Naturii*, **2/1**, 271–8.

References

Ion, I. & Valenciuc, N. (1967). Caracteristicile densității păsărilor din livezile cu pomi de la Bacium-Iași. [Les caractéristiques de la densité des oiseaux des vergers de Bucium-Jassy.] *Analele Științifice ale Universității 'Al I. Cuza' Din Iași (Serie Nouă), Sectiunea II, a. Biologie,* 13, 247–53.

Ion, I. & Valenciuc, N. (1969). Date asupra înmulțirii vrăbiei de cîmp (*Passer montanus* (L.)). [Données sur la reproduction du friquet (*Passer montanus* (L.)).] *Analele Științifice ale Universității 'Al. I. Cuza' Din Iași (Serie Nouă), Sectiunea II, a. Biologie,* 15, 335–42.

Izmaïlov, I. V. & Borovitskaya, G. K. (1967). Kharakteristika naseleniya ptits listvennichnoï taïgi Vitimskogo Ploskogor'ya. [Bird populations in the larch taiga of the Vitimskiï Plateau.] *Ornitologiya,* 8, 192–7.

Jakubiec, Z. (1972). Ptaki rezerwatu Muszkowicki Las Bukowy. [Birds of the reserve Muszkowicki Las Bukowy.] *Ochrona Przyrody,* 37, 135–52.

Janzen, D. H. (1971). Seed predation by animals. *Annual Review of Ecology and Systematics,* 2, 465–92.

Joensen, A. H. (1965). En undersøgelse af fuglebestanden i fire løvskovsområder på Als i 1962 og 1963. [An investigation on bird populations in four deciduous forest areas on Als in 1962 and 1963.] *Dansk Ornithologisk Forenings Tidsskrift,* 59, 115–86.

Johnson, R. E. (1968). Temperature regulation in the white-tailed ptarmigan, *Lagopus leucurus. Comparative Biochemistry and Physiology,* 24, 1003–4.

Johnson, W. E. & Selander, R. K. (1971). Protein variation and systematics in kangaroo rats (genus *Dipodomys*). *Systematic Zoology,* 20, 377–405.

Johnston, R. F. (1964). The breeding birds of Kansas. *University of Kansas Publications of the Museum of Natural History,* 12, 575–655.

Johnston, R. F. (1969). Taxonomy of house sparrows and their allies in the Mediterranean basin. *Condor,* 71, 129–39.

Johnston, R. F. (1973 a). Evolution in the house sparrow. IV. Replicate studies in phenetic variation. *Systematic Zoology,* 22, 219–26.

Johnston, R. F. (1973 b). Intralocality character dimorphism in house sparrows. *Ornithological Monographs,* 14, 24–31.

Johnston, R. F. (1973 c). Color variation and natural selection in Italian sparrows. *Bolletino di Zoologia,* 39, 351–62.

Johnston, R. F. (1976). Evolution in the house sparrow. V. Covariation of skull and hindlimb sizes. *Occasional Papers of the Museum of Natural History, Kansas University,* 56, 1–8.

Johnston, R. F., Niles, D. M. & Rohwer, S. A. (1972). Hermon Bumpus and natural selection in the house sparrow *Passer domesticus. Evolution,* 26, 20–31.

Johnston, R. F. & Selander, R. K. (1964). House sparrow: rapid evolution of races in North America. *Science,* 144, 548–50.

Johnston, R. F. & Selander, R. K. (1971). Evolution in the house sparrow. II. Adaptive differentiation in North American populations. *Evolution,* 25, 1–28.

Johnston, R. F. & Selander, R. K. (1973 a). Evolution in the house sparrow. III. Variation in size and sexual dimorphism in Europe and North and South America. *American Naturalist,* 107, 373–90.

Johnston, R. F. & Selander, R. K. (1973 b). Variation, adaptation, and evolution in the North American house sparrows. In *Productivity, population dynamics and systematics of granivorous birds,* ed. S. C. Kendeigh & J. Pinowski, pp. 301–26. Warszawa: PWN-Polish Scientific Publishers.

Jolly, G. M. (1965). Explicit estimates from capture–recapture data with both death and immigration–stochastic model. *Biometrika,* 52, 225–47.

398

Jones, H. (1969). The common grackle – a nesting study. *Kentucky Warbler*, **45**, 3–8.

Jones, J. M. (1976). The r–K-selection continuum. *American Naturalist*, **110**, 320–3.

Jones, P. J. & Ward, P. (1976). The level of reserve protein as the proximate factor controlling the timing of breeding and clutch-size in the red-billed quelea *Quelea quelea*. *Ibis*, **118**, 547–74.

Jordania, R. (1970). Über die Vernichtung von Haussperlingen durch Bilche. *Der Falke*, **17**, 103.

Judd, D. B. (1950). *Colorimetry*, iii+56 pp. Washington D.C.: US National Bureau of Standards, Circular 478.

Kaatz, C. & Olberg, S. (1975). Investigations on the breeding biology of *Passer montanus* (L.). *International Studies on Sparrows*, **8**, 107–16.

Kahl, M. P. Jr (1962). Bioenergetics of growth in nestling wood storks. *Condor*, **64**, 169–83.

Kale, H. W. II (1965). *Ecology and bioenergetics of the long-billed marsh wren* (Telmatodytes palustris griseus (*Brewster*) *in Georgia salt marshes. Publications of the Nuttall Ornithological Club*, **5**, xiii+142 pp.

Kalela, O. (1949). Über Feldlemming Invasionen und andere irreguläre Tierwanderungen. *Annales Zoologici Societatis Zoologico-Botanicae, Vanamo*, **3**, 1–90.

Kalmbach, E. R. (1940). *Economic status of the English sparrow in the United States, US Department of Agriculture, Technical Bulletin*, No. 711, 66 pp.

Kare, M. R. (1965). The special senses. In *Avian physiology*, ed. P. D. Sturkie, 2nd edn, pp. 428–42. Ithaca, New York: Comstock Publishing Press.

Karr, J. R. & James, F. C. (1975). Eco-morphological configurations and convergent evolution in species and communities. In *Ecology and evolution of communities*, ed. M. L. Cody & J. M. Diamond, pp. 389–427. Cambridge, Massachusetts: Harvard University Press.

Kashkarov, D. Yu., Foss, L. P., Rusinova, K. I., Sataeva, Z. L. & Zaruba, E. A. (1926). Nablyudeniya nad biologieĭ vorob'ya i nad prinosimym im vredom v Turkestane. [Observations on the biology of the sparrows in Turkestan and on their role in the deterioration of the crop.] *Byulleten' Sredne-Aziatskogo Gosudarstvennogo Universiteta*, **13**, 61–81.

Kashkarov, D. Yu. & Puzankova, R. N. (1974). Ptitsȳ – Aves – tkachikovȳe. [Birds – Aves – Ploceidae.] In *Pozvonochnȳe zhivotnȳe Ferganskoĭ Dolinȳ*, ed. G. S. Sultanov, pp. 93–103. Tashkent: Izdatel'stvo 'FAN' Uzbekskoĭ SSR.

Kashkin, V. V. (1961). Heat exchange of bird eggs during incubation. *Biophysics*, **6**, 57–63.

Katz, P. L. (1974). A long-term approach to foraging optimization. *American Naturalist*, **108**, 758–82.

Kayser, C. (1940). Les échanges respiratoires des hibernants. *Theses, University de Strasbourg*, Sér. E, **61**, 1–364.

Kear, J. (1962). Food selection in finches with special reference to interspecific differences. *Proceedings of the Zoological Society of London*, **138**, 163–204.

Keil, W. (1957). Wendehals (*Jynx torquilla*) tötet nestjunge Feldsperlinge. *Ornithologische Mitteilungen*, **9**, 217.

Keil, W. (1970). Untersuchungen zur Ernährung von Haus- und Feldsperling – *Passer domesticus* und *P. montanus* – in einem Getreidebaugebiet im Winterhalbjahr. *Luscinia*, **41**, 76–87.

Keil, W. (1973). Investigations on food of house- and tree sparrows in a cereal-

References

growing area during winter. In *Productivity, population dynamics and systematics of granivorous birds*, ed. S. C. Kendeigh & J. Pinowski, pp. 253–62. Warszawa: PWN-Polish Scientific Publishers.

Keleïnikov, A. A. (1953). Ékologiya domovogo i polevogo vorob'ev, kak massovȳkh vrediteleï zernovȳkh kul'tur v yuzhnȳkh raïonakh SSSR. [The ecology of the house and tree sparrows as species harmful to agriculture in the southern regions of the Soviet Union.] Avtoreferat dissertatsii na soiskanie uchenoï stepeni kandidata biologicheskikh nauk, pp. 1–9. Moskva: Moskovskiï Ordena Lenina Gosudarstvennȳï Universitet Imeni M. V. Lomonosova.

Kendeigh, S. C. (1939). The relation of metabolism to the development of temperature regulation in birds. *Journal of Experimental Zoology*, **82**, 419–38.

Kendeigh, S. C. (1940). Factors affecting length of incubation. *Auk*, **52**, 499–513.

Kendeigh, S. C. (1941). Length of day and energy requirements for gonad development and egg-laying in birds. *Ecology*, **22**, 237–48.

Kendeigh, S. C. (1944). Effect of air temperature on the rate of energy metabolism in the English sparrow. *Journal of Experimental Zoology*, **96**, 1–16.

Kendeigh, S. C. (1949). Effect of temperature and season on energy resources of the English sparrow. *Auk*, **66**, 113–27.

Kendeigh, S. C. (1952). *Parental care and its evolution in birds. Illinois Biological Monographs*, **22**, x+356 pp.

Kendeigh, S. C. (1963). Thermodynamics of incubation in the house wren, *Troglodytes aedon*. In *The Proceedings of the 13th international ornithological congress*, vol. II, pp. 884–904. Baton Rouge, Louisiana: The American Ornithologists' Union.

Kendeigh, S. C. (1969 a). Energy responses of birds to their thermal environments. *Wilson Bulletin*, **81**, 441–9.

Kendeigh, S. C. (1969 b). Tolerance of cold and Bergmann's rule. *Auk*, **86**, 13–25.

Kendeigh, S. C. (1970). Energy requirements for existence in relation to size of bird. *Condor*, **72**, 60–5.

Kendeigh, S. C. (1972). Energy control of size limits in birds. *American Naturalist*, **106**, 79–88.

Kendeigh, S. C. (1973 a). The natural history of incubation. Discussion. In *Breeding biology of birds*, ed. D. S. Farner, pp. 311–20. Washington D.C.: National Academy of Sciences.

Kendeigh, S. C. (1973 b). Monthly variations in the energy budget of the house sparrow throughout the year. In *Productivity, population dynamics and systematics of granivorous birds*, ed. S. C. Kendeigh & J. Pinowski, pp. 17–44. Warszawa: PWN-Polish Scientific Publishers.

Kendeigh, S. C. (1973 c). A symposium of the house sparrow (*Passer domesticus*) and European tree sparrow (*P. montanus*) in North America: introduction. *Ornithological Monographs*, **14**, 1–2.

Kendeigh, S. C. (1973 d). Opening address. In *Productivity, population dynamics and systematics of granivorous birds*, ed. S. C. Kendeigh & J. Pinowski, pp. 11–13. Warszawa: PWN-Polish Scientific Publishers.

Kendeigh, S. C. (1974). Seasonal allocation of time and energy resources in birds (J. R. King). In *Avian energetics*, ed. R. A. Paynter, Jr, pp. 70–9. *Nuttall Ornithological Club Publication*, No. **15**. Cambridge, Massachusetts.

Kendeigh, S. C. & Blem, C. R. (1974). Metabolic adaptation to local climate in birds. *Comparative Biochemistry and Physiology*, **48A**, 175–87.

Kendeigh, S. C., Kontogiannis, J. E., Masac, A. & Roth, R. R. (1969). Environmental regulation of food intake by birds. *Comparative Biochemistry and Physiology*, **31**, 941–57.

400

References

Kendeigh, S. C., Kramer, T. C. & Hamerstrom, F. (1956). Variations in egg characteristics of the house wren. *Auk*, **73**, 42–65.
Kendeigh, S. C. & Wallin, H. E. (1966). Seasonal and taxonomic differences in the size and activity of the thyroid gland in birds. *Ohio Journal of Science*, **64**, 369–79.
Kendeigh, S. C. & West, G. C. (1965). Caloric values of plant seeds eaten by birds. *Ecology*, **46**, 553–5.
Keskpaïk, Yu. (1968). Teploproduktsiya i puti teplootdachi pri polete u lastochek. [Heat production and heat loss in swallows and martins during flight.] *Eesti NSV Teaduste Akadeemia Toimetised, Biologia*, **17**, 179–91.
Kespaïk, Yu. (1972). Obratimaya gipotermiya u beregovȳkh lastochek *Riparia r. riparia* L. v prirode. [Temporary hypothermy in sand-martins *Riparia r. riparia* L. in natural conditions.] In *Soobshcheniya Pribaltiĭskoĭ Komissii po Izucheniyu Migratsii Ptits*, ed. É, Kumari, No. 7, pp. 176–83. Tartu: Akademiya Nauk Éstonskoĭ SSR.
Keskpaïk, Yu. & Khorma, P. (1972). Temperatura tela i ÉKG u morskoĭ chaĭki (*Larus m. marinus* L.) v polete. [Body temperature and heart rate of flying sea-gulls (*Larus m. marinus* L.).] *Eesti NSV Teaduste Akadeemia Toimetised, Biologia*, **21**, 109–16.
King, J. R. (1964). Oxygen consumption and body temperature in relation to ambient temperature in the white-crowned sparrow. *Comparative Biochemistry and Physiology*, **12**, 13–24.
King, J. R. (1973). Energetics of reproduction in birds. In *Breeding biology of birds*, ed. D. S. Farner, pp. 78–107. Washington D.C.: National Academy of Sciences.
King, J. R. (1974). Seasonal allocation of time and energy resources in birds. In *Avian energetics*, ed. R. A. Paynter, Jr, pp. 4–85. *Nuttall Ornithological Club Publication*, No. **15**, Cambridge, Massachusetts.
King, J. R. & Farner, D. S. (1959). Premigratory changes in body weight and fat in wild and captive white-crowned sparrows. *Condor*, **61**, 315–24.
King, J. R. & Farner, D. S. (1965). Studies of fat deposition in migratory birds. *Annals of the New York Academy of Sciences*, **131**, 422–40.
Kintzel, W. (1972). Der Vogelbestand eines isolierten Feldgehölzes. *Ornithologischer Rundbrief Mecklenburgs, Neue Folge*, **13**, 42–3.
Kleiber, M. (1972). Body size, conductance for animal heat flow and Newton's law of cooling. *Journal of Theoretical Biology*, **37**, 139–50.
Klitz, W. (1972). Genetic consequences of colonization with subsequent expansion: the house sparrow in North America. Ph.D. Thesis, University of Kansas, Lawrence, 37 pp.
Klomp, H. (1970). The determination of clutch-size in birds. A review. *Ardea*, **58**, 1–124.
Klopfer, P. H. (1961). Observational learning in birds: the establishment of behavioural modes. *Behaviour*, **17**, 71–80.
Kluge, A. G. & Kerfoot, W. C. (1973). The predictability and regularity of character divergence. *American Naturalist*, **107**, 426–42.
Kontogiannis, J. E. (1968). Effect of temperature and exercise on energy intake and body weight of the white-throated sparrow, *Zonotrichia albicollis*. *Physiological Zoology*, **41**, 54–64.
Koplin, J. R. (1972). Measuring predator impact of woodpeckers on spruce beetles. *Journal of Wildlife Management*, **36**, 308–20.
Korelus, J. (1947). Study of bird's plumage with special consideration of number and weight of their feathers. *Acta Societatis Zoologicae Čechoslovenicae*, **11**, 218–34.

401

References

Korodi Gál, I. (1957). Studii ornithocenologice în cîteva tipuri de păduri foioase din Transilvania. [Études ornithocaenologiques dans quelques types de forêts feuillées de Transylvanie.] *Studii şi Cercetări de Biologie (Cluj)*, 7, 319–29.

Korodi Gál, I. (1958). Untersuchungen über die Vogelpopulation eines Obstgartens. *Ornithologische Mitteilungen*, 10, 66–9.

Korodi Gál, I. (1960). Compoziţia cantitativă şi calitativă a populaţiilor de păsări din grădina botanică din Cluj între anii 1958–1959. [Composition quantitative et qualitative des populations d'oiseaux du jardin botanique de Cluj en 1958–1959.] *Studia Universitatis Babes-Bolyai*, Ser. II, *Biologia*, 2, 153–70.

Korodi Gál, I. (1964). Vogelzönologische Forschungsergebnisse aus einigen Tieflands Eichen- und Mischwäldern Siebenbürgens. *Vertebrata Hungarica*, 6, 41–71.

Korol'kova, G. E. (1963). *Vliyanie ptits na chislennost' vrednȳkh nasekomȳkh.* [*The effect of birds on the numbers of harmful insects.*] 126 pp. Moskva: Izdatel'stvo Akademii Nauk SSSR.

Korompai, V. (1966). The number of the breeding bird-pairs on the inner territory of the town Gyula in spring 1962. *Aquila*, 71–72, 192–3.

Koskimies, J. (1961). Fakultative Kältelethargie beim Mauersegler (*Apus apus*) im Spätherbst. *Die Vogelwarte*, 21, 161–6.

Kovacs, B. (1955). Untersuchungresultat des Kropfinhaltes der Feld- und Haussperlinge sowie deren Wirtschaftliche Bedeutung auf dem Gebiete der Lehrwirtschaft der Akademie in Debrecen. *Különlenyomat a Debreceni Mezögazdaśagi Akadémie Évkönyvéböl*, 63–93.

Koval', N. F. & Samarskiï, S. L. (1972). Osobennosti razmnozheniya vorob'ya polevogo (*Passer montanus* (L.)) v fruktovȳkh nasazhdeniyakh srednego Pridneprov'ya. [Peculiarities in the breeding of the tree sparrow (*Passer montanus* (L.)) in fruit orchards along the middle reaches of the Dnieper River.] *Vestnik Zoologii*, 6, 62–6.

Krebs, C. J. (1972). Ecology: the experimental analysis of distribution and abundance, 694 pp. New York: Harper & Row.

Krogh, A. (1916). The respiratory exchange of animals and man. In *Monographs in biochemistry*, pp. 1–173. London: Longmans Green and Co.

Krüger, C. (1944). En Undersøgelse af Graaspurvens (*Passer d. domesticus*) og Skovspurvens (*Passer m. montanus*) Træk. [An investigation on the migration of the house sparrow (*Passer d. domesticus*) and the tree sparrow (*Passer m. montanus*).] *Dansk Ornithologisk Forenings Tidsskrift*, 38, 105–14.

Kulczycki, A. (1966). Ptaki parku w Łańcucie. [The birds of the Łańcut Park.] *Acta Zoologica Cracoviensia*, 11, 351–86.

Kuroda, N. h. (1966a). Analysis of banding data (1924–43) of the tree sparrow in Japan. *Miscellaneous Reports of the Yamashina Institute for Ornithology*, 4, 129–34.

Kuroda, N. h. (1966b). A bird census in the Imperial and Akasaka Palaces for 1965. *Miscellaneous Reports of the Yamashina Institute for Ornithology*, 4, 269–79.

Kuroda, N. h. (1967). A bird census in the Imperial and Akasaka Palaces for 1966. *Miscellaneous Reports of the Yamashina Institute for Ornithology*, 5, 1–12.

Kuroda, N. h. (1968). A bird census in the Imperial Palace for 1967. *Miscellaneous Reports of the Yamashina Institute for Ornithology*, 5, 202–13.

Kuroda, N. h. (1969). A bird census in the Imperial Palace for 1968. *Miscellaneous Reports of the Yamashina Institute for Ornithology*, 5, 462–72.

Kuroda, N. h. (1970). A bird census in the Imperial Palace for 1969. *Miscellaneous Reports of the Yamashina Institute for Ornithology*, 6, 16–31.

References

Kuroda, N. h. (1971). A bird census in the Imperial Palace for 1970. *Miscellaneous Reports of the Yamashina Institute for Ornithology*, **6**, 217–30.
Kuroda, N. h. (1973 *a*). Analysis of a bird community in southern Kanto Plain. *Miscellaneous Reports of the Yamashina Institute for Ornithology*, **7**, 118–38.
Kuroda, N. h. (1973 *b*). A bird census in the Imperial Palace for 1972. *Miscellaneous Reports of the Yamashina Institute for Ornithology*, **7**, 34–55.
Lack, D. (1940). Variation in the introduced English sparrow. *Condor*, **42**, 239–41.
Lack, D. (1951). Population ecology in birds. A review. In *Proceedings of the Xth international ornithological congress*, ed. S. Hörstadius, pp. 409–48. Uppsala: Almqvist & Wiksell.
Lack, D. (1954). *The natural regulation of animal numbers*, 343 pp. Oxford: Clarendon Press.
Lack, D. (1966). *Population studies of birds*, 341 pp. Oxford: Clarendon Press.
Lack, D. (1968). *Ecological adaptations for breeding in birds*, xii+409 pp. London: Methuen & Co. Ltd.
Lajeunesse, E. J. (1960). *The Windsor Border region*, 47 pp. Toronto: The Champlain Society.
Lambert, H. (1936). Wiederfunde der durch die Zweigberingungsstelle 'Untermain' der Vogelwarte Helgoland gekennzeichneten Feldsperlinge. *Vogelring*, **8**, 65–7.
Lasiewski, R. C. (1963). Oxygen consumption of torpid, resting, active, and flying hummingbirds. *Physiological Zoology*, **36**, 122–40.
Lasiewski, R. C. & Dawson, W. R. (1964). Physiological responses to temperature in the common nighthawk. *Condor*, **66**, 477–90.
Lasiewski, R. C. & Dawson, W. R. (1967). A re-examination of the relation between standard metabolic rate and body weight in birds. *Condor*, **69**, 13–23.
Lasiewski, R. C., Dawson, W. R. & Bartholomew, G. A. (1970). Temperature regulation in the little Papuan frogmouth, *Podargus ocellatus*. *Condor*, **72**, 332–8.
Lasiewski, R. C., Hubbard, S. H. & Moberly, W. R. (1964). Energetic relationships of a very small passerine bird. *Condor*, **66**, 212–20.
Lasiewski, R. C. & Lasiewski, R. J. (1967). Physiological responses of the blue-throated and Rivoli's hummingbirds. *Auk*, **84**, 34–48.
Lasiewski, R. C. & Seymour, R. S. (1974). Thermoregulatory responses to heat stress in four species of birds weighing approximately 40 grams. *Physiological Zoology*, **45**, 106–18.
Lasiewski, R. C., Weathers, W. W. & Bernstein, M. H. (1967). Physiological responses of the giant hummingbird, *Patagona gigas*. *Comparative Biochemistry and Physiology*, **23**, 797–813.
Laux, L. J. (1970). Non-breeding surplus and population structure of the red-winged blackbird (*Agelaius phoeniceus*). Ph.D. Thesis, University of Michigan, Ann Arbor, 85 pp.
Lawrence, J. M. & Schreiber, R. W. (1974). Organic material and calories in the egg of the brown pelican, *Pelecanus occidentalis*. *Comparative Biochemistry and Physiology*, **47A**, 435–40.
Lebreton, J.-D. (1973). Introduction aux modèlés mathématiques de la dynamique des populations. In *Informatique et biosphère*, pp. 76–116. Paris.
LeFebvre, P. W. & Seubert, J. L. (1970). Surfactants and blackbird stressing agents. In *Proceedings of fourth vertebrate pest conference*, ed. R. H. Dana, pp. 156–61. West Sacramento, California.
Leopold, A. (1939). A biotic view of land. *Journal of Forestry*, **37**, 727–30.

403

References

Lewies, R. W. & Dyer, M. I. (1969). Respiratory metabolism of the red-winged blackbird in relation to ambient temperature. *Condor*, 71, 291-8.

Lewontin, R. C. & Hubby, J. L. (1966). A molecular approach to the study of genic heterozygosity in natural populations. II. Amount of variation and degree of heterozygosity in natural populations of *Drosophila pseudoobscura*. *Genetics*, 54, 595–609.

Lifson, N. & McClintock, R. (1966). Theory of use of the turnover rates of body water for measuring energy and material balance. *Journal of Theoretical Biology*, 12, 46–74.

Ligon, J. D. (1968). The biology of the elf owl, *Micrathene whitneyi*. *Miscellaneous Publications, Museum of Zoology, University of Michigan*, 136, 1–70.

Löhrl, H. & Böhringer, R. (1957). Untersuchungen an einer südwestdeutschen Population des Haussperlings (*Passer d. domesticus*). *Journal für Ornithologie*, 98, 229–40.

Long, C. A. & Long, C. F. (1968). Comments on reproduction of the common grackle in central Illinois. *Wilson Bulletin*, 80, 493–4.

Lugovoï, A. E. & Talposh, V. S. (1968). Ptitsÿ urochishcha Chernÿï Mochar posle ego melioratsii. [The birds of Chernÿï Mochar (Black Peat-bog) after it has been reclaimed.] *Ornitologiya*, 9, 238–42.

Lund, H. M.-K. (1956). Graspurven (*Passer domesticus* (L.)) i Nord-Norge. [The house sparrow (*Passer domesticus* (L.)) in North Norway.] *Dansk Ornithologisk Forenings Tidsskrift*, 50, 67–76.

Luniak, M. (1974). Ptaki biotopów parkowych w małych miastach środkowo-wschodniej Polski. [The birds of park biotopes in small towns of central-eastern Poland.] *Acta Ornithologica*, 14, 99–143.

Lustick, S. (1969). Bird energetics: effects of artificial radiation. *Science*, 163, 387–90.

Lustick, S. (1970). Energy requirements of molt in cowbirds. *Auk*, 87, 742–6.

Lustick, S. (1971). Plumage color and energetics. *Condor*, 73, 121–2.

Lyuleeva, D. S. (1970). Énergiya poleta u lastochek i strizheï. [Energy of flight in swallows and swifts.] *Dokladÿ Akademii Nauk SSSR*, 190, 1467–9.

MacArthur, R. H. & Wilson, E. O. (1967). *The theory of island biogeography*, 203 pp. Princeton, New Jersey: Princeton University Press.

Mackowicz, R., Pinowski, J. & Wieloch, M. (1970). Biomass production by house sparrow (*Passer d. domesticus* (L.)) and tree sparrow (*Passer m. montanus* (L.)) populations in Poland. *Ekologia Polska*, 18, 465–501.

Mackrodt, P. (1967–1968). Zur Sterblichkeit beim Haussperling (*Passer domesticus*). *Bericht des Offenbacher Vereins für Naturkunde*, 75, 12–13.

Maclean, G. L. (1973 a). The sociable weaver. Part 1: description, distribution, dispersion and populations. *Ostrich*, 44, 176–90.

Maclean, G. L. (1973 b). The sociable weaver. Part 2: nest architecture and social organization. *Ostrich*, 44, 191–218.

Maclean, G. L. (1973 c). The sociable weaver. Part 3: breeding biology and moult. *Ostrich*, 44, 219–40.

Maclean, G. L. (1973 d). The sociable weaver. Part 4: predators, parasites and symbionts. *Ostrich*, 44, 241–53.

Maclean, G. L. (1973 e). The sociable weaver. Part 5: food, feeding and general behaviour. *Ostrich*, 44, 254–61.

MacMillen, R. E. (1974). Bioenergetics of Hawaiian honeycreepers: the amakihi (*Loxops virens*) and the anianiau (*L. parva*). *Condor*, 76, 62–9.

Magor, J. I. & Ward, P. (1972). Illustrated descriptions, distribution maps, and bibliography of the species of *Quelea* (weaver birds: Ploceidae). *Tropical Pest Bulletin*, **1**, 1–23.

Maïkhruk, M. I. (1974). K biologii poleznogo vorob'ya. [On the biology of a useful sparrow.] In *Materialȳ VI Vsesoyuznoï Ornitologicheskoï Konferentsii, Moskva – 1–5 fevralya 1974 goda*, ed. R. L. Beme & V. E. Flint, chast' II, pp. 78–9. Moskva: Izdatel'stvo Moskovskogo Universiteta.

Malchevskiĭ, A. S. (1955). Ornitofauna parka Lesotekhnicheskoï Akademii im. S. M. Kirova (g. Leningrad) i ee izmeneniya s 1888 po 1950 god. [Avifauna of the park at the Forestry College named after S. M. Kirov (Leningrad) and its changes from 1888 to 1950.] *Uchenye Zapiski Leningradskogo Ordena Lenina Gosudarstvennogo Universiteta Imenii A. A. Zhdanova, Seriya Biologicheskikh Nauk*, **181** (38), 53–60.

Mansfeld, K. (1950). Beiträge zur Erforschung der wissenschaftlichen Grundlagen der Sperlingsbekämpfung. *Nachrichtenblatt des Deutschen Pflanzenschutzdienstes*, **2**, 1–9.

Marder, J. (1973). Body temperature regulation in the brown-necked raven (*Corvus corax ruficollis*). II. Thermal changes in the plumage of ravens exposed to solar radiation. *Comparative Biochemistry and Physiology*, **45A**, 431–40.

Markert, C. L. & Moller, F. (1959). Multiple forms of enzymes: tissue, ontogenetic, and species specific patterns. *Proceedings of the National Academy of Sciences of the United States of America*, **45**, 753–63.

Markus, M. B. (1964). Premaxillae of the fossil *Passer predomesticus* Tchernov and the extant South African Passerinae. *Ostrich*, **35**, 245–6.

Martin, A. C., Zim, H. S. & Nelson, A. L. (1961). *American wildlife and plants: a guide to wildlife food habits*, 500 pp. New York: Dover Publications, Incorporation.

Martin, S. G. (1971). Polygyny in the bobolink: habitat quality and the adaptive complex. Ph.D. Thesis, Oregon State University, Corvallis, 181 pp.

Martin, S. G. (1974). Adaptations for polygynous breeding in the bobolink, *Dolichonyx oryzivorus*. *American Zoologist*, **14**, 108–19.

Matoušek, B. (1956). Prispevok k oológii slovenskej avifauny. [Beitrag zur Oologie der slovakischen Avifaune.] *Biologické Práce*, **2**, 1–89.

Mattson, W. J. & Addy, N. D. (1975). Phytophagous insects as regulators of forest primary production. *Science*, **190**, 515–22.

Matyushkin, E. N. (1967). O naselenii ptits lesostepi Kazakhstana. [Bird populations in the forest–steppe regions of Kazakhstan.] *Ornitologiya*, **8**, 198–210.

Maxwell, G. R. II (1970). Pair formation, nest building and egg laying of the common grackle in Northern Ohio. *Ohio Journal of Science*, **70**, 284–91.

Maxwell, G. R. II & Putnam, L. S. (1972). Incubation, care of young, and nest success of the common grackle (*Quiscalus quiscula*) in northern Ohio. *Auk*, **89**, 349–59.

Mayfield, H. F. (1975). The numbers of Kirtland's warblers. *Jack-Pine Warbler*, **53**, 38–47.

McCullough, D. R. (1970). Secondary production of birds and mammals. In *Analysis of temperate forest ecosystems*, ed. D. Reichle, pp. 107–30. Berlin, Heidelberg & New York: Springer-Verlag.

McFarland, D. J. & Baher, E. (1968). Factors affecting feather posture in the Barbary dove. *Animal Behaviour*, **16**, 171–7.

McGeen, D. S. (1972). Cowbird–host relationships. *Auk*, **89**, 360–80.

McGeen, D. S. & McGeen, J. J. (1968). The cowbirds of Otter Lake. *Wilson Bulletin*, **80**, 84–93.

References

Meanley, B. I. (1971). *Blackbirds and the southern rice crop. Wildlife Resource Publication*, **100**, 64 pp. United States Department of the Interior, Fish and Wildlife Service, Bureau of Sport Fisheries and Wildlife, Washington D.C.

Meanley, B. I. & Webb, J. S. (1963). Nesting ecology and reproductive rate of the red-winged blackbird in tidal marshes of the Upper Chespeake Bay region. *Chesapeake Science*, **4**, 90–100.

Merikallio, E. (1958). Finnish birds, their distribution and numbers. *Societas pro Fauna et Flora Fennica, Fauna Fennica*, **5**, 1–181.

Mertens, J. A. L. (1969). Influence of brood size on the energy metabolism and water loss of nestling great tits *Parus major major. Ibis*, **111**, 11–16.

Mertens, J. A. L. (1972). A model for the prediction of heat loss of great tit broods. In *Institute for Ecological Research, progress report, Royal Netherlands Academy of Arts and Sciences*, ed. J. W. Woldendorp, pp. 89–90. Arnhem.

Michels, H. (1973). Der Vogelbestand des Landschaftschutzparks Salzuflen. *Ornithologische Mitteilungen*, **25**, 260–2.

Michocki, J. (1974). Dziesięć lat badań wpływu praktycznej ochrony ptaków na skład gatunkowy i liczbowy dziuplaków w parku wiejskim w Siemianicach (1962–1971). [Ten-year examination of the influence of practical bird protection upon number and species composition of hollow birds in the park in Siemianice (1962–1971).] *Roczniki Akademii Rolniczej w Poznaniu, Ornitologia Stosowana*, **70**(7), 101–15.

Mikhel'son, G. A. (1958). Obzor obshchikh rezul'tatov raboty̆ po privlecheniyu melkikh lesny̆kh ptits-duplognezdnikov v Latviĭskoĭ SSR. [A review of general results of activities in attracting small hollow-nesting birds in forests in Latvian SSR.] In *Privlechenie polezny̆kh ptits-duplognezdnikov v lesakh Latviĭskoĭ SSR*, ed. Z. D. Spuris, pp. 5–72. Riga: Izdatel'stvo Akademii Nauk Latviĭskoĭ SSR.

Miller, D. S. (1939). A study of the physiology of the sparrow thyroid. *Journal of Experimental Zoology*, **80**, 259–85.

Mirza, Z. B. (1973). Study on the fecundity, mortality, numbers, biomass and food of a population of house sparrows in Lahore, Pakistan. In *Productivity, population dynamics and systematics of granivorous birds*, ed. S. C. Kendeigh & J. Pinowski, pp. 141–50. Warszawa: PWN-Polish Scientific Publishers.

Misch, M. S. (1960). Heat regulation in the northern blue jay, *Cyanocitta cristata bromia* Oberholser. *Physiological Zoology*, **33**, 252–9.

Mitchell, C. J. & Hayes, R. O. (1973). Breeding house sparrows, *Passer domesticus* in captivity. *Ornithological Monographs*, **14**, 39–48.

Mitchell, C. J., Hayes, R. O., Holden, P. & Hughes, T. B. Jr (1973). Nesting activity of the house sparrow in Hale County, Texas, during 1968. *Ornithological Monographs*, **14**, 49–59.

Moberg, C. (1966). Spread of agriculture in the north European periphery. *Science*, **152**, 315–19.

Montell, J. (1917). Fågelfaunan i Muonio socken och angränsande delar af Enontekis och Kittilä socknar. [Die Vogelfauna des Kirchspiels Muonio und der angrenzenden Teile der Kirchspiele Enontekis und Kittilä (Finnisch–Lappland).] *Acta Societatis pro Fauna et Flora Fennica*, **44**, 1–260.

Moore, A. D. (1945). Winter night habits of birds. *Wilson Bulletin*, **57**, 253–60.

Moore, D. J. (1961). Studies in the energy balance of the versicolor pheasant. M.S. Thesis, Ohio University, Athens, Ohio.

Moreau, R. E. (1931). An Egyptian sparrow roost. *Ibis*, **13**, 204–8.

References

Moreau, R. E. (1950). The breeding seasons of African birds. I. Land birds. *Ibis*, **92**, 223–67.

Morel, G. (1968). L'impact écologique de *Quelea quelea* (L.) sur les savanes sahéliennes; raisons du pullulement de ce plocéide. *La Terre et la Vie*, **22**, 69–98.

Morris, R. F., Cheshire, W. F., Miller, C. A. & Mott, D. G. (1958). The numerical response of avian and mammalian predators during a gradation of the spruce budworm. *Ecology*, **39**, 487–94.

Morrison, D. F. (1967). *Multivariate statistical methods*, 338 pp. New York: McGraw-Hill.

Morse, D. H. (1971). The insectivorous bird as an adaptive strategy. *Annual Review of Ecology and Systematics*, **2**, 177–200.

Morse, D. H. (1975). Ecological aspects of adaptive radiation in birds. *Biological Review*, **50**, 167–214.

Morton, M. L. (1967). The effects of isolation on the diurnal feeding patterns of white-crowned sparrows (*Zonotrichia leucophrys gambelii*). *Ecology*, **48**, 690–4.

Moss, R. (1972). Food selection by red grouse (*Lagopus scoticus* (Lath.)) in relation to chemical composition. *Journal of Animal Ecology*, **41**, 411–28.

Moss, R. (1973). The digestion and intake of winter foods by wild ptarmigan in Alaska. *Condor*, **75**, 293–300.

Mott, D. F. & Stone, C. P. (1973). Predation on corn earworms by red-winged blackbirds. *Murrelet*, **54**, 8–10.

Mugas, J. N. & Templeton, J. B. (1970). Thermoregulation in the red-breasted nuthatch. *Condor*, **72**, 125–32.

Mulsow, R. (1967). Untersuchungen zur Siedlungsdichte der Hamburger Vogelwelt. *Abhandlungen und Verhandlungen des Naturwissenschaftlichen Vereins in Hamburg, Neue Folge*, **12**, 124–88.

Munteanu, D. (1963). Cercetări asupra populațiilor de păsări din bazinul Bistriței. I. Livezi din zona montană. [Recherches sur les populations d'oiseaux des vergers de la vallée de la Bistritza.] *Analele Științifice ale Universității ' Al. I. Cuza' din Iași (Serie Nouă), Sectiunea II (Științe Naturale), a. Biologie*, **9**, 257–72.

Murdoch, W. W. (1969). Switching in general predators: experiments of predator specificity and stability of prey populations. *Ecological Monographs*, **39**, 335–54.

Murrish, D. E. (1970). Responses to temperature in the dipper, *Cinclus mexicanus. Comparative Biochemistry and Physiology*, **34**, 859–69.

Murton, R. K. (1968). Some predator–prey relationships in bird damage and population control. In *The problems of birds as pests*, ed. R. K. Murton & E. N. Wright, pp. 157–80. London & New York: Academic Press.

Murton, R. K. (1972). *Man and birds*, xx+364 pp. New York: Taplinger Publishing Company.

Murton, R. K., Isaacson, A. J. & Westwood, N. J. (1963). The feeding ecology of the wood-pigeon. *British Birds*, **56**, 345–75.

Murton, R. K. & Jones, B. E. (1973). The ecology and economics of damage to Brassicae by wood-pigeons *Columba palumbus. Annals of Applied Biology*, **75**, 107–22.

Murton, R. K., Thearle, R. J. P. & Thompson, J. (1972). Ecological studies of the feral pigeon *Columba livia* var. I. Population, breeding biology and methods of control. *Journal of Applied Ecology*, **9**, 835–74.

Murton, R. K., Westwood, N. J. & Isaacson, A. J. (1974). A study of wood-pigeon

References

shooting: the exploitation of a natural animal population. *Journal of Applied Ecology*, **11**, 61–81.

Mustafaev, G. T. (1969). K ékologii girkanskogo domovogo vorob'ya. [The ecology of the Girkan house sparrow.] In *Ornitologiya v SSSR*, ed. A. K. Rustamov, vol. 2, pp. 433–5. Ashkhabad.

Myrcha, A. & Pinowski, J. (1969). Variations in the body composition and caloric value of the nestling tree sparrows (*Passer m. montanus* (L.)). *Bulletin de l'Academie Polonaise des Sciences, Cl. II, Série des Sciences Biologiques*, **17**, 475–80.

Myrcha, A. & Pinowski, J. (1970). Weights, body composition and caloric value of postjuvenal molting European tree sparrows (*Passer m. montanus* (L.)). *Condor*, **72**, 175–81.

Myrcha, A., Pinowski, J. & Tomek, T. (1973). Energy balance of nestlings of tree sparrows, *Passer m. montanus* (L.), and house sparrows, *Passer d. domesticus* (L.). In *Productivity, population dynamics and systematics of granivorous birds*, ed. S. C. Kendeigh & J. Pinowski, pp. 59–83. Warszawa: PWN-Polish Scientific Publishers.

Naik, R. M. (1974). Recent studies on the granivorous birds in India. *International Studies on Sparrows*, **7**, 21–5.

Naik, R. M. & Mistry, L. (1973). Breeding season and reproductive rate of *Passer domesticus* (L.) in Baroda, India. In *Productivity, population dynamics and systematics of granivorous birds*, ed. S. C. Kendeigh & J. Pinowski, pp. 133–40. Warszawa: PWN-Polish Scientific Publishers.

Nekrasov, E. S. (1970). Razmnozhenie i izmenenie biomassȳ polevogo vorob'ya. [Variations in the breeding and biomass of the tree sparrow.] In *Optimal'naya plotnost' i optimal'naya struktura populyatsiĭ zhivotnȳkh*, ed. L. N. Dobrinskiĭ, vol. 2, pp. 62–4. Sverdlovsk: Akademiya Nauk SSSR, Ural'skiĭ Filial.

Nero, R. W. (1956a). A behavior study of the red-winged blackbird. I. Mating and nesting activities. *Wilson Bulletin*, **68**, 5–37.

Nero, R. W. (1956b). A behavior study of the red-winged blackbird. II. Territoriality. *Wilson Bulletin*, **68**, 129–50.

Newton, I. (1967). The adaptive radiation and feeding ecology of some British finches. *Ibis*, **109**, 33–98.

Newton, I. (1968). The temperatures, weights, and body composition of molting bullfinches. *Condor*, **70**, 323–32.

Newton, I. (1972). *Finches*, 288 pp. London: Collins.

Nice, M. M. (1937). Studies in the life history of the song sparrow. *Transactions of the Linnaean Society of New York*, **4**, 1–247.

Nicholson, E. M. (1951). *Birds and men*. London: Collins. 256 pp.

Niethammer, G. (1937). *Handbuch der deutschen Vogelkunde*. Band I. *Passeres*, xxiv+474 pp. Leipzig: Akademische Verlagsgesellschaft M. B. H.

Niethammer, G. (1953). Gewicht und Flügellänge beim Haussperling (*Passer d. domesticus*). *Journal für Ornithologie*, **94**, 282–9.

Noble, D. L. & Shepperd, W. D. (1973). Gray-headed juncos important in first season mortality of Engelmann spruce. *Journal of Forestry*, **71**, 763–5.

Nordmeyer, A., Oelke, H. & Plagemann, E. (1973). Biometrical studies on house sparrow, *Passer domesticus* (L.), population in northwestern Germany. In *Productivity, population dynamics and systematics of granivorous birds*, ed. S. C. Kendeigh & J. Pinowski, pp. 337–50. Warszawa: PWN-Polish Scientific Publishers.

References

North, C. A. (1968). A study of house sparrow populations and their movements in the vicinity of Stillwater, Oklahoma. Ph.D. Thesis, Oklahoma State University, Stillwater, 83 pp.

North, C. A. (1969). Preliminary report on house sparrow reproductivity and population fluctuations in Coldspring, Wisconsin, 1969. *International Studies on Sparrows*, 3, 43–66.

North, C. A. (1973 a). Population dynamics of the house sparrow, *Passer domesticus* (L.), in Wisconsin, USA. In *Productivity, population dynamics and systematics of granivorous birds*, ed. S. C. Kendeigh & J. Pinowski, pp. 195–210. Warszawa: PWN-Polish Scientific Publishers.

North, C. A. (1973 b). Movement patterns of the house sparrow in Oklahoma. *Ornithological Monographs*, 14, 79–91.

Norton, D. W. (1973). Ecological energetics of calidridine sandpipers breeding in northern Alaska. Ph.D. Thesis, University of Alaska, Fairbanks, 163 pp.

Noskov, G. A. & Gaginskaya, A. R. (1969). Yuvenil'naya lin'ka i migratsii polevogo vorob'ya v usloviyakh Leningradskoï Oblasti. [Juvenile moult and migrations of the tree sparrow in Leningrad District.] *Voprosȳ Ékologii i Biotsenologii, Ékologiya Ptits i Mlekopitayushchikh*, 9, 48–58.

Noskov, G. A., Gaginskaya, E. R., Kamenev, V. M., Khaare, A. O. & Bol'shakov, K. V. (1965). Migratsiya ptits v vostochnoï chasti Finskogo zaliva. [Bird migrations in the eastern part of the Bay of Finland.] In *Soobshcheniya Pribaltiïskoï Komissii po Izucheniyu Migratsiï Ptits*, ed. É. Kumari, vol. 3, pp. 3–27. Tartu: Akademiya Nauk Éstonskoï SSR.

Noskov, G. A., Zimin, V. B. & Rezhȳï, S. P. (1975). Migratsii ptits na Ladozhskom ozere. [Bird migrations on Lake Ladoga.] In *Soobshcheniya Pribaltiïskoï Komissii po Izucheniyu Migratsiï Ptits*, ed. É. Kumari, vol. 8, pp. 3–48. Tartu: Akademiya Nauk Éstonskoï SSR.

Nottebohm, F. & Selander, R. K. (1972). Vocal dialects and gene frequencies in the chingolo sparrow (*Zonotrichia capensis*). *Condor*, 74, 137–43.

Novotný, I. (1963). Vliv tvaru hnizda na průběh hnizděni u vrabce domáciho (*Passer domesticus*). [Influence of the nest form on the breeding process of the house sparrow, *Passer domesticus*.] *Zpravý Geografického Ústavu Československé Akademie Ved, Opava*, 4, 4–8.

Novotný, I. (1970). Breeding bionomy, growth and development of young house sparrow (*Passer domesticus* (L.), 1758). *Acta Scientiarum Naturalium Academiae Scientiarum Bohemoslovacae, Brno*, 4, 1–57.

Novotný, I. (1973). Vztah mezi mezoklimatem a průběhem rozmnožování u ptáků. [Correlation between mesoclimate and the course of nesting of birds.] In *Problémy Modernej Bioklimatológie*, pp. 213–18. Bratislava: Vydavatelstvo Slovenskej Akadémie Vied.

Noy-Meir, I. (1973). Desert ecosystems: environment and producers. *Annual Review of Ecology and Systematics*, 4, 25–51.

Oaten, A. & Murdoch, W. W. (1975). Predator switching, functional response, and stability. *American Naturalist*, 109, 299–318.

O'Connor, R. J. (1973). Patterns of weight change in the house sparrow, *Passer domesticus* (L.). In *Productivity, population dynamics and systematics of granivorous birds*, ed. S. C. Kendeigh & J. Pinowski, pp. 111–25. Warszawa: PWN-Polish Scientific Publishers.

O'Connor, R. J. (1975). The influence of brood size upon metabolic rate and body temperature in nestling blue tits *Parus caeruleus* and house sparrows *Passer domesticus. Journal of Zoology*, 175, 391–403.

References

Odum, E. P. (1960). Premigratory hyperphagia in birds. *American Journal of Clinical Nutrition*, **8**, 621–9.

Odum, E. P. (1961). Excretion rate of radio-isotopes as indices of metabolic rates in nature: biological half-life of zinc-65 in relation to temperature, food consumption, growth, and reproduction in arthropods. *Biological Bulletin*, **121**, 371–2.

Oelke, H. (1963). Die Vogelwelt des Peiner Moränen- und Lössgebietes. Ph.D. Thesis, University of Göttingen, Göttingen, 672 pp.

Ohmart, R. D. & Lasiewski, R. C. (1971). Roadrunners: energy conservation by hyperthermia and absorption of sunlight. *Science*, **172**, 67–9.

Olson, J. B. (1965). Effect of temperature and season on the bioenergetics of the eastern field sparrow, *Spizella pusilla pusilla*. Ph.D. Thesis, University of Illinois, Urbana, 86 pp.

Orians, G. H. (1960). Autumnal breeding in the tricolored blackbird. *Auk*, **77**, 379–98.

Orians, G. H. (1961 a). Social stimulation within blackbird colonies. *Condor*, **63**, 330–7.

Orians, G. H. (1961 b). The ecology of blackbird (*Agelaius*) social systems. *Ecological Monographs*, **31**, 285–312.

Orians, G. H. (1969). On the evolution of mating systems in birds and mammals. *American Naturalist*, **103**, 589–603.

Orians, G. H. (1973). The red-winged blackbird in tropical marshes. *Condor*, **75**, 28–42.

Orians, G. H. & Christman, G. M. (1968). A comparative study of the behavior of red-winged, tricolored and yellow-headed blackbirds. *University of California Publications in Zoology*, **84**, 1–81.

Orians, G. H. & Horn, H. S. (1969). Overlap in foods and foraging of four species of blackbirds in the potholes of central Washington. *Ecology*, **50**, 930–8.

Orr, Y. (1970). Temperature measurements at the nest of the desert lark (*Ammomanes deserti deserti*). *Condor*, **72**, 476–8.

Owen, R. B. Jr (1969). Heart rate, a measure of metabolism in blue-winged teal. *Comparative Biochemistry and Physiology*, **31**, 431–6.

Owen, R. B. Jr (1970). The bioenergetics of captive blue-winged teal under controlled and outdoor conditions. *Condor*, **72**, 153–63.

Owre, O. T. (1973). A consideration of the exotic avifauna of southeastern Florida. *Wilson Bulletin*, **85**, 491–500.

Packard, G. C. (1967). House sparrows: evolution of populations from the Great Plains and Colorado Rockies. *Systematic Zoology*, **16**, 73–89.

Payne, B. R. & DeGraaf, R. M. (1975). Economic values and recreational trends associated with human enjoyment of nongame birds. In *Proceedings of the symposium on management of forest and range habitats for nongame birds, USDA Forest Service, general technical report*, **WO-1**, 6–10. Washington D.C.

Payne, R. B. (1965). Clutch size and numbers of eggs laid by brown-headed cowbirds. *Condor*, **67**, 44–60.

Payne, R. B. (1969). Breeding seasons and reproductive physiology of tricolored blackbirds and redwinged blackbirds. *University of California Publications in Zoology*, **90**, 1–137.

Payne, R. B. (1973). The breeding season of a parasitic bird, the brown-headed cowbird, in central California. *Condor*, **75**, 80–99.

Pearson, K. (1901). On lines and planes of closest fit to systems of points in space. *Philosophical Magazine*, Ser. 6, **2**, 559–72.

410

References

Pearson, K. (1902). Variation of egg of the sparrow (*Passer domesticus*). *Biometrika*, **1**, 256–7.

Peiponen, V. (1957). Wechselt der Birkenzeisig, *Carduelis flammea* (L.), sein Brutgebiet während des Sommers? *Ornis Fennica*, **34**, 41–64.

Penney, J. G. & Bailey, E. D. (1970). Comparison of the energy requirements of fledgling black ducks and American coots. *Journal of Wildlife Management*, **34**, 105–14.

Pennycuick, C. J. (1969). The mechanics of bird migration. *Ibis*, **111**, 525–56.

Perkins, S. E. iii (1928). City park nests of red-winged blackbirds. *Bird-Lore*, **30**, 393–4.

Perrins, C. M. (1965). Population fluctuations and clutch-size in the great tit, *Parus major* L. *Journal of Animal Ecology*, **34**, 601–47.

Perrins, C. M. (1970). The timing of birds' breeding seasons. *Ibis*, **112**, 242–55.

Peterson, A. & Young, H. (1950). A nesting study of the bronzed grackle. *Auk*, **67**, 466–76.

Petrusewicz, K. (1967). Suggested list of more important concepts in productivity studies (definitions and symbols). In *Secondary productivity of terrestrial ecosystems*, ed. K. Petrusewicz, vol. 1, pp. 51–8. Warszawa, Kraków: PWN-Polish Scientific Publishers.

Phillips, A. R., Marshall, J. T. Jr & Monson, G. W. (1964). *The birds of Arizona*, xviii+212 pp. Tucson: University of Arizona Press.

Pianka, E. R. (1972). *r* and *K* selection or *b* and *d* selection? *American Naturalist*, **106**, 581–8.

Pianka, E. R. (1974). *Evolutionary ecology*, viii+356 pp. New York, Evanston, San Francisco & London: Harper & Row.

Pianka, E. R. & Parker, W. S. (1975). Age-specific reproductive tactics. *American Naturalist*, **109**, 453–64.

Piggott, S. (1965). *Ancient Europe from the beginnings of agriculture to classical antiquity: a survey*, xxiii+343 pp. Chicago, Illinois: Aldine Publishing Company.

Pikula, J. (1971). Die Variabilität der Eier der Population *Turdus philomelos*, Brehm 1831 in der ČSSR. *Zoologické Listy*, **20**, 69–83.

Pinowska, B. (1975). Foods of female house sparrows (*Passer domesticus* (L.)) in relation to stages of the nesting cycle. *Polish Ecological Studies*, **1** (3), 211–25.

Pinowska, B. (1976). The effect of body composition of female house sparrows, *Passer domesticus* (L.), on the clutch size and the number of broods (preliminary report). *International Studies on Sparrows*, **9**, 55–71.

Pinowska, B., Chyliński, G. & Gondek, B. (1976). Studies on the transmitting of Salmonellae by house sparrows (*Passer domesticus* (L.)) in the region of Żuławy. *Polish Ecological Studies*, **2** (1), 113–21.

Pinowska, B. & Pinowski, J. (1977). Fecundity, mortality, numbers and biomass dynamics of a population of the house sparrow, *Passer d. domesticus* (L.). *International Studies on Sparrows*, **10**, 26–41.

Pinowski, J. (1964). Der Feldsperling (*Passer m. montanus* (L.)) als potentieller Überträger von Krankheitserregern. In *Schriftenreihe der Landesstelle für Naturschutz und Landschaftspflege in Nordrhein-Westfalen, Festschrift zum 25 Jährigen Bestehen der nordrhein-westfälischen Vogelschutzwarte Essen-Altenhundem*, ed. W. Przygodda, vol. 1, pp. 109–14. Recklinghausen: Verlag Aurel Bongers.

Pinowski, J. (1965 a). Dispersal of young tree sparrows (*Passer m. montanus* (L.)). *Bulletin de l'Academie Polonaise des Sciences, Cl. II, Série des Sciences Biologiques*, **13**, 509–14.

411

References

Pinowski, J. (1965 b). Overcrowding as one of the causes of dispersal of young tree sparrows. *Bird Study*, **12**, 27–33.

Pinowski, J. (1966). Der Jahreszyklus der Brutkolonie beim Feldsperling (*Passer m. montanus* (L.)). *Ekologia Polska*, Ser. A, **14**, 145–72.

Pinowski, J. (1967 a). Die Auswahl des Brutbiotops beim Feldsperling (*Passer m. montanus* (L.)). *Ekologia Polska*, Ser. A, **15**, 1–30.

Pinowski, J. (1967 b). Estimation of the biomass produced by a tree sparrow (*Passer m. montanus* (L.)) population during the breeding season. In *Secondary productivity of terrestrial ecosystems*, ed. K. Petrusewicz, vol. 1, pp. 357–67. Warszawa, Kraków: PWN-Polish Scientific Publishers.

Pinowski, J. (1968). Fecundity, mortality, numbers and biomass dynamics of a population of the tree sparrow (*Passer m. montanus* (L.)). *Ekologia Polska*, Ser. A, **16**, 1–58.

Pinowski, J. (1973). The problem of protecting crops against harmful birds in Poland. *OEPP/EPPO Bulletin*, **3**, 107–10.

Pinowski, J. & Myrcha, A. (1970). Winter fat deposition in the tree sparrow (*Passer m. montanus* (L.)). *Bulletin de l'Academie Polonaise des Sciences, Cl. II, Série des Sciences Biologiques*, **18**, 457–63.

Pinowski, J., Pinowska, B. & Truszkowski, J. (1973). Escape from the nest and brood desertion by the tree sparrow, *Passer m. montanus* (L.), the house sparrow, *Passer d. domesticus* (L.), and the great tit, *Parus m. major* L. In *Productivity, population dynamics and systematics of granivorous birds*, ed. S. C. Kendeigh & J. Pinowski, pp. 397–406. Warszawa: PWN-Polish Scientific Publishers.

Pinowski, J., Tomek, T. & Tomek, W. (1973). Food selection in the tree sparrow, *Passer m. montanus* (L.). Preliminary report. In *Productivity, population dynamics and systematics of granivorous birds*, ed. S. C. Kendeigh & J. Pinowski, pp. 263–73. Warszawa: PWN-Polish Scientific Publishers.

Pinowski, J. & Wieloch, M. (1973). Energy flow through nestlings and biomass production of house sparrow, *Passer d. domesticus* (L.), and tree sparrow, *Passer m. montanus* (L.), populations in Poland. In *Productivity, population dynamics and systematics of granivorous birds*, ed. S. C. Kendeigh & J. Pinowski, pp. 151–63. Warszawa: PWN-Polish Scientific Publishers.

Pinowski, J. & Wójcik, Z. (1968). Produkcja chwastów na polach i stopień wyżerowania ich nasion przez wróble polne (*Passer montanus* (L.)). [Production of weeds in fields and degree to which their seeds are consumed by the tree sparrow (*Passer montanus* (L.)).] *Ekologia Polska*, Ser. B, **14**, 297–301.

Pinowski, J. & Wójcik, Z. (1969). Die Unkrautproduktion auf den Feldern und die Ausnutzung des Unkrautsamens durch die Feldsperlinge (*Passer montanus* (L.)). *Der Falke*, **16**, 256–61.

Pitelka, F. A., Holmes, R. T. & Maclean, S. F. Jr (1974). Ecology and evolution of social organization in arctic sandpipers. *American Zoologist*, **14**, 185–204.

Pohl, H. (1971). Seasonal variation in metabolic functions of bramblings. *Ibis*, **113**, 185–93.

Pohl, H. & West, G. C. (1973). Daily and seasonal variation in metabolic response to cold during rest and forced exercise in the common redpoll. *Comparative Biochemistry and Physiology*, **45A**, 851–67.

Porter, W. P. & Gates, D. M. (1969). Thermodynamic equilibria of animals with environment. *Ecological Monographs*, **39**, 227–44.

Potapov, R. L. & Andreev, A. V. (1973). K bioénergetike tetereva *Lyrurus tetrix*

References

(L.) v zimnii period. [On bioenergetics of *Lyrurus tetrix* (L.) in the winter period.] *Dokladȳ Akademii Nauk SSSR, Seriya Biologiya*, **210**, 499–500.

Power, D. M. (1970). Geographic variation of red-winged blackbirds in central North America. *University of Kansas Publications of the Museum of Natural History*, **19**, 1–83.

Prange, H. D. & Schmidt-Nielsen, K. (1970). The metabolic cost of swimming in ducks. *Journal of Experimental Biology*, **53**, 763–77.

Preiser, F. (1957). Untersuchungen über die Ortsstetigkeit und Wanderung der Sperlinge (*Passer domesticus domesticus* (L.)) als Grundlage für die Bekämpfung. Ph.D. Thesis, Institut für Pflanzenschutz-Landwirtschaftliche Hochschule, Hohenheim, 57 pp.

Preston, F. W. (1948). The commonness and rarity of species. *Ecology*, **29**, 254–83.

Przygodda, W. (1960). Beringung von Haussperlingen in Bonn. *Ornithologische Mitteilungen*, **12**, 21–5.

Ptushenko, E. S. & Inozemtsev, A. A. (1968). *Biologiya i khozyaïstvennoe znachenie ptits Moskovskoï Oblasti i sopredel'nykh territorii.* [*Biology and agricultural significance of birds in the Moscow District and surrounding regions.*] 461 pp. Moskva: Izdatel'stvo Moskovskogo Universiteta.

Pulliam, H. R. (1973). On the advantages of flocking. *Journal of Theoretical Biology*, **38**, 419–22.

Pulliam, H. R. & Enders, F. (1971). The feeding ecology of five sympatric finch species. *Ecology*, **52**, 557–66.

Quick, H. F. (1963). Animal population analysis. In *Wildlife investigational techniques*, ed. H. S. Mosby, 2nd edn, pp. 190–228. Washington D.C.: The Wildlife Society.

Rademacher, B. (1951). Beringungsversuche über die Ortstreue der Sperlinge (*Passer d. domesticus* (L.) und *Passer m. montanus* (L.)). *Zeitschrift für Pflanzenkrankheiten (Pflanzenpathologie) und Pflanzenschutz*, **58**, 416–26.

Rahn, H., Paganelli, C. V. & Ar, A. (1975). Relation of avian egg weight to body weight. *Auk*, **92**, 750–65.

Ranoszek, E. (1969). Ilościowe obserwacje ptaków w grądzie nadodrzańskim. [Quantitative observations on breeding population of birds in Querceto-Carpinetum forest at the Odra River.] *Notatki Ornitologiczne*, **10**, 10–20.

Rasmussen, G. & Brander, R. (1973). Standard metabolic rate and lower critical temperature for the ruffed grouse. *Wilson Bulletin*, **85**, 223–9.

Ravkin, Yu. S. (1973). *Ptitsȳ severo-vostochnogo Altaya.* [*The birds of the northeastern Altaï.*] 374 pp. Novosibirsk: Izdatel'stvo 'Nauka', Sibirskoe Otdelenie.

Rékási, J. (1968a). Data on the food biology of *Passer d. domesticus* (L.). *International Studies on Sparrows*, **2**, 25–39.

Rékási, J. (1968b). Zur Ernährungsbiologie des Haussperlings (*Passer domesticus* (L.)). *Aquila*, **75**, 111–25.

Rékási, J. (1969–1970). Hamster (*Cricetus cricetus*) destroyer of the young of tree sparrows. *Aquila*, **76/77**, 197.

Renfrew, J. M. (1973). *Palaeoethnobotany: the prehistoric food plants of the Near East and Europe*, xviii+248 pp. New York: Columbia University Press.

Richet, C. (1885). Recherches de calorimetrie. *Archives de Physiologie*, Third Ser., **6**, 237–91, 450–97.

Ricklefs, R. E. (1967). Relative growth, body constituents, and energy content of nestling barn swallows and red-winged blackbirds. *Auk*, **84**, 560–70.

Ricklefs, R. E. (1968). Patterns of growth in birds. *Ibis*, **110**, 419–51.

References

Ricklefs, R. E. (1969a). An analysis of nesting mortality in birds. *Smithsonian Contributions to Zoology*, **9**, 1–48.

Ricklefs, R. E. (1969b). Natural selection and the development of mortality rates of young birds. *Nature, London*, **223**, 922–5.

Ricklefs, R. E. (1972). Patterns of growth in birds. II. Growth rate and mode of development. *Ibis*, **115**, 177–201.

Ricklefs, R. E. (1973). Fecundity, mortality, and avian demography. In *Breeding biology of birds*, ed. D. S. Farner, pp. 366–435. Washington D.C.: National Academy of Sciences.

Ricklefs, R. E. (1974). Energetics of reproduction in birds. In *Avian energetics*, ed. R. A. Paynter, Jr, pp. 152–297. *Nuttall Ornithological Club, Publication*, No. 15. Cambridge, Massachusetts.

Rising, J. D. (1968). The effect of temperature variation on the metabolic activity of the Harris sparrow. *Comparative Biochemistry and Physiology*, **25**, 327–33.

Rising, J. D. & Hudson, J. W. (1974). Seasonal variation and thyroid activity of the black-capped chickadee (*Parus atricapillus*). *Condor*, **76**, 198–203.

Risser, P. G. (ed.) (1972). A preliminary compartment model of a tallgrass prairie, Osage site, 1970. *US IBP Grassland Biome Technical Report*, No. **159**, 21 pp.

Robbins, C. S. (1973). Introduction, spread, and present abundance of the house sparrow in North America. *Ornithological Monographs*, **14**, 3–9.

Robbins, C. S. & Van Velzen, W. T. (1974). Progress report on the North American breeding bird survey. In *Proceedings of the fourth meeting of the International Bird Census Committee and second meeting of the European Ornithological Atlas Committee*, ed. J. Pinowski & K. Williamson. *Acta Ornithologica*, **14**, 170–91.

Robertson, R. J. (1972). Optimal niche space of the red-winged blackbird (*Agelaius phoeniceus*). I. Nesting success in marsh and upland habitat. *Canadian Journal of Zoology*, **50**, 247–63.

Robertson, R. J. (1973). Optimal niche space of the red-winged blackbird. III. Growth rate and food of nestlings in marsh and upland habitat. *Wilson Bulletin*, **85**, 209–22.

Robson, D. S. & Youngs, W. D. (1971). *Statistical analysis of reported tag-recaptures in the harvest from an exploited population. Cornell Biometrics Unit Publications*, **BU–369–M**, 15 pp.

Robson, R. W. & Williamson, K. (1972). The breeding birds of a Westmorland farm. *Bird Study*, **19**, 202–14.

Rogers, J. G. Jr (1974). Responses of caged red-winged blackbirds to two types of repellents. *Journal of Wildlife Management*, **38**, 418–23.

Rolnik, V. V. (1970). *Bird embryology*, vi+379 pp. Springfield, Virginia: US Department of Commerce. (Translated from Russian.)

Root, R. B. (1967). The niche exploitation pattern of the blue-gray gnatcatcher. *Ecological Monographs*, **37**, 317–50.

Rothstein, S. I. (1973). The niche variation model – is it valid? *American Naturalist*, **107**, 598–620.

Roughgarden, J. (1972). Evolution of niche width. *American Naturalist*, **106**, 683–718.

Rowley, I. (1973a). The comparative ecology of Australian corvids. II. Social organization and behaviour. *CSIRO Wildlife Research*, **18**, 25–65.

Rowley, I. (1973b). The comparative ecology of Australian corvids. IV. Nesting and the rearing of young to independence. *CSIRO Wildlife Research*, **18**, 91–129.

414

Rowley, I., Braithwaite, L. W. & Chapman, G. S. (1973). The comparative ecology of Australian corvids. III. Breeding seasons. *CSIRO Wildlife Research*, **18**, 67–90.

Rowley, I. & Vestjens, W. J. M. (1973). The comparative ecology of Australian corvids. V. Food. *CSIRO Wildlife Research*, **18**, 131–55.

Royama, T. (1970). Factors governing the hunting behaviour and selection of food by the great tit (*Parus major* L.). *Journal of Animal Ecology*, **39**, 619–68.

Rubner, M. (1883). Ueber den Einfluss der Körpergrösse auf Stoff- und Kraftwechsel. *Zeitschrift für Biologie*, **19**, 535–62.

Rubner, M. (1910). Über Kompensation und Summation von funktionellen Leistungen des Körpers. *Sitzungsberichte der Königlichen Preussischen Akademie der Wissenschaften*, **16**, 316–24.

Ruprecht, A. L. (1968). Morphological variability of the *Passer domesticus* (L.) skull in postnatal development. *Acta Ornithologica*, **11**, 27–43.

Rustamov, A. K. (1958). *Ptitsÿ Turkmenistana*. [*Birds of Turkmenistan*], vol. II, 252 pp. Ashkhabad: Izdatel'stvo Akademii Nauk Turkmenskoï SSR.

Rydzewski, W. (1949a). Sprawozdanie z działalności Stacji Badania Wędrówek Ptaków za rok 1938. [Compte rendu de l'activité de la station pour l'étude des migrations des oiseaux pour l'année 1938.] *Acta Ornithologica*, **4**, 1–113.

Rydzewski, W. (1949b). Sprawozdanie z działalności Stacji Badania Wędrówek Ptaków za rok 1939. [Compte rendu de l'activité de la station pour l'étude des migrations des oiseaux pour l'année 1939.] *Acta Ornithologica*, **4**, 115–221.

Savidge, I. R. (1974). A model for management predictions of territorial bird populations. *Ohio Journal of Science*, **74**, 301–12.

Saxena, B. B. (1957). Unterschiede physiologischer Konstanten bei Finkenvögel aus verschiedenen Klimazonen. *Zeitschrift für Vergleichende Physiologie*, **40**, 376–96.

Schafer, E. W. Jr, Brunton, R. B. & Cunningham, D. J. (1973). A summary of the acute oral toxicity of 4-aminopyridine to birds and mammals. *Applied Pharmacology*, **26**, 532–8.

Schafer, E. W. Jr, Brunton, R. B. & Lockyer, N. F. (1974). Hazards to animals feeding on blackbirds killed with 4-aminopyridine baits. *Journal of Wildlife Management*, **38**, 424–6.

Schartz, R. L. & Zimmerman, J. L. (1971). The time and energy budget of the male dickcissel (*Spiza americana*). *Condor*, **73**, 65–76.

Scherner, E. R. (1972). Untersuchungen zur Ökologie des Feldsperlings *Passer montanus*. *Vogelwelt*, **93**, 41–68.

Scherner, E. R. (1974). Untersuchungen zur populären Variabilität des Haussperlings (*Passer domesticus*). *Vogelwelt*, **95**, 41–60.

Schlegel, R. (1966). Betrachtungen über Ergebnisse von Vogelschutzmassnahmen und Siedlungsdichteermittelungen im Auenwald Laske. *Aufsätze zu Vogelschutz und Vogelkunde*, **2**, 12–18.

Schoener, T. W. (1965). The evolution of bill size differences among sympatric congeneric species of birds. *Evolution*, **19**, 189–213.

Schönwetter, M. (1960–72). *Handbuch der Oölogie*, ed. W. Meise, 448 pp. Berlin: Akademie-Verlag.

Schreiber, R. W. & Lawrence, J. M. (1976). Organic matter and calories in laughing gull eggs. *Auk*, **93**, 46–52.

Sealy, S. G. (1971). The irregular occurrences of the dickcissel in Alberta, Manitoba and Saskatchewan. *Blue Jay*, **29**, 12–16.

Seber, G. A. F. (1970). Estimating time-specific survival and reporting rates for adult birds from band returns. *Biometrika*, **57**, 313–18.

References

Seel, D. C. (1964). An analysis of the nest cards of the tree sparrow. *Bird Study*, **11**, 265–71.

Seel, D. C. (1968*a*). Breeding seasons of the house sparrow and the tree sparrow *Passer* spp. at Oxford. *Ibis*, **110**, 129–44.

Seel, D. C. (1968*b*). Clutch size, incubation and hatching success in the house sparrow and tree sparrow *Passer* spp. at Oxford. *Ibis*, **110**, 270–82.

Seel, D. C. (1969). Food, feeding rates and body temperature in the nestling house sparrow, *Passer domesticus*, at Oxford. *Ibis*, **111**, 36–47.

Seel, D. C. (1970). Nestling survival and nestling weights in the house sparrow and tree sparrow *Passer* spp. at Oxford. *Ibis*, **112**, 1–14.

Seibert, H. C. (1949). Differences between migrant and non-migrant birds in food and water intake at various temperatures and photoperiods. *Auk*, **66**, 128–53.

Seibert, H. C. (1963). Metabolizable energy in the Reeves pheasant. *Game Research in Ohio*, **2**, 185–200.

Selander, R. K. (1950). The birds of Utah. M.A. Thesis, University of Utah, Logan, Utah, pp. 1–455.

Selander, R. K., Hunt, W. G. & Yang, S. Y. (1969). Protein polymorphism and genic heterozygosity in two European subspecies of the house mouse. *Evolution*, **23**, 379–90.

Selander, R. K. & Johnson, W. E. (1973). Genetic variation among vertebrate species. *Annual Review of Ecology and Systematics*, **4**, 75–92.

Selander, R. K. & Johnston, R. F. (1967). Evolution in the house sparrow. I. Intrapopulation variation in North America. *Condor*, **69**, 217–58.

Shake, W. F. & Mattsson, J. P. (1975). Three years of cowbird control: an effort to save the Kirtland's warbler. *Jack-Pine Warbler*, **53**, 48–53.

Sharipov, M. M. (1974). Gorod kak sreda obitaniya. [The town as an ecosystem.] In *Pozvonochnÿe Zhivotnÿe Ferganskoĭ Doliny*, ed G. S. Sultanov, pp. 33–42. Tashkent: Izdatel'stvo 'FAN' Uzbekskoĭ SSR.

Sharrock, J. T. R. (1974). The ornithological Atlas project in Britain and Ireland. Methods and preliminary results. In *Proceedings of the fourth meeting of the International Bird Census Committee and second meeting of the European Ornithological Atlas Committee*, ed. J. Pinowski & K. Williamson. *Acta Ornithologica*, **14**, 412–28.

Shilov, I. A. (1968). *Heat regulation in birds*, x+279 pp. Springfield, Virginia: US Department of Commerce. (Translated from Russian, 1973.)

Shott, A. R. & Preston, F. W. (1975). The surface area of an egg. *Condor*, **77**, 103–4.

Shtegman, B. K. (1956). Vorob'i v Kazakhstane i izÿskanie mer bor'bÿ s nimi. [The sparrows in Kazakhstan and searching for methods of their control.] *Zoologicheskiĭ Zhurnal*, **35**, 1203–13.

Siegfried, W. R. (1973). Breeding success and reproductive potential in the Cape sparrow, *Passer melanurus* (Müller). In *Productivity, population dynamics and systematics of granivorous birds*, ed. S. C. Kendeigh & J. Pinowski, pp. 167–80. Warszawa: PWN-Polish Scientific Publishers.

Simeonov, S. D. (1963). Prouchvane v''rkhu khranata na polskoto vrabche (*Passer montanus* (L.)) v Sofiĭsko. [Untersuchung der Nahrungszusammenstellung des Feldsperlings (*Passer montanus* (L.)) im Sofioter Bezirk.] *Izvestiya na Zoologicheskiya Institut s Muzeĭ*, **14**, 93–109.

Simeonov, S. D. (1964). Prouchvane v''rkhu khraneneto na domashnoto vrabche (*Passer domesticus* (L.)) v Sofiĭsko. [Über die Nahrung des Haussperlings in der Umgebung von Sofia.] *Godishnik na Sofiĭskiya Universitet, Biologo–Geologo–Geografski Fakultet, Biologiya (Zoologiya)*, **56**, 239–75.

416

Sims, R. W. (1955). The morphology of the head of the hawfinch. *Bulletin of the British Museum of Natural History*, **2**, 369–93.

Skead, C. J. (1964). The ecology of the ploceid weavers, widows and bishop-birds in the southeastern Cape Province, South Africa. In *Ecological studies in Southern Africa*, ed. D. H. S. Davis. *Monographiae Biologicae*, **14**, 219–32. The Hague: Dr W. Junk Publishers.

Smith, H. M. (1943). Size of breeding populations in relation to egg-laying and reproductive success in the eastern red-wing (*Agelaius p. phoeniceus*). *Ecology*, **24**, 183–207.

Smith, K. G. & Prince, H. H. (1973). The fasting metabolism of subadult mallards acclimatized to low ambient temperatures. *Condor*, **75**, 330–5.

Smith, N. J. H. (1973). House sparrows (*Passer domesticus*) in the Amazon. *Condor*, **75**, 242–3.

Smith, W. K., Roberts, S. W. & Miller, P. C. (1974). Calculating the nocturnal energy expenditure of an incubating Anna's hummingbird. *Condor*, **76**, 176–83.

Smithies, O. (1955). Zone electrophoresis in starch gels: group variation in serum proteins of normal human adults. *Biochemical Journal*, **61**, 629–41.

Sneath, P. & Sokal, R. R. (1973). *Numerical taxonomy*, xv+573 pp. San Francisco: Freeman.

Snelling, J. C. (1968). Overlap in feeding habits of redwinged blackbirds and common grackles nesting in a cattail marsh. *Auk*, **85**, 560–85.

Snow, B. K. (1960). The breeding biology of the shag *Phalacrocorax aristotelis* on the island of Lundy, Bristol Channel. *Ibis*, **102**, 554–75.

Sokołowski, J. (1929). Gniazda ptasie w skrzynkach. [Nests of the birds in boxes.] *Ochrona Przyrody*, **8**, 9–13.

Soulé, M. & Stewart, B. R. (1970). The 'niche-variation' hypothesis: a test and alternatives. *American Naturalist*, **104**, 85–97.

Southern, H. N. (1945). The economic importance of the house sparrow, *Passer domesticus* (L.), a review. *Annals of Applied Biology*, **32**, 57–62.

Spellerberg, I. F. (1969). Incubation temperatures and thermoregulation in the McCormick skua. *Condor*, **71**, 59–67.

Speyer, W. (1956). Beringungsversuche mit Sperlingen in den Jahren 1951 bis 1954. *Nachrichtenblatt des Deutschen Pflanzenschutzdienstes*, **8**, 27–9.

Šťastný, K. (1973). Vyuzití ptáků a savců pro charakterizaci hrází Třeboňska z hlediska krajinné ekologie. [Characterisation of Třeboňska dike by birds and mammals from point of view of landscape ecology.] Ph.D. Thesis, Praha University, Praha, 168 pp.

Steen, J. (1958). Climatic adaptation in some small northern birds. *Ecology*, **39**, 625–9.

Stewart, P. A. (1975). Development of roosting congregations of common grackles and associated species. *Bird-Banding*, **46**, 213–16.

Stickley, A. R. Jr & Guarino, J. L. (1972). A repellent for protecting corn seed from blackbirds and cows. *Journal of Wildlife Management*, **36**, 150–2.

Stickley, A. R. Jr, Mitchell, R. T., Heath, R. G., Ingram, C. R. & Bradley, E. L. Jr (1972). A method for appraising the bird repellency of 4-aminopyridine. *Journal of Wildlife Management*, **36**, 1313–16.

Stickley, A. R. Jr, Mitchell, R. T., Seubert, J. L., Ingram, C. R. & Dyer, M. I. (1976). Large-scale evaluation of blackbird frightening agent 4-aminopyridine in corn. *Journal of Wildlife Management*, **40**, 126–31.

Stiven, A. E. (1961). Food energy available for and required by the blue grouse chick. *Ecology*, **42**, 547–53.

Stohn, H. (1971). Amsel plündert Haussperlingsnest. *Der Falke*, **18**, 138.

417

References

Stone, C. P. Jr (1973). Phenetic variation of breeding red-winged blackbirds in Ohio. Ph.D. Thesis, Ohio State University, Columbus, 301 pp.

Stone, C. P. Jr, Mott, D. F., Besser, J. F. & DeGrazio, J. W. (1972). Bird damage to corn in the United States in 1970. *Wilson Bulletin*, **84**, 101–5.

Strawiński, S. & Wieloch, M. (1972). Charakterystyka populacji *Passer domesticus* (L.) i *P. montanus* (L.) jako podłoże do badań aktywności. [The characteristics of the populations of *Passer domesticus* (L.) and *P. montanus* (L.) as a background to activity studies.] *Zeszyty Naukowe, Instytut Ekologii PAN*, **5**, 329–40.

Sturges, F. W., Holmes, R. T. & Likens, G. E. (1974). The role of birds in nutrient cycling in a northern hardwoods ecosystem. *Ecology*, **55**, 149–55.

Sudilovskaya, A. M. (1954). Semeïstvo tkachikovȳe Ploceidae (Family Ploceidae). In *Ptitsȳ Sovetskogo Soyuza*, ed. G. P. Dement'ev & N. A. Gladkov, vol. 5, pp. 306–74. Moskva: Gosudarstvennoe Izdatel'stvo 'Sovetskaya Nauka'.

Sugden, L. G. & Harris, L. E. (1972). Energy requirements and growth of captive lesser scaup. *Poultry Science*, **51**, 625–33.

Sulkava, S. (1969). On small birds spending the night in the snow. *Aquilo Series Zoology*, **7**, 33–7.

Summers-Smith, D. (1954). The communal display of the house sparrow. *Ibis*, **96**, 116–28.

Summers-Smith, D. (1956). Movements of the house sparrow. *British Birds*, **49**, 465–88.

Summers-Smith, D. (1959). The house sparrow, *Passer domesticus*: population problem. *Ibis*, **101**, 449–54.

Summers-Smith, D. (1963). *The house sparrow*, xvi+269 pp. London: Collins Clear-Type Press.

Szczepski, J. B. (1951). Sprawozdanie z działalności Stacji Ornitologicznej w latach 1945–1948. [Compte rendu de l'activité de la station ornithologique pour les années 1945–1948.] *Acta Ornithologica*, **4**, 237–72.

Szczepski, J. B. & Szczepska, M. W. (1953). Sprawozdanie z działalności Stacji Ornitologicznej za rok 1949. [Compte rendu de l'activité de la station ornithologique pour l'année 1949.] *Acta Ornithologica*, **4**, 273–310.

Tangl, F. (1903). Beiträge zur Energetik der Ontogenese. I Mitteilung. Die Entwicklungsarbeit in Vogelei. *Pfluger's Archiv für Gesamte Physiologie des Menschen und der Tiere*, **93**, 327–76.

Tatum, J. B. (1975). Egg volume. *Auk*, **92**, 576–80.

Taylor, C. R., Dmi'el, R., Fedak, M. & Schmidt-Nielsen, K. (1971). Energetic cost of running and heat balance in a large bird, the rhea. *American Journal of Physiology*, **221**, 597–601.

Tchernov, E. (1962). Paleolithic avifauna in Palestine. *Bulletin of the Research Council of Israel*, **11**, 95–131.

Teal, J. M. (1969). Direct measurements of CO_2 production during flight in small birds. *Zoologica*, **54**, 17–23.

Terroine, E. F. & Trautmann, S. (1927). Influence de la température extérieure sur la production calorique des homéothermes et loi des surfaces. *Annales de Physiologie et de Physicochimie Biologique*, **3**, 422–57.

Thomas, J. W., Crouch, G. L., Bumstead, R. S. & Bryant, L. D. (1975). Silvicultural options and habitat values in coniferous forests. In *Proceedings of the symposium on management of forest and range habitats for nongame birds*, *USDA Forest Service, general technical report*, **WO-1**, 272–87. Washington D.C.

418

References

Thurber, W. A. (1972). House sparrows in Guatemela. *Auk*, **89**, 200.
Tinbergen, L. (1960). The natural control of insects in pine woods. *Archives Néerlandaises de Zoologie*, **13**, 265–379.
Tomek, T. (1975). Elementy bilansu energetycznego piskląt gawrona *Corvus f. frugilegus* L. [Elements of energy balance of nestlings of rooks, *Corvus f. frugilegus* L.] Ph.D. Thesis, Jagiellonian University, Kraków, 64 pp.
Tomek, W. (1973). Ptaki zachodniej części Pogórza Ciężkowickiego. [Birds of the western part of the Ciężkowice Uplands.] *Acta Zoologica Cracoviensia*, **18**, 529–82.
Tomiałojć, L. (1970). Badania ilościowe nad synantropijną awifauną Legnicy i okolic. [Quantitative studies on the synanthropic avifauna of Legnica and its environs.] *Acta Ornithologica*, **12**, 293–392.
Tomiałojć, L. (1974). Charakterystyka ilościowa lęgowej i zimowej awifauny lasów okolic Legnicy (Śląsk Dolny). [The quantitative analysis of the breeding and winter avifauna of the forests in the vicinity of Legnica, Lower Silesia.] *Acta Ornithologica*, **14**, 59–97.
Trivers, R. L. (1972). Parental investment and sexual selection. In *Sexual selection and the descent of man 1871–1971*, ed. R. Campbell, pp. 136–79. Chicago, Illinos: Aldine Publishing Co.
Trost, C. H. (1972). Adaptations of horned larks (*Eremophila alpestris*) to hot environments. *Auk*, **89**, 506–27.
Truszkowski, J. (1963). Ptaki parku miejskiego w Pruszkowie. [Birds of the town park in Pruszków.] *Przegląd Zoologiczny*, **7**, 62–71.
Tucker, V. A. (1968). Respiratory exchange and evaporative water loss in the flying budgerigar. *Journal of Experimental Biology*, **48**, 67–87.
Tucker, V. A. (1973). Bird metabolism during flight: evaluation of a theory. *Journal of Experimental Biology*, **58**, 689–709.
Tucker, V. A. (1974). Energetics of natural avian flight. In *Avian energetics*, ed. R. A. Paynter, Jr, pp. 298–333. *Nuttall Ornithological Club Publication*, No. 15. Cambridge, Massachusetts.
Turček, F. J. (1966). On plumage quantity in birds. *Ekologia Polska*, Ser. A, **14**, 617–34.
Turček, F. J. (1968). Some suggestions on sparrows as biological indicators. *International Studies on Sparrows*, **2**, 16–19.
Turček, F. J. (1969). On some functional aspects of biological production. *Ekologia Polska*, Ser. B, **15**, 31–5.
Ulfstrand, S., Roos, G., Alerstam, T. & Österdahl, L. (1974). Visible bird migration at Falsterbo, Sweden. *Vår Fågelvärld*, Supplement, **8**, 1–245.
Van Valen, L. (1965). Morphological variation and width of ecological niche. *American Naturalist*, **94**, 377–90.
Van Valen, L. & Grant, P. R. (1970). Variation and niche width re-examined. *American Naturalist*, **104**, 589–90.
Varley, G. C. (1967). The effects of grazing by animals on plant productivity. In *Secondary productivity of terrestrial ecosystems (principles and methods)*, ed. K. Petrusewicz, vol. 2, pp. 773–8. Warszawa, Kraków: PWN-Polish Scientific Publishers.
Varney, J. R. & Ellis, D. H. (1974). Telemetering egg for use in incubation and nesting studies. *Journal of Wildlife Management*, **38**, 142–8.
Veghte, J. H. (1964). Thermal and metabolic responses of the gray jay to cold stress. *Physiological Zoology*, **37**, 316–28.
Verheyen, R. (1957). Over de Verplaatsingen van de Boommus, *Passer montanus*

419

References

(L.), in en door Belgie. [About the displacements of the tree sparrow, *Passer montanus* (L.), in and through Belgium.] *Die Giervalk*, **47**, 161–70.

Verheyen, R. K. (1967). *Oologia Belgica*, 331 pp. Bruxelles: Patrimoine de l'Institut royal des Sciences naturelles de Belgique.

Vermeer, K. (1969). Egg measurements of California and ring-billed gull eggs at Miquelon Lake, Alberta, in 1965. *Wilson Bulletin*, **81**, 102–3.

Verner, J. (1965). Breeding biology of the long-billed marsh wren. *Condor*, **67**, 6–30.

Verner, J. & Willson, M. F. (1966). The influence of habitats on mating systems of North American passerine birds. *Ecology*, **47**, 143–7.

Vernon, C. J. (1973). Avian biomass in a suburb of Pietermaritzburg. *Ostrich*, **44**, 142–3.

Vierke, J. (1970). Die Besiedlung Südafrikas durch den Haussperling (*Passer domesticus*). *Journal für Ornithologie*, **111**, 94–103.

Vik, R. (1962). Bird observations in the North Atlantic. *Sterna*, **5**, 15–23.

Vine, I. (1973). Detection of prey flocks by predators. *Journal of Theoretical Biology*, **40**, 207–10.

Vladÿshevskiï, D. V. (1972). Naselenie ptits sosnovÿkh lesov Kievshchinÿ. [Bird populations in the pine forests of Kievshchina (Kiev Province).] *Ornithologiya*, **10**, 130–8.

Volchanetskiï, I. B., Lisetskiï, A. S. & Kholupyak, Yu. K. (1970). O formirovanii faunÿ ptits iskusstvennÿkh nasazhdeniï yuga Ukrainÿ za period s 1936 po 1967 g. [On succession in the avifauna of forest plantations in the south of Ukraine for the period from 1936 to 1967.] *Vestnik Zoologii*, **4**(1), 39–48.

Volkov, N. I. (1968). Éksperimental'noe izuchenie temperaturnÿkh usloviï v snezhnÿkh norakh teterevinÿkh ptits. [An experimental study of thermal conditions in snow burrows of tetraonid birds.] *Zoologicheskiï Zhurnal*, **47**, 283–6.

Vtorov, P. P. (1962). K landshaftnoï ornitogeografii Tsentral'nogo Kavkaza. [On the landscape avian geography of the Central Caucasus.] *Ornitologiya*, **4**, 218–33.

Wagner, H. O. (1959). Die Einwanderung des Haussperlings in Mexiko. *Zeitschrift für Tierpsychologie*, **16**, 584–92.

Wallgren, H. (1954). Energy metabolism of two species of the genus *Emberiza* as correlated with distribution and migration. *Acta Zoologica Fennica*, **84**, 5–110.

Walter, H. W. & Mocci Demartis, A. (1972). Brutdichte und oekologische Nische sardischer Stadtvögel. *Journal für Ornithologie*, **113**, 391–406.

Ward, P. (1963). Lipid levels in birds preparing to cross the Sahara. *Ibis*, **105**, 109–11.

Ward, P. (1964). A suggested relationship between wing shape of migrants and migratory fat. *Ibis*, **106**, 256–7.

Ward, P. (1965 a). Feeding ecology of the black-faced dioch *Quelea quelea* in Nigeria. *Ibis*, **107**, 173–214.

Ward, P. (1965 b). The breeding biology of the black-faced dioch *Quelea quelea* in Nigeria. *Ibis*, **107**, 326–49.

Ward, P. (1966). Distribution, systematics, and polymorphism of the African weaver-bird *Quelea quelea*. *Ibis*, **108**, 34–40.

Ward, P. (1971 a). The migration patterns of *Quelea quelea* in Africa. *Ibis*, **113**, 275–97.

Ward, P. (1971 b). *Manual of techniques used in research on quelea birds*, 70 pp. Rome: FAO.

Ward, P. (1972). New views on controlling queleas. *Span*, **15**, 136–7.

Ward, P. (1973). A new strategy for the control of damage by queleas. *PANS*, **19**, 97–106.

Ward, P. & Jones, P. J. (1977). Pre-migratory fattening in three races of the red-billed quelea *Quelea quelea*. *Journal of Zoology*, **181**, 43–56.

Ward, P. & Poh, G. E. (1968). Seasonal breeding in an equatorial population of the tree sparrow *Passer montanus*. *Ibis*, **110**, 359–63.

Ward, P. & Zahavi, A. (1973). The importance of certain assemblages of birds as 'information-centres' for food-finding. *Ibis*, **115**, 517–34.

Weaver, R. L. (1943). Reproduction in English sparrows. *Auk*, **60**, 62–74.

Webb, J. S. & Royall, W. C. Jr (1970). National survey of blackbird-starling roosts. In *Proceedings of fifth bird control seminar*, ed. D. L. Rintamaa & W. B. Jackson, pp. 134–5. Bowling Green, Ohio: Bowling Green State University.

Weiner, J. (1973). Energy requirements of house sparrow, *Passer d. domesticus* (L.), in southern Poland. In *Productivity, population dynamics and systematics of granivorous birds*, ed. S. C. Kendeigh & J. Pinowski, pp. 45–58. Warszawa: PWN-Polish Scientific Publishers.

Weiner, J. & Głowaciński, Z. (1975). Energy flow through a bird community in a deciduous forest in southern Poland. *Condor*, **77**, 233–42.

Wentworth, B. C. (1968). Avian birth control potentialities with synthetic grit. *Nature, London*, **220**, 1243–5.

Wentworth, B. C., Hendricks, B. G. & Sturtevant, J. (1968). Sterility induced in Japanese quail by spray treatment of eggs with mestranol. *Journal of Wildlife Management*, **32**, 879–87.

West, G. C. (1960). Seasonal variation in the energy balance of the tree sparrow in relation to migration. *Auk*, **77**, 306–29.

West, G. C. (1968). Bioenergetics of captive willow ptarmigan under natural conditions. *Ecology*, **49**, 1035–45.

West, G. C. (1972 a). The effect of acclimation and acclimatization on the resting metabolic rate of the common redpoll. *Comparative Biochemistry and Physiology*, **43A**, 293–310.

West, G. C. (1972 b). Seasonal differences in resting metabolic rate of Alaskan ptarmigan. *Comparative Biochemistry and Physiology*, **42A**, 867–76.

West, G. C. (1973). Foods eaten by tree sparrows in relation to availability during summer in northern Manitoba. *Arctic*, **26**, 7–21.

West, G. C. & Hart, J. S. (1966). Metabolic responses of evening grosbeaks to constant and to fluctuating temperatures. *Physiological Zoology*, **39**, 171–84.

West, G. C. & Meng, M. S. (1966). Nutrition of willow ptarmigan in northern Alaska. *Auk*, **83**, 603–15.

Westerterp, K. (1973). The energy budget of the nestling starling *Sturnus vulgaris*, a field study. *Ardea*, **61**, 137–58.

White, F. N., Bartholomew, G. A. & Howell, T. R. (1975). The thermal significance of the nest of the sociable weaver, *Philetairus socius*: winter observations. *Ibis*, **117**, 171–9.

White, F. N. & Kinney, J. L. (1974). Avian incubation: interactions among behavior, environment, nest, and eggs result in regulation of egg temperature. *Science*, **186**, 107–15.

Whittaker, R. H. (1965). Dominance and diversity in land plant communities. *Science*, **147**, 250–60.

Wieloch, M. (1975). Food of nestling house sparrows *Passer domesticus* (L.) and tree sparrows *Passer montanus* (L.) in agrocoenoses. *Polish Ecological Studies*, **1** (3), 227–42.

References

Wieloch, M. & Fryska, A. (1975). Biomass production and energy requirements in populations of the house sparrow (*Passer d. domesticus* (L.)) and tree sparrow (*Passer m. montanus* (L.)) during the breeding season. *Polish Ecological Studies*, **1** (3), 243–52.

Wieloch, M. & Strawiński, S. (1976). Produkcja populacji wróbli *Passer domesticus* (L.) i *Passer montanus* (L.) i próba oceny roli wróbli domowych w agrocenozach jako konsumentów w okresie lęgowym. [Production of the populations of the sparrows *Passer domesticus* (L.) and *Passer montanus* (L.) and a trial at assessing the role of the house sparrows as consumers during their breeding season in agrocenoses.] In *Ekologia ptaków wybrzeża*, ed. S. Strawiński, pp. 7–15. Gdańsk: Gdańskie Towarzystwo Naukowe.

Wiens, J. A. (1963). Aspects of cowbird parasitism in southern Oklahoma. *Wilson Bulletin*, **75**, 130–9.

Wiens, J. A. (1965). Behavioral interactions of red-winged blackbirds and common grackles on a common breeding ground. *Auk*, **82**, 356–74.

Wiens, J. A. (1973). Pattern and process in grassland bird communities. *Ecological Monographs*, **43**, 237–70.

Wiens, J. A. (1974 a). Habitat heterogeneity and avian community structure in North American grasslands. *American Midland Naturalist*, **91**, 195–213.

Wiens, J. A. (1974 b). Climatic instability and the 'ecological saturation' of bird communities in North American grasslands. *Condor*, **76**, 385–400.

Wiens, J. A. (1975). Avian communities, energetics, and functions in coniferous forest habitats. In *Proceedings of the symposium on management of forest and range habitats for nongame birds, USDA Forest Service, general technical report*, **WO–1**, 226–65. Washington D.C.

Wiens, J. A. (1976). Population responses to patchy environments. *Annual review of Ecology and Systematics*, **7**, 81–120.

Wiens, J. A. & Dyer, M. I. (1975 a). Simulation modelling of red-winged blackbird impact on grain crops. *Journal of Applied Ecology*, **12**, 63–82.

Wiens, J. A. & Dyer, M. I. (1975 b). Rangeland avifaunas: their composition, energetics, and role in the ecosystem. In *Proceedings of the symposium on management of forest and range habitats for nongame birds, USDA Forest Service, general technical report*, **WO–1**, 146–82. Washington D.C.

Wiens, J. A. & Emlen, J. T. (1966). Post-invasion status of the dickcissel in southern Wisconsin. *Passenger Pigeon*, **28**, 63–9.

Wiens, J. A. & Innis, G. S. (1973). Estimation of energy flow in bird communities. II. A simulation model of activity budgets and population bioenergetics. In *Proceedings of the 1973 summer computer simulation conference*, pp. 739–52. La Jolla, California: Simulation Councils, Inc.

Wiens, J. A. & Innis, G. S. (1974). Estimation of energy flow in bird communities: a population bioenergetics model. *Ecology*, **55**, 730–46.

Wiens, J. A. & Nussbaum, R. A. (1975). Model estimation of energy flow in northwestern coniferous forest bird communities. *Ecology*, **56**, 547–61.

Wiens, J. A. & Scott, J. M. (1975). Model estimation of energy flow in Oregon coastal seabird populations. *Condor*, **77**, 439–52.

Wilkie, D. R. (1959). The work output of animals: flight by birds and by man-power. *Nature, London*, **183**, 1515–16.

Will, R. L. (1969). Fecundity, density, and movements of a house sparrow population in southern Illinois,. Ph.D. Thesis, University of Illinois, Urbana, 67 pp.

422

References

Will, R. L. (1973). Breeding success, numbers, and movements of house sparrows at McLeansboro, Illinois. *Ornithological Monographs*, **14**, 60–78.

Williams, C. B. (1964). *Patterns in the balance of nature*, 324 pp. New York: Academic Press.

Williams, G. C. (1975). *Sex and evolution*, 200 pp. *Monographs in Population Biology*, No. **8**. Princeton, New Jersey: Princeton University Press.

Williams, J. E. (1965). Energy requirements of the Canada goose in relation to distribution and migration. Ph.D. Thesis, University of Illinois, Urbana, 86 pp.

Williams J. F. (1940). The sex ratio in nestling eastern redwings. *Wilson Bulletin*, **52**, 267–77.

Williamson, K. (1967). The bird community of farmland. *Bird Study*, **14**, 210–26.

Williamson, K. (1975). The breeding bird community of chalk grassland scrub in the Chiltern Hills. *Bird Study*, **22**, 59–70.

Willson, M. F. (1969). Avian niche size and morphological variation. *American Naturalist*, **103**, 531–42.

Willson, M. F. (1971). Seed selection in some North American finches. *Condor*, **73**, 415–29.

Willson, M. F. & Harmeson, J. C. (1973). Seed preferences and digestive efficiency of cardinals and song sparrows. *Condor*, **75**, 225–34.

Willson, M. F., John, R. D. St., Lederer, R. J. & Muzos, S. J. (1971). Clutch size in grackles. *Bird-Banding*, **42**, 28–35.

Winkel, K. (1951). Vergleichende Untersuchungen einiger physiologischer Konstanten bei Vögeln aus verschiedenen Klimazonen. *Zoologische Jahrbücher, Abteilung für Systematik, Ökologie und Geographie der Tiere*, **80**, 256–76.

Winkel, W. (1970). Experimentelle Untersuchungen zur Brutbiologie von Kohl- und Blaumeise (*Parus major* und *P. caeruleus*). *Journal für Ornithologie*, **111**, 154–74.

Witherby, H. F., Jourdain, F. C. R., Ticehurst, N. F. & Tucker, B. W. (1948). *The handbook of British birds*. Vol. I. *Crows to firecrest*, xl+326 pp. London: H. F. & G. Witherby Ltd.

Wood, H. B. (1938). Nesting of red-winged blackbirds. *Wilson Bulletin*, **50**, 143–4.

Wood, M. (1928). Mortality of young red-winged blackbirds. *Bird-Lore*, **30**, 262.

Woolfenden, G. E. (1975). Florida scrub jay helpers at the nest. *Auk*, **92**, 1–15.

Wright, H. E. Jr (1968). Natural environment of early food production north of Mesopotamia. *Science*, **161**, 334–40.

Yarbrough, C. G. (1971). The influence of distribution and ecology on the thermoregulation of small birds. *Comparative Biochemistry and Physiology*, **39** (2A), 235–66.

Young, H. (1963 a). Breeding success of the cowbird. *Wilson Bulletin*, **75**, 115–22.

Young, H. (1963 b). Age specific mortality in the eggs and nestlings of blackbirds. *Auk*, **80**, 145–55.

Young, T. G. (1962). Unseasonable breeding of house sparrows. *Scottish Birds*, **2**, 102.

Zar, J. H. (1968). Standard metabolism comparisons between orders of birds. *Condor*, **70**, 278.

Zar, J. H. (1974). *Biostatistical analysis*, xiv+620 pp. Englewood Cliffs, New Jersey: Prentice-Hall, Inc.

Zeidler, K. (1966). Untersuchungen über Flügelbefiederung und Mauser des Haussperlings (*Passer domesticus* (L.)). *Journal für Ornithologie*, **107**, 113–45.

Zimmerman, J. L. (1965 a). Bioenergetics of the dickcissel, *Spiza americana*. *Physiological Zoology*, **38**, 370–89.

423

References

Zimmerman, J. L. (1965 b). Digestive efficiency and premigratory obesity in the dickcissel. *Auk*, **82**, 278–9.

Zimmerman, J. L. (1971). The territory and its density dependent effect in *Spiza americana*. *Auk*, **88**, 591–612.

Ziswiler, V. (1965). Zur Kenntnis des Samenöffnens und der Struktur des hörnernen Gaumens bei körnerfressenden Oscines. *Journal für Ornithologie*, **106**, 1–48.

Index

Index

climatic conditions, and the initiation of breeding activity, 320–1, 324

clutch size; environmental influences on, 322–3; and natality rates, 322–3. *See also under individual species*

coal tit, *see Parus ater*

Coccothraustes coccothraustes: diet and feeding behavior, 306, 331–2; flocking and roosting behavior, 332; structure of bill, 306

Coccothraustes vespertinus: basal metabolism, daily rhythm, 130; bill size and structure, 306; diet, 192, 306; metabolizable energy coefficient, 192; as a non-pest species, 337; opportunistic population movements, 263

Colinus virginianus: calorific value, eggs, 167; as a non-pest species, 335, 337

colonies, bird: agricultural habitats, 248–9; breeding, 328–30; community attributes, 249–51, 262; energy demands, 251–3, 259, 262; feeding behavior, 295, 330–2; food consumption, 253–60, 262; migrations, 86–7; in natural ecosystems, 261–3; as pests, 267–8, 295, 327–8, 342; social organization, 327–8, 342

Columba livia: commensalism with man, 48; as a pest species, 275, 335–9

Columba palumbus, as a pest species, 268–9, 271, 274

commensalism with man, *Passer domesticus*, 4, 9, 15–18, 22–8, 47–51, 53, 205–66 *passim*, 314–15

common grackle, *see Quiscalus quiscula*

common redpoll, *see Acanthis flammea*

computer simulation models for estimating: energy requirements, 10, 11, 200–3, 213–21, 264, 343; food consumption, 10, 219–20, 264, 343; potential impact, 10, 264–6, 343–4

cooperative breeding, 327

coot, see *Fulica americana*

Corvus corax tibetanus, size, 146

Corvus corone, as a pest species, 335, 337–40

Corvus frugilegus, energy requirement for growth, 186–7; as a pest species, 335, 337–40

Corvus monedula, as a pest species, 335, 337–40

Corvus spp.: breeding behavior, 320–2; distribution, 313; as pests, 268, 334

Coturnix coturnix: calorific value, lean dry weight, 189; chemical control of, 274

courtship, energy cost of, 174

cowbird, *see Molothrus ater*

crested mynah, *see Acridotheres cristatellus*

critical temperature (defined): lower, 134, 136–7; upper, 134

crop, function of, 303, 307

crossbills, *see Loxia* spp.

crows, *see Corvus* spp.

daily energy budget (DEB), 10, 177–83, 195–6, 342–3; adaptation to local climate, 179–81; components, 198; estimation of, 197–203; and existence metabolism, 181; and food consumed, 191–4, 343; and growth of young, 183–91; and species weight, 182–3

Delichon urbica, nests occupied by *Passer domesticus*, 74

Dendrocygna autumnalis: calorific value, eggs, 167; energy requirement, growth of young, 183–6, 189–91; metabolizable energy coefficient, 192

Dendroica kirtlandii, population decrease caused by *Molothrus ater*, 95

Dendrocopus major, egg predator, 73

Dendropagus obscurus, energy requirement for growth, 187

desert lark, *see Ammomanes deserti*

desert sparrow, *see Passer simplex*

dickcissel, *see Spiza americana*

dietary opportunism, 308–13

digestive tract, structural modifications, 307

distributional opportunism, 313–19

Dolichonyx oryzivorus: brood reduction, 326; 'distant flight' behavior, 319; as a non-pest species, 335, 337; as a pest species, 339

domestic fowl, *see Gallus gallus*

domestic pigeon, *see Columba livia*

double-crested cormorant, *see Phalacrocorax auritis*

Drosophila pseudoobscura, polymorphism, 39

Dryomys nitedula, nestling predator, 78

dunlin, *see Erolia alpina*

eared dove, *see Zenaidura auriculata*

ecosystem functioning, role of birds in, 207–9

Ectopistes migratorius: colonial breeding, 328; eradication, 269

egg laying, energy cost of, 165–6

Eliomys quercinus, egg predator, 73

Emberiza citrinella: egg weight loss during incubation, 112; heat conserved by fluffing plumage, 157; as a non-pest species, 335,337

Emberiza hortulana, heat conserved by fluffing plumage, 157

energy-conserving conditions and activities, 153–9

energy-demanding conditions and activties, 159–77

energy units, conversion of, 14

Eremophila alpestris, 4, 5; basal metabolism, daily rhythm, 130; bill structure and size, 306; breeding behavior, 239, 261, 263; breeding biotopes, 263; diet and feeding behavior, 239, 242, 261, 263, 306; distribution, 7, 9; energy demand, 239, 242; migration, 242; as a non-pest species, 335, 337; population (breeding) density, 239; size, 242

Erithacus rubecula, egg weight loss during incubation, 112

Erolia alpina: calorific value, eggs, 167; energy requirement for growth, 187; metabolizable energy coefficient, 192

Erolia bairdii: calorific value, eggs, 167; temperature regulation on hatching, 189

Euplectes orix, as a pest species, 299

Euplectes spp.: breeding behavior, 334; diet and feeding behavior, 334

European tree sparrow, *see Passer montanus*

evening grosbeak, *see Coccothraustes vespertinus*

evolution, avian energetics and, 194–7

evolutionary history, *Passer* spp., 15–19, 341

426

Index

Melospiza melodia, as a non-pest species, 335, 337
metabolizable energy coefficient (MEC), 191–4
migration: energy cost of, 164–5, 195–6; of pest species, 295–6. *See also under individual species*
Molothrus ater, 4–5; breeding behavior, 86; breeding biotopes, 86; diet and feeding behavior, 293; distribution, 6–7, 53, 207; flocking and roosting behavior, 86–7, 104, 207, 293; migration, 86–7; nest parasitism, 86, 95, 326; as a pest species, 95, 104, 207, 268, 272, 277, 293; production values, 95–9, (clutch size) 95, 97, (hatching success) 95, 97, (fledgling and juvenile survival) 95, 97, (adult survival) 97–9, (life expectancy) 93, 97
monogamous mating systems, 328, 333
morphological adaptations to seed diet, 306–8; bill size and structure, 306–7; digestive tract modifications, 307; leg structure, 307–8; sexual dimorphism, 308
morphological relationships, *Passer* spp., 16–18
mortality and survivorship, 324–6; factors, 324–5; of young, 325. *See also under individual species*
moulting, 65; energy cost of, 175–7, 196
Mus musculus, polymorphism, 39
Mycteria americana, energy requirement for growth, 188

natality potential, 322–4; brood number, 323; clutch size, 322–4
nests: energy cost of building, 167; insulation, 172–3, 196; location, 171–2, 196–7; parasitism, 86, 326
net energy cost of moulting (NEM), 175–7
Nyctea scandiaca, heat loss increased by wind, 159
Nyroca affinis, energy requirement for growth, 187–8

olive-crowned sparrow, *see Passer flaveolus*
opportunism: dietary, 308–13; distributional, 313–19
optimization models: dickcissel feeding patterns, 311; quelea feeding patterns, 311
ortolan bunting, see *Emberiza hortulana*
Ouum-Qatafa Cave (Israel), fossils, 19
overwintering, energy stress of, 165, 195

parakeets, *see Aratinga* spp.
Parus ater: diet, 193; roosting behavior, 156; metabolizable energy coefficient, 193
Parus atricapillus, energy conserved by hypothermy, 158
Parus caeruleus: diet, 193; egg weight and clutch size, 108; metabolizable energy coefficient, 193
Parus cinctus, energy conserved by hypothermy, 158
Parus major: calorific value, eggs, 167; diet and feeding behavior, 193, 312; egg weight and clutch size, 108; energy cost of incubating eggs, 170; metabolizable energy coefficient, 193
Parus montanus, energy conserved by hypothermy, 158
Parus spp., diet and feeding behavior, 312
passenger pigeon, *see Ectopistes migratorius*

Passer ammodendri: morphological relationship with other *Passer* spp., 16–18; as a non-pest species, 335, 337–9
Passer castanopterus: commensalism with man, 48; morphological relationship with other *Passer* spp., 16–18
Passer domesticus, 4; ancestors, 19, 22, 24; biomass production, 107–22, 125–6, (eggs) 107–12, (young) 112–18, (adults) 118–22; breeding behavior, 53, 61–5, (egg laying and altitude 62–3, (egg laying and latitude) 62, 64, 355; breeding biotopes, 55, 60–1, 99–100; calorific value, eggs, 167; diet and feeding behavior, 53–5, 99, 210–11, 232–8, 242–6, 248, 271, 315; dispersal, 9, 25–8, 49–50, 53, 55–6; distribution, 3, 18, 53, 99–100 (in North America) 15, 24–6, 99, 314; energy demand, 221–38; evolutionary biology, 24–5; evolutionary history, 15–19, 22, 24; flocking and roosting behavior, 53–5, 210–11, 248, 338; migration, 55; morphological relationship with other *Passer* spp., 16–18; niche partitioning, variable, 49; as a pest species, 103, 210–11, 245, 268, 271, 283, 355–8; population (breeding) density, 55, 57–8, 60–1, 83–6, 100, 102–3, 105, 345–8; production values, 65–86, (brood number) 65–6, 126, 323–4, (clutch size) 67–70, 101, 323–4, (number of eggs laid) 65–70, 83, 323–4, 361, (egg loss) 70–4, 83, 100–1, 325, 358–9, 361, (nestling mortality) 74–8, 83, 85, 100–1, 325, 358–9, 361, (fledgling and juvenile survival) 78–9, 83, 85, 101, 325, 361, (adult survival) 79–82, 85, (life expectancy) 82, 84; sex ratio, 90; sexual dimorphism, 32–4, 37, 308; *see also* commensalism with man, genetic variation, phenetic variations
Passer domesticus bactrianus, 18; breeding behavior, 54; clutch size, 67; diet and feeding behavior, 54; egg weight, 110; flocking and roosting behavior, 54; migration, 54, 57; as a non-pest species, 335, 337–9; poisoning program against, 104; population (breeding) density, 57; weight change, 153
Passer domesticus griseogularis, egg weight, 110, 112
Passer domesticus hyrcanus, egg weight, 110, 112
Passer domesticus indicus, egg weight, 107, 110, 112
Passer eminibey, morphological relationship with other *Passer* spp., 16–18
Passer flaveolus, morphological relationship with other *Passer* spp., 16–18
Passer griseus: commensalism with man, 48; feeding behavior, 293–4; morphological relationship with other *Passer* spp., 16–18
passer hispaniolensis, 4; breeding behavior, 239, 321, 328; breeding biotopes, 263; commensalism with man, 48; diet and feeding behavior, 239, 240, 263, 267; dispersal, 9, 23; distribution, 9, 18, 23, 53; energy demand, 239, 240; evolutionary history, 16–19; flocking and roosting behavior, 60, 210–11, 267, 328, 338; migration, 9, 22; morphological relationship with other *Passer* spp., 16–18; as a pest species, 104, 210–11, 267–9, 335, 337–9; population (breeding)

428

429

Index